Mathematical Physics

Proceedings of the XI Regional Conference

Mathematical Physics

Proceedings of the XI Regional Conference

Tehran, Iran 3 – 6 May 2004

editors

S Rahvar
N Sadooghi

Sharif University of Technology and Institute for Theoretical Physics & Mathematics (IPM), Iran

F Shojai

University of Tehran and Institute for Theoretical Physics & Mathematics (IPM), Iran

World Scientific

NEW JERSEY • LONDON • SINGAPORE • BEIJING • SHANGHAI • HONG KONG • TAIPEI • CHENNAI

Published by

World Scientific Publishing Co. Pte. Ltd.
5 Toh Tuck Link, Singapore 596224
USA office: 27 Warren Street, Suite 401-402, Hackensack, NJ 07601
UK office: 57 Shelton Street, Covent Garden, London WC2H 9HE

British Library Cataloguing-in-Publication Data
A catalogue record for this book is available from the British Library.

MATHEMATICAL PHYSICS
Proceedings of the XI Regional Conference

ISBN 981-256-334-2

Printed in Singapore by Mainland Press

Preface

The XIth Regional Conference on Mathematical Physics was organized by the Institute for Studies in Theoretical Physics and Mathematics (IPM) from May 3 to 6, 2004 in Tehran, Iran. This year's conference was dedicated to the memory of the victims of the earthquake in Bam, Iran in December 26, 2003.

The aim of this series of Regional Conferences which was originally initiated by a group of physicists from Iran, Pakistan and Turkey, is to encourage research in Theoretical Physics in the region and to facilitate the regional and international contact of young scientists. In this conference 93 lecturers from the region and outside were invited. The lectures included 12 plenary and 81 parallel talks covering a wide range of topics in Theoretical Physics such as Astrophysics, Condensed Matter and Statistical Physics, High Energy Physics, General Relativity, Classical and Quantum Gravity, Mathematical Physics, Noncommutative Field Theory, Plasma Physics and String Theory.

We wish to acknowledge financial support from the Abdus Salam International Centre for Theoretical Physics (ICTP); the Centre of Excellence (CEP) of the Physics Department of the Sharif University of Technology, Iran; the Centre for International Research and Collaboration (ISMO), Iran; Atmospheric Science and Meteorological Research Centre (ASMERC) and Iranian Meteorological Organization (IRIMO). The participation of our Pakistani colleagues was financed by COMSTECH, the Higher Education Commission and the Science Foundation of Pakistan.

Our special thanks on the Rahbaran Petrochemical Co. providing the conference with three lectures halls, computer facilities and accommodation for some of our guests.

We would like to express our gratitude to Shirin Davarpanah, the Secretary of the Conference, Maryam Soltani, the Computer Administrator, Nassim Bagheri and Niloufar Pileroudi for secretarial services. We are also grateful to the officers from the Petrochemical Co. for their logistics help.

The Editorial Board
S. Rahvar, N. Sadooghi* and F. Shojai

September 2004
Tehran, Iran

For more information about the XIth Regional Conference on Mathematical Physics, please check the conference homepage:
http://physics.ipm.ac.ir/conferences/regconf11/index.htm

*Electronic Address: sadooghi@sharif.edu.

Contents

III. High Energy Physics; Phenomenology

IV. General Relativity and Quantum Gravity

V. Mathematical Physics

VI. Noncommutative Field Theory and String Theory

VII. Plasma Physics

CHAPTER 1: ASTROPHYSICS & COSMOLOGY

DEGENERATE FERMIONIC DARK MATTER IN GALAXIES

G. KUPI

Konkoly Observatory Budapest, Hungary

E-mail: kupiga@ludens.elte.hu

This work was motivated by the attempts to describe the spatial distribution of dark matter in galaxies by means of neutrinos. Their distribution is ruled by degenerate fermion distribution now. We give first a general description of this kind of dark matter. Then we study a special case. The size of such a dark matter cloud depends on rest mass of fermions. If this mass is high enough, the dark matter can be very dense in the galactic nuclei. We study the interaction between this dense phase and a black hole.

1 General description of degenerate fermionic dark matter in galaxies

[a]Gravitation manifests itself in the observed rotational velocities of astronomical objects orbiting around the center of galaxies: $v(r)^2/r = GM(r)/r^2$. We would expect that the velocity $v(r)$ decreases outside the central mass like $\frac{1}{r}$. We do not observe this decreasing. $v(r)$ is approximately constant far behind the center of galaxies. It means that there must be some dark (more exactly invisible) matter there. There are lots of idea what this matter can be, but, we do not have a perfect candidate.

We know that $kT \ll m_\nu c^2$ is valid for cosmological neutrinos with some eV rest mass ($T = 2$ K today). Degenerate Fermi distribution is a good approximation for them: $16\pi p_F^3/3h^3 = \rho_\nu/m_\nu$, where p_F is the Fermi momentum, m_ν and ρ_ν are mass and mass density of neutrinos, respectively. We can obtain from it for non-relativistic case the equation of state by means of $\epsilon = \frac{8\pi}{h^3 m_\nu}\int_0^{p_F} p^4 dp = \frac{8\pi}{5h^3 m_\nu} p_F^5$ and $P = \frac{2}{3}\epsilon$, that is $P = \frac{h^2}{5m_\nu^{8/3}}\left(\frac{3}{16\pi}\right)^{2/3}\rho_\nu^{5/3}$ $(\equiv K\rho_\nu^{5/3})$. We have two other relatively simple equations if we assume hydrostatic equilibrium and spherical symmetry. Euler's equation $\frac{dP}{dr} = -\frac{dU}{dr}\rho_\nu$ and Poisson's equation $\frac{d^2U}{dr^2} + \frac{2}{r}\frac{dU}{dr} = 4\pi G(\rho_\nu + \rho_b)$. ρ_b is the baryonic component.

Combining all these facts we get the following equation which leads to the saptial distribution of degenerate fermionic dark matter in galaxies:

$$\frac{d^2(U_0 - U)}{dr^2} + \frac{2}{r}\frac{d(U_0 - U)}{dr} = -4\pi G\left[\left(\frac{2}{5K}\right)^{3/2}(U_0 - U)^{3/2} + \rho_b\right],$$

where U_0 is the value of gravitational potential at which ρ_ν is 0^1.

The solutions of this equation for different m_ν and total mass of dark matter (the latter depends on $\frac{dU}{dr}|_{r=0}$) show that if we increase m_ν and/or total mass of dark matter, this matter spreads over a decreasing region around the center of the galaxy. It forms a thin halo around the whole galaxy for m_ν in some eV range. What can happen if m_ν is larger?

2 Degenerate fermionic dark matter in galactic nuclei

There are more and more observations concerning the centers of the galaxies. The stellar-kinematical data indicate dark compact objects of $10^{6.5}$ to $10^{9.5}$ solar masses there. From the time variability of the energy output, a few light days or light hours can be obtained for the size of these objects. These observations are mostly explained by supermassive black holes. However, the size of these objects can be larger than their Schwarzschild radius.

In Ref. [2] Viollier *et al.* suggest an alternative to black holes in galactic nuclei. They state that these objects could be neutrino or neutralino clouds. The relic light, neutral, weakly interacting fermions may have been collected into the centers of the galaxies. If this self-graviting degenerate fermionic matter consisted of light particles with about 17 keV rest mass, then it could mimic phenomena that are expected around supermassive black holes. The direct laboratory data are for neutrinos $m_{\nu_\mu} \leq 190$ keV and $m_{\nu_\tau} \leq 18.2$ MeV.[3] These values allow the 17 keV rest mass. However, we have to note that most of our result remains valid even if we do not specify the fermions. This matter can be dense enough, if

[a]Talk presented at the XIth Regional Conference on Mathematical Physics, May 2004, Tehran, Iran.

the gravitational phase transition of fermionic matter is conseidered[4,5,6,7]. Hence it cannot extend over the whole galaxy but only a few light days.

If we assume a dark matter star at the center of a galaxy then it is evidently important to study the interaction of such an object and a black hole. A simple model is presented here: let a non-rotating dark matter sphere and a non-rotating black hole in its center. The accretion flow is stationary. These are not too strict constraints because of strong degeneracy of fermionic gas. Let the mass of the whole system 10^9 solar mass.

We need the following equations to describe this object:

- Schwarzschild line element:

$$ds^2 = (1 - 2m/r)c^2dt^2 - \frac{dr^2}{1 - 2m/r} - r^2 d\Omega^2.$$

- Continuity equation:

$$(nU^i)_{;i} = 0.$$

- Relativistic Euler's equation:

$$(\epsilon + P)U^k U_{i;k} = -\frac{\partial P}{\partial x^i} - U_i U^k \frac{\partial P}{\partial x^k}.$$

- First law of thermodynamics:

$$\frac{d\epsilon}{dr} - \frac{\epsilon}{n}\frac{dn}{dr} - \frac{P}{n}\frac{dn}{dr} = \frac{\Lambda}{u}.$$

- The equation of state:

$$P = P(n).$$

It is proportional to $n^{5/3}$ for non-relativistic case and to $n^{4/3}$ for relativistic case.

Here $m = GM/c^2$, $U^i = dx^i/ds$ is the four-velocity, n is the scalar number density of particles, $\omega = \epsilon + P$ is the internal enthalpy per unit proper volume, ϵ is the proper internal energy density, and $\Lambda = \bar{v}\sigma n\epsilon$ is the energy loss function. σ is the cross section of the annihilation of a fermion pair into two photons.

The continuity equation and Euler's equation are simpler in our case:

$$4\pi n u r^2 = A.$$

$$\frac{1}{2}\frac{d}{dr}\left(\frac{u^2}{c^2}\right) = -\frac{1}{(P+\epsilon)}\frac{dP}{dr}\left(\frac{u^2}{c^2} - 1 + \frac{2m}{r}\right) - \frac{m}{r^2},$$

where $u = c|U^1|$, $A = \frac{\pi (GM)^2 n_0}{c_{s0}^3}$ and c_s is the velocity of sound. 0 subscript denotes the value in the infinity[8,9].

We can compute from these equations the number density of particles. Next figure shows the results. The mass of black hole is in solar mass unit.

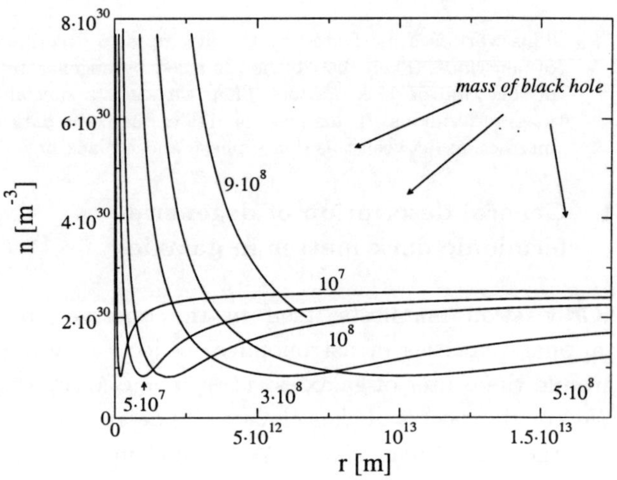

If we combine these data with in fall velocities and specify σ, we can compute the X-ray spectrum originated from annihilation. We can see it in the next figure.

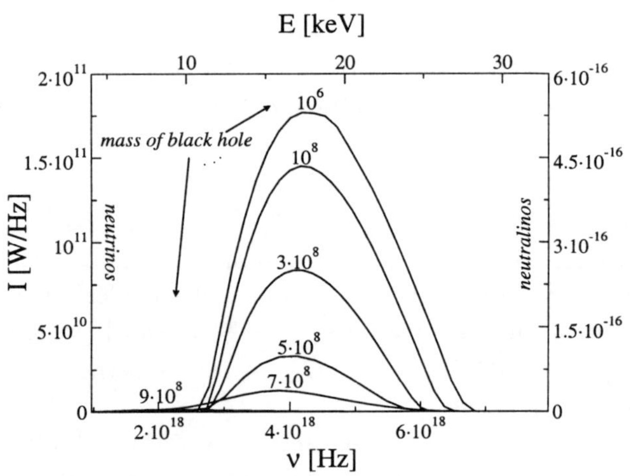

For neutrinos σ describes the $\nu + \tilde{\nu} \rightarrow 2\gamma$ process. For neutralinos σ of the following annihilation process $2\chi \rightarrow 2\gamma$ is calculated[10]. The used cross section is valid in pure photino limit of minimal supersymmetric extension of the Standard Model and it is only an upper limit, because its value depends on masses of supersymmetrical particles like $\sim 1/m^4$.

3 Time evolution of the system

One can estimate the lifetime of this system. The accretion rate is $dM/dt = 4\pi\rho_0 G^2 M^2 / c_{s0}^3 \equiv KM^2$, where M is the mass of the black hole. After integration we get for its growing $M(t) = 1/[1/M(0) - Kt]$. This gives some ten million years to the black hole to engulf the whole cloud of mass of some million solar mass if it starts growing with mass of a supernova remnant black hole. It is a faster process for more massive clouds. Such a neutrino or neutralino star radiates for long time with the same luminosity but the radiation decreases fast at the end of its life and stops in some years because the black hole grows hyperbolically. This can be seen in the table.

M [$M_\odot$]	t	E_{max}/E_{max0}	I/I_0 [%]
10	$2.2 \cdot 10^5$ y	1.00	100
10^3	2209 y	1.00	100
10^6	2.2 y	1.00	100
10^7	79.8 d	1.00	98
10^8	7.25 d	1.00	74
$5 \cdot 10^8$	19.35 h	0.97	14
$7 \cdot 10^8$	8.29 h	0.94	6.2
$9 \cdot 10^8$	2.15 h	0.54	0.7
$9.5 \cdot 10^8$	60.6 min	0.20	$3 \cdot 10^{-2}$
$9.9 \cdot 10^8$	11.7 min	0.14	$3 \cdot 10^{-4}$

The first column shows the mass of the central black hole. The second contains the time to the final stage of the system when the black hole remains only. E_{max}/E_{max0} is the ratio of the location of the peak of the spectrum to the original location. I/I_0 is the ratio of the intensity to the original intensity.

4 Conclusion

We can see the main properties of the spectrum: the Doppler broadening and the gravitational redshift of the line corresponding to the rest mass of light particles. It can be seen that the shapes of the two spectra are the same. They differ only in the amount of radiated energy. It can be imagined that they can have different shapes because of different energy dependence of the two annihilation processes but in the studied ranges of the parameters we have not experienced it. The difference is huge: $I_\nu / I_\chi \geq 10^{26}$. This is an inequality because we can give only an upper limit for the cross section of the neutralino annihilation. So we can hope to observe such dense neutrino stars but we do not have any chances of observing a neutralino star, which radiates some thousands of watts in the X-ray range maximum.

If we accept this simple model as a basic case, one may see that the black hole grows hyperbolically. This decrease ends in a short time. So the neutrino/neutralino quasars work mostly in the early Universe and we see that the big dark matter clouds have very little chance of existing now in galactic nuclei.

Black holes in galactic nuclei are derived usually from baryonic hyperstars. These objects radiate only for some decades. The heavy nuclei ($70 \leq A \leq 209$) are synthesized by the r process, that is by *rapid* process. This can happen only if the neutron flux is very high. Many neutrons can be captured subsequently and many nucleides have not enough time to decay. Conditions for the r process do not exist in normal stars but only in hyperstars at the end of their lives (and in supernovae). So these heavy nuclei can be found in bulk only in galaxies that contained hyperstars. Galaxies comprised of neutrino or neutralino quasars contain fewer heavy nuclei now.

References

1. G. Marx and G. Kupi, *Acta Physica Hungarica* N.S. **15**, 81 (2002).
2. R.D. Viollier, D. Trautmann and G.B. Tupper, *Phys. Lett.* B **306**, 79 (1993).
3. D. Groom *et al.*, *Particle Data Group*, *European Phys. J.* C **15 (1-4)**, 350 (2000).
4. N. Bilić and R.D. Viollier, *Phys. Lett.* B **408**, 75 (1997).
5. J. Messer, *J. Math. Phys.* 22 **No. 12**, 2910 (1981).
6. P. Hertel and W. Thirring, *Commun. Math. Phys.* **24**, 22 (1971).
7. P. Hertel, H. Narnhofer and W. Thirring, *Commun. Math. Phys.* **28**, 159 (1972).
8. S.L. Shapiro, *ApJ* **180**, 531 (1973).
9. H. Bondi, *MNRAS* **112**, 195 (1952).
10. G. Kupi, *Phys. Rev.* D **64**, 103507 (2001).

DOES TRANSPARENT HIDDEN MATTER GENERATE OPTICAL SCINTILLATION?

M. MONIEZ

Laboratoire de l'Accélérateur Linéaire
IN2P3-CNRS, *Université de Paris-Sud*
B.P. 34, 91898 Orsay Cedex, France
E-mail: moniez@lal.in2p3.fr

It is proposed to search for scintillation of extragalactic sources through the last unknown baryonic structures. Stars twinkle because their light goes through the atmosphere. The same phenomenon is expected when the light of extragalactic stars goes through a Galactic – disk or halo – refractive medium. Because of the large distances involved here, the length and time scales of the optical intensity fluctuations resulting from the wave distortions are accessible to the current technology. In this paper, we discuss the different possible scintillation regimes and we focus on the so-called strong diffractive regime that is likely to produce large intensity contrasts. Appropriate observation of the scintillation process described here should allow one to detect column density stochastic variations in cool Galactic molecular clouds of order of $\sim 3\times10^{-5}\,\mathrm{g/cm^2}$ – that is 10^{19} molecules/cm^2 – per $\sim 10\,000$ km transverse distance.

1 Introduction

[a]Considering the results of baryonic compact massive objects searches[1,2,3], it seems that the only constituent that could contribute quite significantly to the Galactic baryonic hidden matter is the cool molecular hydrogen (H_2). It has been suggested that a hierarchical structure of cold H_2 could fill the Galactic thick disk[4] or halo[5], providing a solution for the Galactic dark matter problem. This gas could form undetectable "clumpuscules" of 10 AU size at the smallest scale, with a column density of $10^{24-25}\,\mathrm{cm}^{-2}$, and a surface filling factor less than 1%. Light propagation is delayed through such a structure and the *average* transverse gradient of optical path variations is of order of 1λ *per* 10 000 km at $\lambda = 500$ nm. These H_2 clouds could then be detected through their diffraction and refraction effects on the propagation of the light of background stars (for a more detailed paper on this proposal see Ref. [6]).

2 Detection Mode of Extended H_2 Clouds: Principle

Due to index refraction effects, an inhomogeneous transparent medium (hereafter called screen) distorts the wave-fronts of incident electromagnetic waves (see Fig. 1). The extra optical path induced by a screen at distance z_0 can be described by a phase delay $\delta(x_1, y_1)$ in the plane transverse to the observer-source line. The amplitude A_0 in the observer's plane results from the subsequent propagation, that is described by the Huygens-Fresnel diffraction theory:

$$A_0(x_0, y_0) = \frac{Ae^{2i\pi z_0/\lambda}}{2i\pi R_F^2} \times \int\int_{-\infty}^{+\infty} e^{\frac{2i\pi\delta(x_1,y_1)}{\lambda}} e^{i\frac{(x_0-x_1)^2+(y_0-y_1)^2}{2R_F^2}} dx_1 dy_1, \quad (1)$$

where A is the incident amplitude (before the screen), taken as a constant for a very distant on-axis point-source, and $R_F = \sqrt{\lambda z_0/2\pi}$ is the Fresnel radius. R_F is of order of 1500 km to 15 000 km at $\lambda = 500$ nm, for a screen distance z_0 between 1 kpc to 100 kpc. This length scale characterizes the (x_1, y_1) domain that contributes significantly to the integral (a few Fresnel radii). The resulting intensity in the observer's plane is affected by strong interferences (the so-called speckle) if $\delta(x_1, y_1)$ varies stochastically of order of λ within the Fresnel radius domain. This is precisely the same order of magnitude as the average gradient that characterizes the hypothetic H_2 structures.

As for radio-astronomy[7,8], the stochastic variations of $\delta(x_1, y_1)$ can be characterized by the diffusion length scale R_{diff}, defined as the transverse separation for which the root mean square of the optical path difference is $\lambda/2\pi$.

- If $R_{diff} \gg R_F$, the screen is weakly diffusive. The wavefront is weakly corrugated, producing weakly contrasted patterns with length scale R_F in the observer's plane (Fig. 1).

[a]Invited plenary talk presented at the XIth Regional Conference on Mathematical Physics, May 2004, Tehran, Iran.

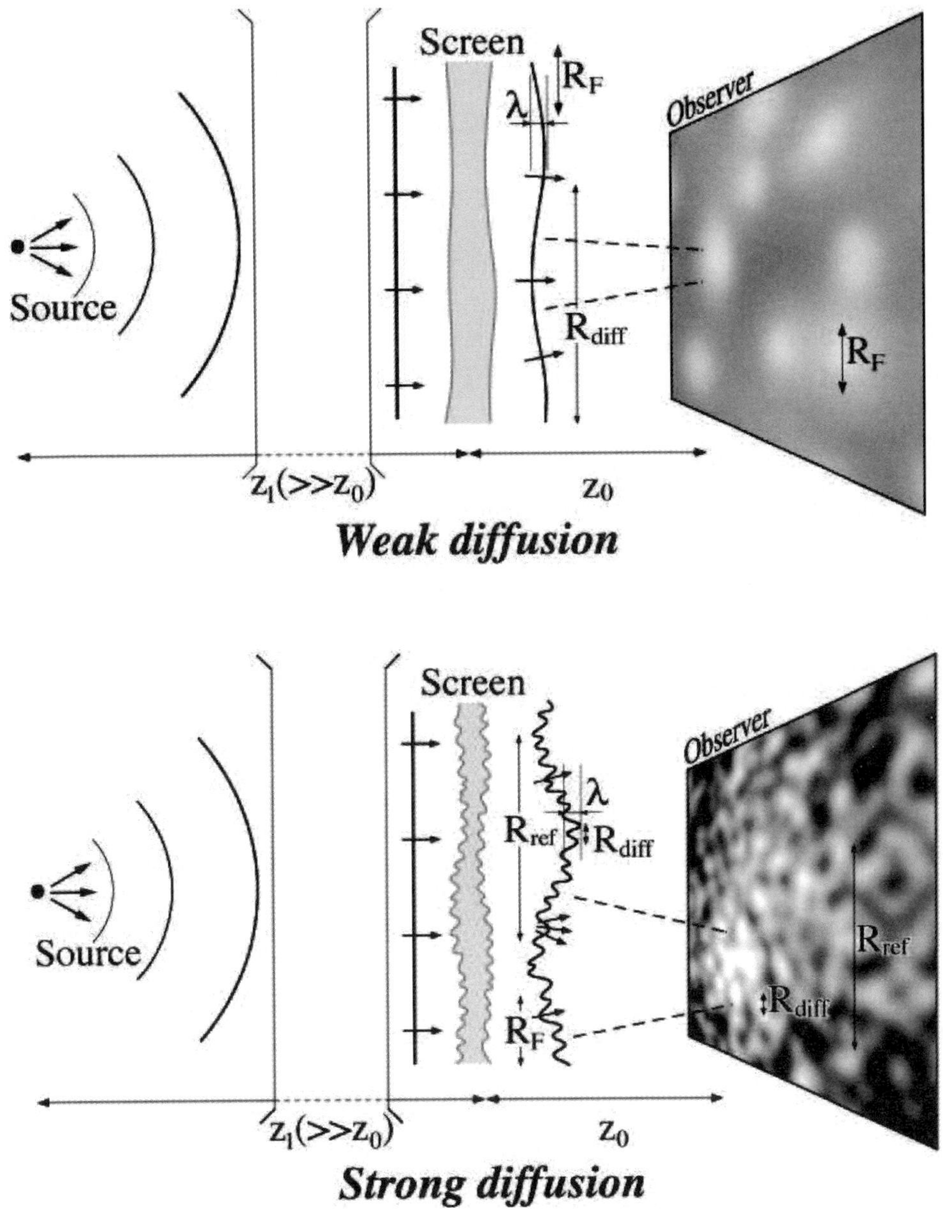

Figure 1. The two scintillation regimes: Up: $R_{diff} \gg R_F$. The weakly distorted wavefront produces a weak scintillation at scale R_F in the observer's plane. Down: $R_{diff} \ll R_F$. The strongly distorted wavefront produces strong scintillation at scales R_{diff} (diffractive mode) and R_{ref} (refractive mode).

- If $R_{diff} \ll R_F$, the screen is strongly diffusive. Two modes occur, the diffractive scintillation – producing strongly contrasted patterns characterized by the length scale R_{diff} – and the refractive scintillation – giving less contrasted patterns and characterized by the large scale structures of the screen R_{ref} –.

We focus here on the strong diffractive mode, which should produce the most contrasted patterns and is easily predictable.

3 Basic Configurations

Fig. 2 (left) displays the expected intensity variations in the observer's plane for a point-like monochromatic source observed through a transparent screen with a step of optical path $\delta = \lambda/4$ and through a prism edge. The inter-fringe is – in a natural way – close to the length scale defined by $\sqrt{\pi}R_F$. Such configurations model the edge of a H_2 structure, because they have the same average gradient of optical path. More generally, contrasted patterns take place as soon as the second derivative of the optical path is different from zero.

4 Limitations from Spatial and Temporal Coherences

At optical wavelengths, diffraction pattern contrast is severely limited by the size of the source r_S. Fig. 3 shows the diffraction patterns for different reduced source radii, defined as $R_S = r_s/(\sqrt{\pi}R_F) \times z_0/z_1$, where z_1 is the distance from the source to the screen.

In return, temporal coherence with the standard UBVRI filters is sufficiently high to enable the formation of contrasted interferences in the configurations considered here. The fringe jamming induced by the wavelength dispersion is also much smaller than the jamming due to the source extension.

5 What is to See?

An interference pattern with inter-fringe of $\sim R_F$ ($1000 - 10\,000$ km at $\lambda = 500$ nm) is expected to sweep across the Earth when the line of sight of a sufficiently small astrophysical source crosses an inhomogeneous transparent Galactic structure (see Table 1). This pattern moves at the relative transverse velocity V_T of the diffusive screen. Its shape may also evolve, due to random turbulence in the scattering medium. In the present study, we assume that the scintillation is mainly due to pattern motion rather than pattern instability, as it is usually the case in radioastronomy observations[7]. For the Galactic H_2 clouds we are interested in, we expect a typical modulation index m_{scint} (or contrast) ranging between 1% and $\sim$ 20%, critically depending on the source apparent size. The time scale $t_{scint} = R_F/V_T$ of the intensity variations is of order of 30 s. As the inter-fringe scales with $\sqrt{\lambda}$, one expects a significant difference in the time scale t_{scint} between the red side of the optical spectrum and the blue side. This property might be used to sign the diffraction phenomenon at the R_F natural scale.

6 How to See it?

Minimal hardware: from Table 1, we deduce that the minimal magnitude of extragalactic stars that are likely to undergo a few percent modulation index from a Galactic molecular cloud is $M_V \sim 20.5$ (A5V star in LMC or B0V in M31). Therefore, the search for diffractive scintillation needs the capability to sample every ~ 10 s (or faster) the luminosity of stars with $M_V > 20.5$, with a point-to-point precision better than a few percent. This performance can be achieved using a telescope diameter larger than two meters, with a high quantum efficiency detector allowing a negligible dead time between exposures (like frame-transfer CCDs). Multi-wavelength detection capability is highly desirable to exploit the dependence of the diffractive scintillation pattern with the wavelength.

Chances to see something? The 1% surface filling factor predicted for gaseous structures is also the maximum optical depth for *all* the possible refractive (weak or strong) and diffractive scintillation regimes. Under the pessimistic hypothesis that strong diffractive regime occurs only when a Galactic structure enters or leaves the line of sight, the duration for this regime is of order of $\sim$ 5 minutes (time to cross a few fringes) over a total typical crossing time of $\sim$ 400 days. Then the diffractive regime optical depth τ_{scint} should be at least of order of 10^{-7} and the average exposure needed to observe one event of $\sim$ 5 minute

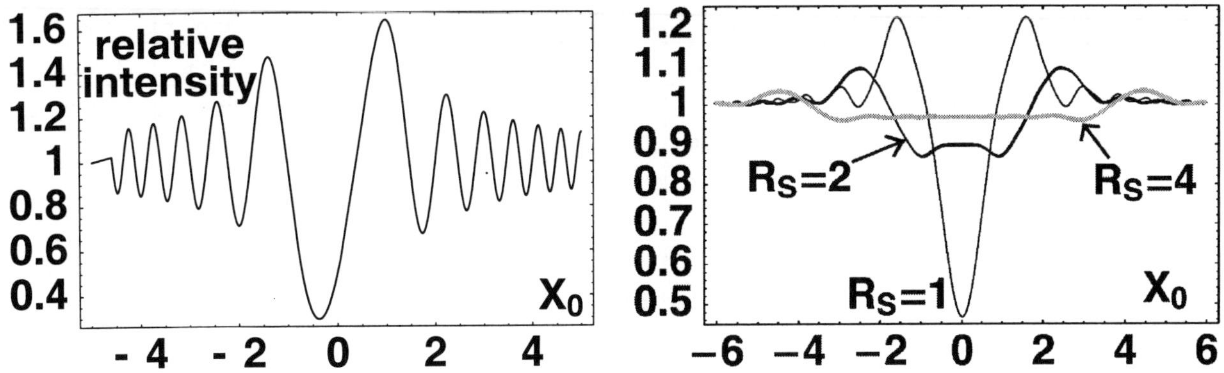

Figure 2. Relative intensity diffraction patterns produced in the observer's plane, perpendicularly to:
- a step of optical path $\delta = \lambda/4$ (left).
- a prism of optical path with $\delta = X_0 \times \lambda/2$ for $X_0 > 0$ (right).
The X_0 − axis origin ($X_0 = x_0/\sqrt{\pi}R_F$) is the intercept of the source-step line with the observer's plane.

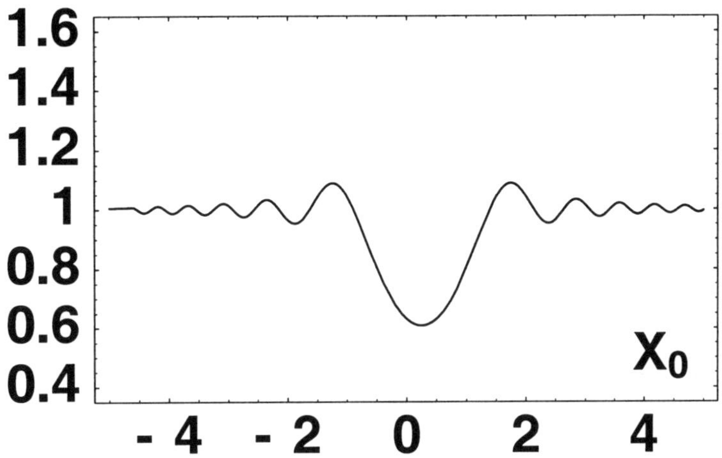

Figure 3. Relative intensity diffraction pattern produced in the observer's plane, perpendicularly to a step of optical path $\delta = \lambda/2$, for extended disk-sources of reduced radius R_S = 1, 2 and 4. The X_0 − axis origin ($X_0 = x_0/\sqrt{\pi}R_F$) is the intercept of the source-step line with the observer's plane.

Table 1. Configurations leading to strong diffractive scintillation. Here we assume a regime characterized by $R_{diff} \leq R_F$, or at least a transitory regime as described in Section 3 – characterized by R_F –, for example if an inhomogeneity due to a turbulent mechanism crosses the line of sight. Numbers are given for $\lambda = 500\,\text{nm}$.

		SCREEN					
		atmosphere	solar system	solar suburbs	Galactic thin disk	Galactic thick disk	Galactic halo
Distance		10 km	1 AU	10 pc	300 pc	1 kpc	10 kpc
R_F to × by $\left[\frac{\lambda}{500\,\text{nm}}\right]^{\frac{1}{2}}$		2.8 cm	109 m	157 km	860 km	1570 km	5000 km
Transverse speed V_T		1 m/s	10 km/s	20 km/s	30 km/s	40 km/s	200 km/s
t_{scint} to × $\left[\frac{\lambda}{500\,\text{nm}}\right]^{\frac{1}{2}} \left[\frac{R_{diff}}{R_F}\right]$		0.03 s	0.01 s	8 s	29 s	40 s	25 s
Optical depth τ_{scint}		1	total $> 10^{-7}$				
m_{scint} in % to multiply by $\left[\frac{\lambda}{500\,\text{nm}}\right]^{\frac{1}{2}} \left[\frac{R_{diff}}{R_F}\right] \left[\frac{r_S}{r_\odot}\right]^{-1}$		100%	32%× $\left[\frac{d}{10\text{pc}}\right]$	2.2%× $\left[\frac{d}{1\text{kpc}}\right]$	4.1%× $\left[\frac{d}{10\text{kpc}}\right]$	2.2%× $\left[\frac{d}{10\text{kpc}}\right]$	7.1%× $\left[\frac{d}{100\text{kpc}}\right]$
SOURCE		**DIFFRACTIVE MODULATION INDEX** m_{scint} **(to multiply by** $\sqrt{\lambda/500\,\text{nm}} \times R_{diff}/R_F$**)**					
Location	Type						
nearby 10pc	any star	$\ll 1\%$ in a telescope $> 1\,\text{m}$	$< 100\%$				
Galactic 8kpc	star		100%	1-70%	1-10%		
LMC 55kpc	A5V ($r_S = 1.7 r_\odot$)		100%	70%	13%	7%	2%
M31 725kpc	B0V ($r_S = 7.4 r_\odot$)		100%	100%	40%	22%	7%
z=0.2–0.9Gpc	SNIa		100%	70%	13%	7%	2%
z=1.7–1.7Gpc	Q2237+0305		100%	$> 45\%$	$> 8\%$	$> 4\%$	$> 1.4\%$

duration is 10^6 star × hr [b]. It follows that a wide field detector is necessary to monitor a large number of stars.

7 Foreground Effects, Background to the Signal

Atmospheric effects: Surprisingly, atmospheric intensity scintillation is negligible through a large telescope ($m_{scint} \ll 1\%$ for a $> 1\,\text{m}$ diameter telescope[9]). Any other long time scale atmospheric effect such as absorption variations at the sub-minute scale (due to fast cirruses for example) should be easy to recognize as long as nearby stars are monitored together.

The solar neighborhood: Overdensities at 10 pc could produce a signal very similar to the one expected from the Galactic clouds. But in this case, even big stars should undergo a contrasted diffractive scintillation; the distinctive feature of scintillation through more distant screens ($> 300\,\text{pc}$) is that *only the smallest stars* are expected to scintillate. It follows that simultaneous monitoring of various types of stars at various distances should allow one to discriminate effects due to solar neighborhood gas and due to more distant gaseous structures.

Sources of background? Physical processes such as asterosismology, granularity of the stellar surface, spots or eruptions produce variations of very different amplitudes and time scales. A few categories of recurrent variable stars exhibit important emission variations at the minute time scale[10], but their types are easy to identify from spectrum.

8 Conclusions and Perspectives

Structuration of matter is perceptible at all scales, and the eventuality of stochastic fluctuations producing diffractive scintillation is not rejected by observations. In this paper, I showed that there is an observational opportunity resulting from the subtle compromise between the arm-lever of interference patterns due to hypothetic diffusive objects in the Milky-Way and the size of the extra-galactic stars. The hardware and software techniques required for scintillation searches are currently available. Tests are under way to validate some of the ideas discussed here.

[b]Turbulence or any process creating filaments, cells, bubbles or fluffy structures should increase these estimates.

If some indications are discovered with a single telescope, one will have to consider a project involving a 2D array of telescopes, a few hundred and/or thousand kilometers apart. Such a setup would allow to temporally and spatially sample an interference pattern, unambiguously providing the diffusion length scale R_{diff}, the speed and the dynamics of the scattering medium.

References

1. T. Lasserre *et al.* (EROS Collab.) *A&A* **L39**, 355 (2000).
2. C. Alfonso *et al.* (EROS collab.) *A&A* **400**, 951 (2003).
3. C. Alcock *et al.* (MACHO collab.) *ApJ* **542**, 281 (2000).
4. D. Pfenniger and F. Combes, *A&A* **285**, 94 (1994).
5. F. De Paolis *et al.* *Phys. Rev. Lett.* **74**, 14 (1995).
6. M. Moniez, *A&A* **412**, 105 (2003).
7. A.G. Lyne and F. Graham-Smith in *Pulsar Astronomy*, (Cambridge University Press 1998).
8. R. Narayan, *Phil. Trans. R. Soc. Lond. A* **341**, 151 (1992).
9. D. Dravins *et al.* *Pub. of the Ast. Soc. of the Pacific* **109**, (I, II) (1997), **110**, (III) (1998).
10. C. Sterken and C. Jaschek in *Light Curves of Variable Stars, a Pictorial Atlas*, (Cambridge University Press 1996).

GALACTIC MACHO BUDGET: PROBLEMS AND POSSIBLE SOLUTION WITH THE ABUNDANT BROWN DWARFS

S. RAHVAR
Department of Physics, Sharif University of Technology
P.O. Box 11365-9161, Tehran, Iran
and
Institute for Theoretical Physics and Mathematics (IPM), School of Physics
P.O. Box 19395-5531, Tehran, Iran
E-mail: rahvar@sharif.edu

The gravitational microlensing experiments in the direction of Large Magellanic Cloud (LMC) predict a large amount of white dwarfs ($\sim 20\%$) filling the galactic halo. The main contradiction of this result with the other astrophysical observations is (i) the predicted white dwarfs are not observed at the galactic halo and (ii) the evidence of the existence of white dwarfs is the heavy metals whose signature have not been detected. To interpret the microlensing results and resolving the mentioned problems, we use the hypothesis of spatially varying mass function of MACHOs, proposed by Kerins and Evans[1] (hereafter KE). However the KE model is not compatible with the duration distribution of the events. To have a better parameters for the model, we do a likelihood analysis and show that in contrast to the abundant brown dwarfs of the halo, heavy MACHOs reveal themselves frequently as the microlensing events.

1 Introduction

[a]The rotation curves of spiral galaxies and the Milky Way as well, show that these type of galaxies have dark halo component. The most trivial candidate for the dark halo structure is the baryonic matter that can be in the gaseous or Massive Compact Halo Objects (MACHOs) forms. Since MACHOs are expected to be too light to be luminous and difficult to be detectable, Paczyński[2] proposed an indirect method so-called gravitational microlensing to observe them indirectly. Following his suggestion several experiments such as EROS and MACHO started monitoring LMC stars for one decade and observed less than 20 events in this direction[3,4].

Due to the degeneracy nature of gravitational microlensing problem, it is impossible to obtain the mass, distance and transverse velocity of a lens by measuring one parameter of duration of events. The only way with the results of present experiments is statistical studying of the microlensing events with the models and the results are the mean mass of MACHOs and their mass fraction in the halo.

This study has been done for a category of Galactic models, called power law halo and almost in all the models the derived mean mass of MACHOs is large enough to detect them directly. For instance, in the case of standard model, halo is comprised by 20% MAHCOs with about 0.5 solar mass. Comparing luminous mass of Milky Way with that of the halo, we can conclude that white dwarfs of the halo should be two times as much as the ordinary stars of galactic disk and bulge.

In the present study we use the hypothesis of spatially varying Mass Function[1] (MF) of MACHOs as a possible solution for the interpretation of microlensing events. In this model, in contrast to the traditional Dirac-Delta MF used in the microlensing analysis, the MF changes monotonically from the center of galaxy to the outer parts of halo.

The latest MACHO experiment microlensing events is used in the framework of power-law halo model[5] for our analysis. A likelihood analysis is performed to obtain the best parameter for the MF to be compatible with the duration distribution of observed data. The advantage of using spatially varying MF is that in spite of dominant brown dwarfs of the halo, heavy MACHOs are responsible to the microlensing events. To quantify this effect, we define two mean masses for the MACHOs as the active and passive mean masses. The former is the mean mass of observed events while the latter is the mean mass of overall lenses in the halo. By a Monte-Carlo simulation we show that active mean mass is always larger than the passive one.

[a]Talk presented at the XIth Regional Conference on Mathematical Physics, May 2004, Tehran, Iran.

2 Spatially Varying MF and Power-Law Halo Model

This section contains the physical motivation of using spatially varying MF for the MACHOs and introducing the power-law Galactic halo model.

2.1 *Spatially Varying MF*

The tradition in gravitational microlensing analysis is using the non-realistic Dirac-Delta MF. The star formation theories predicts that the MF of stars should depend on the density of interstellar medium[6,7]. Kerins and Evans (1998) proposed a non-homogeneous MF as follows:

$$MF(r) = \delta[M - M(r)], \tag{1}$$

where the mass scale is $M(r) = M_U(\frac{M_L}{M_U})^{r/R_{halo}}$, M_L and M_U are the mass scales which represent the lower and upper limits of the mass function and R_{halo} is the galactic halo size filled with MACHOs. The parameters of KE model are indicated in Table 1.

2.2 *Power-Law Galactic Models*

In the direction of LMC, the galactic disk, spheroid, halo and LMC disk can contribute to the gravitational microlensing events. The galactic components can be combined to build various galactic models. The tradition is using S, A, B, C, D, E, F and G for calling them. Here we use the components of galactic models in simulating the microlensing events. The Galactic halo is chosen as the power-law model[5] and the Galactic disk[9] and matter distribution of LMC disk[8] as the double exponential[9].

Table 1. M_L, M_U are the lower and upper bounds for the mass of MACHOs in terms of solar mass and R_{halo} in terms of kilo parsec is the length scale where the MACHOs extended to that size. The first I, second II and third III MF models are for medium, small and large halo models, respectively.

Model	M_L	M_U	$R_{halo}(kpc)$	Description
I	10^{-3}	3	100	isotropic dispersion
II	10^{-3}	10	50	small halo
III	10^{-3}	2	200	large halo

3 Comparing Observed Data with the Theory

Using the numerical method, we obtain the theoretical duration distribution of microlensing events in power-law halo models[10]. Eight galactic models are used for generating the microlensing events. The observational efficiency of experiment is multiplied to the theoretical distributions to obtain the expected one to compare with the MACHO data. The duration distribution for eight Galactic models are shown in Fig. 1.

Table 2. The result of likelihood analysis: the first column indicates the name of eight galactic models. The second column shows the size of halo that MACHOs are extended. The third column is the lower limit for the mass of MACHOs that are located at the edge of halo and the fourth column is the upper limit for the mass of MACHOs that resides at the center of halo. The fifth column is the mean mass of the MACHOs in each model, the so-called passive mean mass of the lenses and the sixth column is the active mean mass of the observed lenses by the experiment. The seventh column shows the halo fraction made by MACHOs in each model.

Model	R_{halo}	M_L	M_U	$< M_{ml} >$	$< \tilde{M}_{ml} >$	f_{ML}
S	126	10^{-3}	1	0.16	0.26	0.2
A	177	10^{-3}	0.5	0.16	0.31	0.14
B	163	10^{-3}	0.9	0.22	0.5	0.1
C	85	10^{-3}	0.5	0.04	0.21	0.25
D	103	10^{-3}	0.4	0.06	0.2	0.12
E	87	10^{-3}	0.5	0.04	0.15	0.8
F	96	10^{-3}	0.4	0.04	0.13	0.3
G	110	10^{-3}	0.3	0.05	0.13	0.18

The next step is to compare the duration distribution of MACHO candidates with the theoretical distributions of events. Two statistical parameters of the mean and the width of duration of events are used for this comparison[11,12]. The width of distribution of observed events is the difference between the maximum and the minimum values of the duration of events. Comparing these parameters with Fig. 1 shows that unlike the Dirac-Delta MF, KE model is not compatible with the observed data.

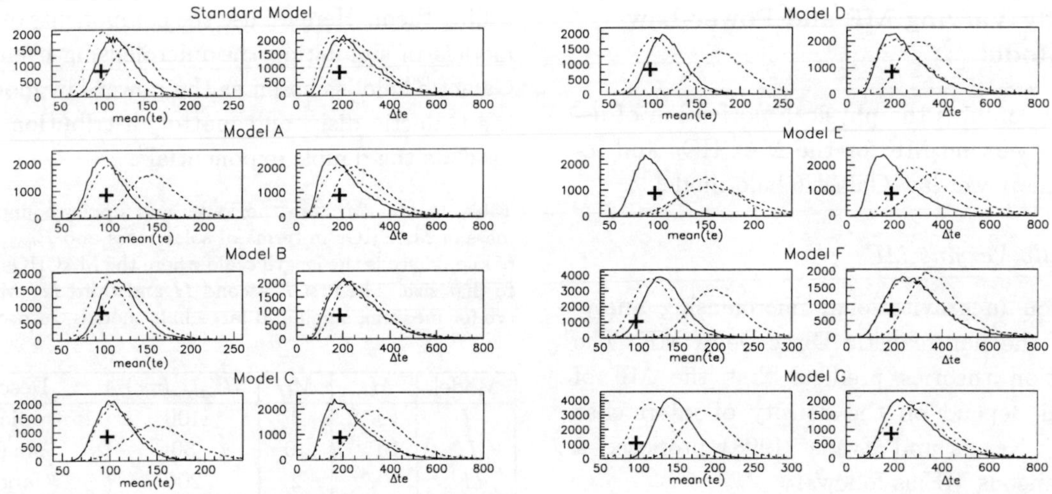

Figure 1. The expected distributions of the mean and the width of the duration of events are shown for the S, A, B, C, D, E, F and G Galactic models. The cross indicates the position of the observed value by the MACHO experiments. The uniform Dirac Delta MF, KE model and the result of likelihood analysis are shown by the solid, dashed and dash-dotted lines.

To find the compatible parameters for the KE model we do a likelihood analysis to find the halo size and the upper limit for the mass of MACHOs to be in agreement with data. The results are shown in Table 2. To quantify the effect of spatially varying MF on the interpretation of the microlensing results we define two masses of active and passive as the mean mass of observed events and the mean mass of overall lenses in the halo. The passive mean mass of lenses can be obtain directly from the MF and the halo model. The active mean mass is obtained by a Monte-Carlo simulation according to the distribution of lenses in the line of sight, using the MF and producing the duration of events. The duration of events are compared with the observational efficiency to be selected or rejected. The selected microlensing events are used to obtain the active mean mass of lenses.

The advantage of using the spatially varying MF is that in spite of dominant brown dwarfs in the halo, heavy MACHOs are responsible for the microlensing events. Table 2 shows that in all the cases, the active mean mass is always larger than the passive one and this point may resolve interpretation of microlensing data.

References

1. K. Kerins and N.W. Evans, *ApJ* **503**, 75 (1998).
2. B. Paczyński, *ApJ* **304**, 1 (1986).
3. C. Alcock *et al.* (MACHO Collab.) *ApJ* **542**, 281 (2000).
4. T. Lasserre *et al.* (EROS Collab.) *A&A* **355**, L39 (2000).
5. N.W. Evans, *MNRAS* **267**, 333 (1994).
6. S.M. Fall and M. Rees, *ApJ* **298**, 18 (1985).
7. K. Ashman, *MNRAS* **247**, 662 (1990).
8. G. Gyuk, N. Dalal and K. Griest *ApJ* **535**, 90 (2000).
9. S. Binney and S. Tremaine in *Galactic Dynamics*, (Princeton University Press 1987).
10. C. Alcock *et al.* (MACHO Collab.) *ApJ* **461**, 84 (1996).
11. A.M. Green and K. Jedamzik *A&A* **395**, 31 (2002).
12. S. Rahvar, *MNRAS* **347**, 213 (2004).

THE MYSTERIOUS NATURE OF DARK ENERGY

V. SAHNI

Inter-University Centre for Astronomy & Astrophysics (IUCAA)
Post Bag 4, Ganeshkhind, Pune 411 007, India
E-mail: varun@iucaa.ernet.in

I briefly review some theoretical models which give rise to an accelerating Universe. These "Dark Energy" models include the cosmological constant, scalar field models (quintessence) as well as braneworld models. Recent diagnostic tools for studying the properties of dark energy, such as the Statefinders, are also discussed.

1 The Accelerating Universe and Dark Energy

1.1 *The Cosmological Constant*

[a]Type Ia supernovae, when treated as standardized candles, indicate that our Universe is accelerating[15,16,19], and that this acceleration is fueled by a form of matter having large negative pressure and called "dark energy"[25,23].[b]

The simplest example of dark energy is a cosmological constant, introduced by Einstein in 1917. The Einstein equations, in the presence of the cosmological constant, acquire the form

$$R_{ik} - \frac{1}{2}g_{ik}R = \frac{8\pi G}{c^4}T_{ik} + \Lambda g_{ik}. \quad (1)$$

Although Einstein originally introduced the cosmological constant (Λ) into the left hand side of his field equations, it has now become conventional to show the Λ-term in the RHS, treating it as an effective form of matter. This indeed has been the rationale behind most dark energy models (though not all – see for instance the discussion of braneworld dark energy models in this section). In a homogeneous and isotropic Friedmann-Robertson-Walker (FRW) universe consisting of pressureless dust (dark matter) in addition to Λ the Raychaudhury equation, which follows from (1), takes the form

$$\ddot{a} = -\frac{4\pi G}{3}a\rho_m + \frac{\Lambda}{3}. \quad (2)$$

Equation (2) can be rewritten in the form of a force law:

$$\mathcal{F} = -\frac{GM}{R^2} + \frac{\Lambda}{3}R, \quad (R \equiv a), \quad (3)$$

which demonstrates that the cosmological constant gives rise to a *repulsive* force whose value increases with distance, and which could therefore be responsible for the current acceleration of the universe.

Although introduced into physics in 1917, the physical basis for a cosmological constant was a bit of a mystery until the 1960's when it was realized that zero-point vacuum fluctuations must respect Lorenz invariance and therefore have the form $\langle T_{ik}\rangle = \Lambda g_{ik}$. As it turns out, the vacuum expectation value of the energy momentum is divergent both for bosonic and fermionic fields, and this gives rise to what is known as "the cosmological constant problem". Indeed the effective cosmological constant generated by vacuum fluctuations is

$$\frac{\Lambda}{8\pi G} = \langle T_{00}\rangle_{\rm vac} \propto \int_0^\infty \sqrt{k^2+m^2}k^2dk, \quad (4)$$

this integral diverging like k^4 implies an infinite value for the vacuum energy[36,34]. Even if one chooses to "regularise" $\langle T_{ik}\rangle$ by invoking an ultraviolet cutoff at the Planck scale, one is still left with an enormously large value for the vacuum energy $\langle T_{00}\rangle_{\rm vac} \simeq c^5/G^2\hbar \sim\sim 10^{76}{\rm GeV}^4$ which is 123 orders of magnitude larger than the currently observed $\rho_\Lambda \simeq 10^{-47}{\rm GeV}^4$. A smaller ultraviolet cut-off does not fare much better since a cutoff at the QCD scale results in $\Lambda^4_{QCD} \sim 10^{-3}{\rm GeV}^4$, which is still forty orders of magnitude larger than observed.

1.2 *Dynamical Dark Energy Models – Quintessence*

The cosmological constant is but one example of a form of matter (dark energy) which can drive an accelerated phase in the expansion history of our uni-

[a]Invited plenary talk presented at the XIth Regional Conference on Mathematical Physics, May 2004, Tehran, Iran.
[b]This invited review talk is largely based on my lectures on 'Dark Matter and Dark Energy' delivered in Greece in 2003 [24].

verse. Indeed, (2) is easily generalized to

$$\frac{\ddot{a}}{a} = -\frac{4\pi G}{3}\sum_i(\rho_i + 3p_i) = -\frac{4\pi G}{3}\sum_i \rho_i(1+3w_i)\ , \tag{5}$$

where the summation is over all forms of matter present in the universe and $w = p/\rho$ is their equation of state. Equation (5) together with its companion equation

$$H^2 \equiv \left(\frac{\dot{a}}{a}\right)^2 = \frac{8\pi G}{3}\sum_i \rho_i - \frac{k}{a^2}, \tag{6}$$

completely describes the dynamics of a FRW universe (k/a^2 is the Gaussian curvature of space).

Clearly a universe consisting of only a single component will accelerate if $w < -1/3$.

Fluids satisfying $\rho + 3p \geq 0$ or $w \geq -1/3$ are said to satisfy the 'strong energy condition' (SEC). We therefore find that, in order to accelerate, "dark energy" must violate the SEC. Another condition which is usually assumed to be sacrosanct, but has recently been called into question is the "weak energy condition" (WEC) according to which $\rho + p \geq 0$ or $w \geq -1$.

It is interesting that the fine tuning problem facing dark energy models with a constant equation of state can be alleviated if we assume that the equation of state is time dependent. An important class of models having this property are scalar fields which couple minimally to gravity and whose energy momentum tensor is

$$\rho \equiv T_0^0 = \frac{1}{2}\dot{\phi}^2 + V(\phi),\ P \equiv -T_\alpha^\alpha = \frac{1}{2}\dot{\phi}^2 - V(\phi). \tag{7}$$

Potentials which are sufficiently steep in order to satisfy $\Gamma \equiv V''V/(V')^2 \geq 1$ have the interesting property that scalar fields approach a common evolutionary path from a wide range of initial conditions[38]. In these so-called "tracker" models the scalar field density (and its equation of state) remains close to that of the dominant background matter during most of cosmological evolution.

An excellent example is provided by $V(\phi) = V_0/\phi^\alpha$.[18] During tracking the ratio of the energy density of the scalar field (quintessence) to that of radiation/matter gradually increases $\rho_\phi/\rho_B \propto t^{4/(2+\alpha)}$, while its equation of state remains marginally smaller than the background value $w_\phi = (\alpha w_B - 2)/(\alpha + 2)$. For large values of ϕ this potential becomes flat and this ensures that the scalar field rolls slower ($\dot{\phi}^2 \ll V(\phi)$) thereby allowing the universe to accelerate.

The scalar field therefore eventually dominates the density of the universe and generates a late-time epoch of accelerated expansion. Current observations place the strong constrain $\alpha < 2$.

A an extreme example of quintessence is provided by the exponential potential $V(\phi) = V_0 \exp\left(-\sqrt{8\pi}\lambda\phi/M_{\rm Pl}\right)$[18,35] for which

$$\frac{\rho_\phi}{\rho_B + \rho_\phi} = \frac{3(1+w_B)}{\lambda^2} = \text{constant} < 0.2. \tag{8}$$

ρ_B is the background energy density while w_B is the associated equation of state. The lower limit $\rho_\phi/\rho_{\rm total} < 0.2$ arises because of nucleosynthesis constraints which prevent the energy density in quintessence from being large initially (at $t \sim few$ sec.). Equation (8) suggests that the exponential potential will remain subdominant if it was so initially. An interesting potential which interpolates between an exponential and a power law can however give rise to late time acceleration from tracker-like initial conditions[20]

$$V(\phi) = V_0[\cosh\lambda\phi - 1]^p, \tag{9}$$

has the property that $w_\phi \simeq w_B$ at early times whereas $\langle w_\phi \rangle = (p-1)/(p+1)$ at late times. Consequently (9) describes *quintessence* for $p \leq 1/2$ and pressureless "cold" dark matter (CDM) for $p = 1$. Thus the cosine hyperbolic potential (9) is able to describe both dark matter and dark energy within a common framework (also see Refs. [3,33]).

Remarkably, quintessence can even accommodate a constant equation of state (w = constant) by means of the potential[25,26,32]

$$V(\phi) \propto \sinh^{\frac{2(1+w)}{w}}(C\phi + D), \tag{10}$$

with suitably chosen values of C, D.

1.3 Braneworld Models of Dark Energy

Inspired by the Randall-Sundrum[17] scenario, braneworld cosmology suggests that we could be living on a three dimensional "brane" which is embedded in a higher (usually four) dimensional bulk. According to such a scheme, all matter fields are confined to the brane whereas the graviton if free to propagate in the brane as well as in the bulk[12]. Within the RS setting the equation of motion of a scalar field propagating on the brane is

$$\ddot{\phi} + 3H\dot{\phi} + V'(\phi) = 0, \tag{11}$$

where [28]

$$H^2 = \frac{8\pi}{3m^2}\rho(1+\frac{\rho}{2\sigma}) + \frac{\Lambda_4}{3} + \frac{\mathcal{E}}{a^4}\,,$$
$$\rho = \frac{1}{2}\dot{\phi}^2 + V(\phi)\,. \tag{12}$$

$\mathcal{E}$ is an integration constant which transmits bulk graviton influence onto the brane. The brane tension σ provides a relationship between the four dimensional Planck mass (m) and the five-dimensional Planck mass (M)

$$m = \sqrt{\frac{3}{4\pi}}(\frac{M^3}{\sqrt{\sigma}})\,. \tag{13}$$

σ also relates the four-dimensional cosmological constant Λ_4 on the brane to the five-dimensional (bulk) cosmological constant Λ_b through

$$\Lambda_4 = \frac{4\pi}{M^3}\left(\Lambda_b + \frac{4\pi}{3M^3}\sigma^2\right). \tag{14}$$

Note that (12) contains an additional term ρ^2/σ whose presence can be attributed to junction conditions imposed at the bulk-brane boundary. Because of this term the damping experienced by the scalar field as it rolls down its potential *dramatically increases* so that inflation can be sourced by potentials which are normally too steep to produce slow-roll.

Indeed the slow-roll parameters in braneworld models (for $V/\sigma \gg 1$) are[11]

$$\epsilon \simeq 4\epsilon_{\rm FRW}(V/\sigma)^{-1},\ \eta \simeq 2\eta_{\rm FRW}(V/\sigma)^{-1}, \tag{15}$$

illustrating that slow-roll ($\epsilon, \eta \ll 1$) is easier to achieve when $V/\sigma \gg 1$. Inflation can therefore arise for the very steep potentials associated with quintessence such as $V \propto e^{-\lambda\phi}$, $V \propto \phi^{-\alpha}$, etc. This gives rise to the intriguing possibility that both inflation and quintessence may be sourced by one and the same scalar field. Termed "quintessential inflation", these models have been examined in Refs. [4,6,10,13,14,21,29].

Other examples of Braneworld dark energy will be provided by Y. Shtanov in this meeting; see also Refs. [1,5,22].

2 Dark Energy and the Statefinder Diagnostic

In view of the considerable number of dark energy models suggested in the literature, it becomes meaningful to ask whether we can reconstruct the properties of DE from observations in a model independent manner. This indeed may be possible if one notices that the the Hubble parameter is related to the luminosity distance[30,27]

$$H(z) = \left[\frac{d}{dz}\left(\frac{d_L(z)}{1+z}\right)\right]^{-1}, \tag{16}$$

and that, in the case of quintessence, the scalar field potential as well as its equation of state can be directly expressed in terms of the Hubble parameter and its derivative[30,27]

$$\frac{8\pi G}{3H_0^2}V(x) = \frac{H^2}{H_0^2} - \frac{x}{6H_0^2}\frac{dH^2}{dx} - \frac{1}{2}\Omega_m x^3, \tag{17}$$
$$\frac{8\pi G}{3H_0^2}\left(\frac{d\phi}{dx}\right)^2 = \frac{2}{3H_0^2 x}\frac{d\ln H}{dx} - \frac{\Omega_m x}{H^2},\ x = 1+z, \tag{18}$$

$$w_\phi(x) \equiv \frac{p}{\varepsilon} = \frac{(2x/3)d\ln H/dx - 1}{1-(H_0^2/H^2)\,\Omega_m x^3}. \tag{19}$$

Both the quintessence potential $V(\phi)$ as well as the equation of state $w_\phi(z)$ may therefore be reconstructed provided the luminosity distance $d_L(z)$ is known to reasonable accuracy from observations.

The Sn inventory is increasing dramatically every year and so are increasingly precise measurements of galaxy clustering and the CMB. To keep pace with the better quality observational data which will soon become available and the increasing sophistication of theoretical modeling, a new diagnostic of DE called "Statefinder" was introduced in Ref. [26].

The Statefinder probes the expansion dynamics of the universe through higher derivatives of the expansion factor $\dddot{a}$ and is a natural companion to the deceleration parameter q which depends upon $\ddot{a}$. The Statefinder pair $\{r, s\}$ is defined as

$$r \equiv \frac{\dddot{a}}{aH^3} = 1 + \frac{9w}{2}\Omega_X(1+w) - \frac{3}{2}\Omega_X\frac{\dot{w}}{H}\,, \tag{20}$$
$$s \equiv \frac{r-1}{3(q-1/2)} = 1 + w - \frac{1}{3}\frac{\dot{w}}{wH}\,, \tag{21}$$

where q is the deceleration parameter

$$q = -\frac{\ddot{a}}{aH^2}. \tag{22}$$

Inclusion of the Statefinder pair $\{r, s\}$, increases the number of cosmological parameters to four: H, q, r and s. The Statefinder is a "geometrical" diagnostic in the sense that it depends upon the expansion factor and hence upon the metric describing space-time. An important property of the Statefinder is that spatially flat LCDM corresponds to the fixed point

$$\{r, s\}\Big|_{\text{LCDM}} = \{1, 0\} \ . \qquad (23)$$

Departure of a given DE model from this fixed point provides a good way of establishing the "distance" of this model from LCDM[2]. As demonstrated in Refs. [2,9,26,27] the Statefinder can successfully differentiate between a wide variety of DE models including the cosmological constant, quintessence, the Chaplygin gas, braneworld models and interacting DE models.

References

1. U. Alam and V. Sahni, arXiv: astro-ph/0209 443.
2. U. Alam, V. Sahni, T.D. Saini and A.A. Starobinsky *MNRAS* **344**, 1057 (2003), arXiv: astro-ph/0303009.
3. A. Arbey, J. Lesgourgues and P. Salati, *Phys. Rev.* D **68**, 023511 (2003).
4. E.J. Copeland, A.R. Liddle and J.E. Lidsey, *Phys. Rev.* D **64**, 023509 (2001).
5. C. Deffayet, G. Dvali and G. Gabadadze, *Phys. Rev.* D **65**, 044023 (2002), arXiv: astro-ph/0105068; C. Deffayet, S.J. Landau, J. Raux, M. Zaldarriaga and P. Astier, *Phys. Rev.* D **66**, 024019 (2002), arXiv: astro-ph/0201164.
6. K. Dimopoulos, *Phys. Rev.* D **68**, 123506 (2003), arXiv: astro-ph/0212264.
7. P.G. Ferreira and M. Joyce, *Phys. Rev. Lett.* **79**, 4740 (1997); P.G. Ferreira and M. Joyce, *Phys. Rev.* D **58**, 023503 (1998).
8. W. Fischler, A. Kashani-Poor, R. McNees and S. Paban, *JHEP* **0107**, 003 (2001), arXiv: hep-th/0104181.
9. V. Gorini, A. Kamenshchik and U. Moschella, arXiv: astro-ph/0209395.
10. G. Huey and J. Lidsey, *Phys. Lett.* B **514**, 217 (2001).
11. R. Maartens, D. Wands, B.A. Bassett and I.P.C. Heard *Phys. Rev.* D **62**, 041301 (2000).
12. R. Maartens, arXiv: gr-qc/0312059.
13. A.S. Majumdar, *Phys. Rev.* D **64**, 083503 (2001).
14. P.J.E. Peebles and A. Vilenkin, *Phys. Rev.* D **59**, 063505 (1999).
15. S.J. Perlmutter *et al.*, *Nature* **391**, 51 (1998).
16. S.J. Perlmutter *et al.*, *ApJ* **517**, 565 (1999).
17. L. Randall and R. Sundrum, *Phys. Rev. Lett.* **83**, 4690 (1999).
18. B. Ratra and P.J.E. Peebels, *Phys. Rev.* D **37**, 3406 (1988).
19. A.G. Riess *et al.*, *Astron. J.* **116**, 1009 (1998).
20. V. Sahni and L. Wang, *Phys. Rev.* D **62**, 103517 (2000).
21. V. Sahni, M. Sami and T. Souradeep, *Phys. Rev.* D **65**, 023518 (2002).
22. V. Sahni and Yu.V. Shtanov, *JCAP* **0311**, 014 (2003), arXiv: astro-ph/0202346.
23. V. Sahni, *Class. Quant. Grav.* **19**, 3435 (2002), arXiv: astro-ph/0202076.
24. V. Sahni, arXiv: astro-ph/0403324.
25. V. Sahni and A.A. Starobinsky, *IJMP* D **9**, 373 (2000).
26. V. Sahni, T.D. Saini, A.A. Starobinsky and U. Alam, *JETP* **77**, 201 (2003), arXiv: astro-ph/0201498.
27. T.D. Saini, S. Raychaudhury, V. Sahni and A.A. Starobinsky, *Phys. Rev. Lett.* **85**, 1162 (2000).
28. T. Shiromizu, K. Maeda and M. Sasaki, *Phys. Rev.* D **62**, 024012 (2000).
29. T. Shiromizu, T. Torii and T. Uesugi, arXiv: hep-th/0302223.
30. A.A. Starobinsky, *JETP* **68**, 757 (1998).
31. A.A. Starobinsky, *Grav. Cosmol.* **6**, 157 (2000).
32. L.A. Ureña-López and T. Matos, *Phys. Rev. D* **62**, 081302 (2000).
33. L.A. Ureña-López and A. Liddle, *Phys. Rev.* D **66**, 083005 (2002), arXiv: astro-ph/0207493.
34. S. Weinberg, *Rev. Mod. Phys.* **61**, 1 (1989).
35. C. Wetterich, *Nucl. Phys.* B **302**, 668 (1988).
36. Ya.B. Zel'dovich, *Sov. Phys. – Uspekhi* **11**, 381 (1968).
37. W. Zimdahl and D. Pavon, arXiv: gr-qc/0311067.
38. I. Zlatev, L. Wang and P.J. Steinhardt, *Phys. Rev. Lett.* **82**, 896 (1999).

CHAPTER 2: CONDENSED MATTER & STATISTICAL PHYSICS

ARNOLD TONGUES IN ONE-, AND MULTI-DIMENSIONAL MAPPINGS OF PHYSICAL SYSTEMS

N. S. ANANIKIAN and L. N. ANANIKYAN

Yerevan Physics Institute
Alikhanian Brothers 2, 375036 Yerevan, Armenia
E-mail: ananik@yerphi.am

We have obtained Arnold tongues for Ising and Potts models with winding number $w = \frac{1}{2}$ in one-dimensional mapping. Using the dynamical systems approach, we have got the Yang-Lee zeros for classical picture of hydrogen bond between $N - H \cdots O = C$ of helix-coil phase transition for polypeptides and proteins in thermodynamic limit on recursive zigzag ladder. We used a model of α-helical proteins or polypeptides systems to evaluate the energetics of organic solvents exposed $C - H \cdots O$ interactions known as non-classical helix-stabilizing ones. Applying multi-dimensional mapping on zigzag ladder, we got Arnold tongues for non-classical helix-coil phase transition for neutral points of mapping with angle $\varphi = \frac{5}{6}\pi$, winding number $w = \frac{5}{12}$ and $\varphi = \frac{3}{4}\pi$, winding number $w = \frac{3}{8}$ with $Q = 50$.

1 Introduction

[a]It is well established that the thermodynamic properties of a physical system can be derived from a knowledge of the partition function. Since the discovery of statistical mechanics, it has been a central theme to understand the mechanism how the analytic partition function for a finite-size system acquires a singularity in the thermodynamic limit when the system undergoes a phase transition. The answer to this question was given in 1952 by Lee and Yang in their seminal papers[1].

It was shown that phase transitions occur in the systems in which the continuous distribution of zeros of the partition function intersects the real axis in the thermodynamic limit.

For anti-ferromagnetic Potts models, by contrast, there are some tantalizing conjectures concerning the critical loci, but many aspects still remain obscure[2].

It is interesting to regard the line between paramagnetic and modulated phases. This line is defined by neutral points of dynamical mapping. Neutral points are defined as eigenvalues of Jacobian mapping with modules equal to one ($\lambda = e^{i\varphi}$). There are two types of modulated phases: commensurate and incommensurate ones. For commensurate phases, when $\varphi = \frac{p}{q}\pi$, there exist Arnold tongues. Typically, for multi-dimensional maps, the border of such regions (Arnold tongues) split into two branches in parameter space. Consequently Arnold tongues can be crossed, leading to a situation in which two or more different periodic orbits associated with different rotation (winding) numbers are found with the same parameter values[3]. These techniques for multi-dimensional mapping are used for non-classical helix-coil phase transition of anti-ferromagnetic Potts model for biopolymer, regarding on the recursive zigzag ladder.

The advantage of recursive lattices is that for the models formulated on them, the exact recurrence relations for branches of the partition function can be derived and the thermodynamic properties of ferromagnetic and anti-ferromagnetic ones may be studied in terms of dynamical systems. Recently, the investigation of the partition function zeros has become a powerful tool for studying phase transition and critical phenomena. Particularly, much attention is being attached to the study of zeros of partition function of helix-coil transition of biological macromolecules[4]. The key step in the folding of proteins is a formation of secondary structure elements such as α-helices. Traditionally, the transition from random coiled conformation to the helical state in DNA, RNA or proteins are described in the framework of Zimm-Bragg[5] type Ising model. But this type of one-dimensional model cannot account for non-trivial topology of hydrogen bonds.

As the simple example, in Sec. 2 we regard Arnold tongues of one-dimensional mapping in Ising and Potts models on recursive Bethe lattice. In Sec. 3 we have obtained Yang-Lee zeros for classical picture of hydrogen bond between $N - H \cdots O = C$ of helix-coil phase transition for polypeptides and pro-

[a]Talk presented by N. S. A. at the XIth Regional Conference on Mathematical Physics, May 2004, Tehran, Iran.

teins in thermodynamic limit on recursive zigzag ladder. In Sec. 4 we have obtained Arnold tongues for non-classical helix-coil phase transition.

2 Arnold Tongues in Ising and Potts Models

Let us regard the anti-ferromagnetic Ising and Potts models on the recursive Bethe lattice connected through sites. For Ising model the partition function can be written as

$$Z = \sum_{\{\sigma_0\}} \exp\{h\sigma_0\} g_n^q(\sigma_0), \tag{1}$$

where σ_0 is the central spin, $g_n(\sigma_0)$ the contribution of each lattice branch, h magnetic field and q coordination number[6]. $g_n(\sigma_0)$ is obviously expressed through $g_{n-1}(\sigma_1)$

$$g_n(\sigma_0) = \sum_{\sigma_1} \exp\{\frac{h\sigma_1 - \sigma_0\sigma_1}{T}\} g_{n-1}^2(\sigma_1), \tag{2}$$

for $q = 3$ and interaction between the spins is constant $J = -1$. Introducing the notation

$$x_n = \frac{g_n(+)}{g_n(-)}, \tag{3}$$

the recursion relation (2) can be rewritten in the form

$$x_n = f(x_{n-1}, T, h). \tag{4}$$

As is known, if the derivative of $f(x, T, h)$ is equal to -1, we will have a bifurcation point, corresponding to the second order phase transition for anti-ferromagnetic model. We define $v = e^{-\frac{2}{T}}$ and after a simple calculation we get the following system of equations

$$\begin{cases} x = \frac{v^h + vx^2}{v^{h+1} + x^2}, \\ \frac{2vx - 2x^2}{v^{h+1} + x^2} = -1. \end{cases} \tag{5}$$

Eliminating x we obtain

$$4v^2(v^{h+1} + 1)(v^h + v) = v^h(1 - v^2)^2. \tag{6}$$

Solving this equation we get

$$\begin{aligned} -\tfrac{2h}{T} &= -3\ln 2 + \tfrac{6}{T}, \\ &+ \ln\Big\{1 - 6v^2 - 3v^4 \\ &\pm\sqrt{(1 - 6v^2 - 3v^4)^2 - 64v^6}\Big\}. \end{aligned} \tag{7}$$

This equation define Arnold tongues between paramagnetic and modulated phases with winding number $w = 1/2$.

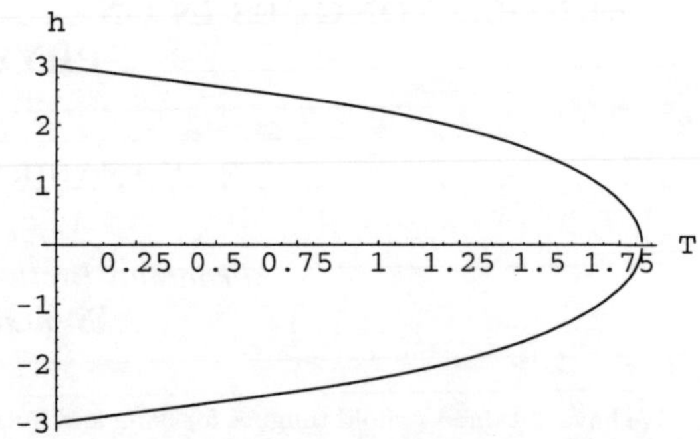

Figure 1. Arnold tongue for anti-ferromagnetic Ising model on recursive Bethe lattice with coordination number q=3.

The Arnold tongue begins at the temperature of $T = \frac{2}{\ln 3}$, when the external magnetic field $h = 0$, and ends ($T = 0$) at $h = \pm 3$ (see Fig. 1).

The same procedure we can perform for anti-ferromagnetic Potts model on recursive Bethe lattice with Hamiltonian

$$-\beta H = -\sum_{<i,j>} \delta(\sigma_i, \sigma_j) + h\sum_i \delta(\sigma_i, 1), \tag{8}$$

where σ_i takes the values 1, 2 and 3. Introducing the notation

$$x_n = \frac{g_n(*)}{g_n(1)}, \tag{9}$$

where $g_n(1)$ is the branch of partition function with central spin $\sigma = 1$ and $g_n(*)$ is the branch of partition function with central spin $\sigma \neq 1$. For the coordination number of the Bethe lattice $q = 3$, we obtain the following system of equations

$$\begin{cases} x = \frac{z^{-h} + (z+1)x^2}{z^{-h+1} + 2x^2}, \\ \frac{2(z+1)x - 4x^2}{z^{-h+1} + 2x^2} = -1. \end{cases} \tag{10}$$

Here again the derivative of $f(x, T, h)$ is equal to -1 which corresponds to the second order phase transition of anti-ferromagnetic model and where $z = e^{-\frac{1}{T}}$. The Arnold tongue begins at $z = \frac{1}{6}(\sqrt{17} - 3)$, when external magnetic field $h = 1.5$, and ends ($T = 0$) at $h = 0$ and $h = 3$ (see Fig. 2).

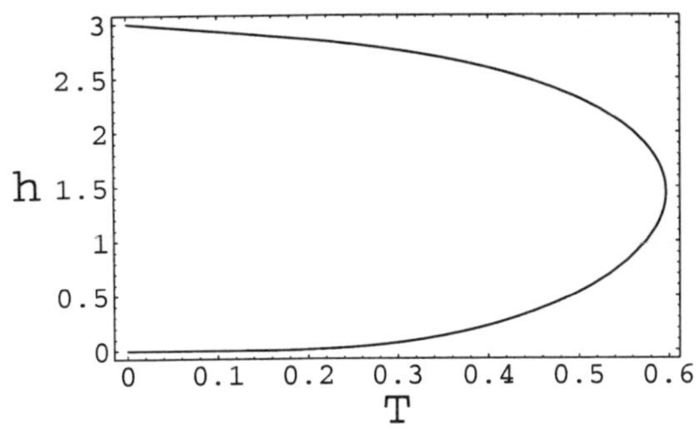

Figure 2. Arnold tongue for anti-ferromagnetic Potts model on recursive Bethe lattice with coordination number $q = 3$.

3 Multi-Dimensional Mapping and Yang-Lee Zeros for Biological Macromolecules

The helix-coil transition of DNA, RNA or proteins are described by Zimm-Bragg or Lifson-Roig type Ising model[5]. The relationship between the Lifson-Roig and Zimm-Bragg parameters can be established by using both theories to set up the statistical weight of an entire segment, consisting of a helix embedded between coil regions and equating the results[7]. But this type of one-dimensional model can not account for non-trivial topology of hydrogen bonds[11].

The backbone chain of the polypeptide molecule is shown (see Fig. 3). **R**(amino acid residue) denotes the side chain. Because of the planar structure of the amide group, almost whole conformational flexibility of the polypeptide backbone chain is determined by the rotation angles around the single bonds $N - C_\alpha$ and $C_\alpha - C$ which are usually denoted as φ and ψ, respectively. We formulate the three-site interaction Potts model for zigzag ladder[11]. To each pair of rotation angles (φ_i, ψ_i) a spin variable s_i is assigned which can obtain Q discrete values.

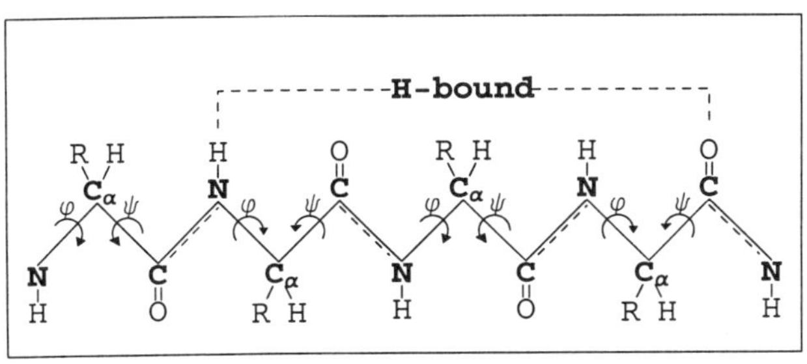

Figure 3. Zigzag ladder with H-bound.

When 3 successive rotation pairs (spins) are zero $(s_i = s_{i+1} = s_{i+2} = 0)$, an intramolecular H-bond appears which leads to some energy gain. This model can be described by the terms of zigzag ladder lattice model. The Hamiltonian of the system is written as

$$-\beta H = J \sum_{\Delta^i} \delta(s_{i-1}, 0)\delta(s_i, 0)\delta(s_{i+1}, 0)$$
$$+K \sum_{\Delta^i} [1 - \delta(s_{i-1}, 0)\delta(s_i, 0)\delta(s_{i+1}, 0)], \quad (11)$$

where $\beta = 1/kT$, J is the energy of intramolecular hydrogen bond, K is the energy of protein-solvent hydrogen bond, s_i denotes the Potts variable at the site i and takes the values $0, 1, 2, \cdots, Q-1$.

In the one-dimensional case we got one recursion relation. In this multi-dimensional case we obtain more than one recursion relations. Let us introduce the notations

$$z \equiv \alpha = \exp \frac{J}{kT},$$
$$\mu \equiv \beta = \exp \frac{K}{kT}, \quad (12)$$

$$z_n(0,0) = z_1^n; \quad z_n(0,*) = z_2^n;$$
$$z_n(*,0) = z_3^n; \quad z_n(*,*) = z_4^n. \quad (13)$$

Since $z_2^n = z_4^n$, then in fact we have only 3 equations instead of 4. We further introduce

$$x_n = \frac{z_1^n}{z_4^n},$$
$$y_n = \frac{z_3^n}{z_4^n}, \quad (14)$$

and obtain the two dimensional rational mapping relation for x_n and y_n

$$x_n = f_1(x_{n-1}, y_{n-1}),$$
$$y_n = f_2(x_{n-1}, y_{n-1}), \quad (15)$$

where

$$f_1(x,y) = \frac{\gamma x + (Q-1)}{y + (Q-1)},$$
$$f_2(x,y) = \frac{x + (Q-1)}{y + (Q-1)}, \quad (16)$$

with $\gamma = \alpha/\beta = z/\mu$.

In case of multi-dimensional rational mapping the fixed point x^*, y^* is attracting when the eigenvalues of Jacobian $|\lambda| < 1$, repelling, when $|\lambda| > 1$, and neutral, when $|\lambda| = 1$. So the system undergoes a phase transition when

$$\begin{vmatrix} \frac{\partial f_1}{\partial x} - \exp(\imath\varphi) & \frac{\partial f_1}{\partial y} \\ \frac{\partial f_2}{\partial x} & \frac{\partial f_2}{\partial y} - \exp(\imath\varphi) \end{vmatrix} = 0. \quad (17)$$

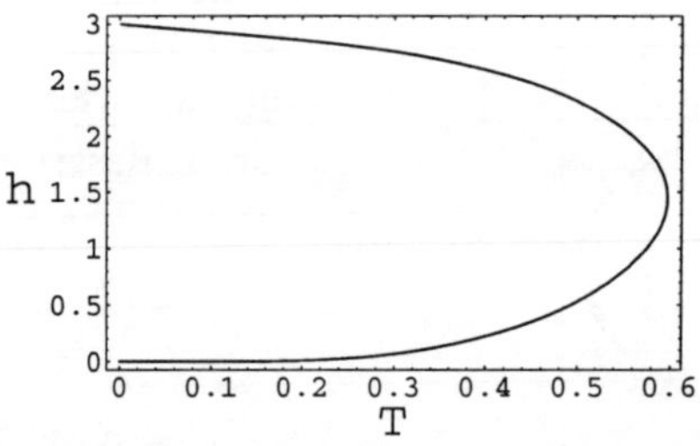

Figure 4. The Yang-Lee zeros: a) Q=9, J=2.0, T=0.56, K=0.76; b) Q=30, J=2.1, T=0.54, K=0.26; c) Q=40, J=2.4, T=0.56, K=0.334; d) Q=50, J=2.5, T=0.6, K=0.1523.

Using the same techniques as in the previous section for multi-dimensional rational mapping, after eliminating x and y from (16) and (17), we obtain the following equation for the partition function zeros

$$b_0 + b_1\cos(\varphi) + b_2\cos^2(\varphi) + b_3\cos^3(\varphi)[\cos(3\varphi) + \imath\sin(3\varphi)] = 0, \quad (18)$$

where

$$\begin{aligned} b_0 &= Q^2(\gamma-1) + (\gamma-1)\gamma^2 + Q^3(\gamma+1) + Q\gamma(\gamma^2+\gamma-2),\\ b_1 &= 2\{Q^3 + 2Q(\gamma-1) - Q^2(\gamma^2-1) + \gamma(\gamma^2+\gamma-2)\},\\ b_2 &= -4(Q-1)(\gamma-1)(2+Q+\gamma),\\ b_3 &= -8(Q-1)(\gamma-1). \end{aligned} \quad (19)$$

One can solve (18) for μ and find the Yang-Lee zeros of partition function with different parameters Q, J and T like in previous papers[12]. These parameters are different for each polypeptides and proteins. After making discrete values of Q and comparing with Ramachardan and Shceraga[13], we confirm that the circle in classical helix-coil transition does not cut the real axis. So we have not a real phase transition in polypeptides (proteins). According to phenomenological theory of Zimm-Bragg or Lifson-Roig, there is only pseudo-phase transition in 2-site (Ising) model. Our results describe the microscopic theory of helix-coil transition of polypeptides or proteins with non-trivial topology of hydrogen bonds and find Yang-Lee zeros of pseudo phase transition.Yang-Lee zeros of helix-coil transition for polyalanine, polyvaline and polyglysine was regarded[4]. The authors made Monte Carlo simulation technique and considered polypeptide chain up to $N = 30$ monomers and determine the (pseudo-)critical temperatures of the helix-coil transition in all-atom model of polypeptides. In this paper we find Yang-Lee zeros in microscopic theory for biological macromolecules in thermodynamic limit.

4 Arnold Tongues in Multi-Dimensional Mapping for Biological Macromolecules

Recently the thermodynamic stability in polypeptides can be described by the distance constraint model with many phenomenological parameters taking into account rearrangements of thermally fluctuating constrains that are independent of temperature and chain length[8]. Q-state Potts lattice model can be used[9] for the description of cold denaturation of proteins in solvents[10]. The organic solvents (trifluoroethanol, urea and hexafluoroisopropanol) played an important role in helix-stabilizing of proteins and polypeptides. We use the microscopic model on a zigzag ladder. We have taken into account this non-classical H-bond to each amino acid residue in protein or polypeptide by describing the interaction through discrete Q-state Potts model. Non-classical H-bonds may exist in every $C_\alpha - H$. So in terms of Q-state Potts model, if the value of spin is equal to zero ($s_i = 0$), then the non-classical H-bonds do not exist. Taking into account non-classical H-bonds interaction with solvent, the Hamiltonian may be written in the following way

$$\begin{aligned} -\beta H = J\sum_{\Delta^i}\delta(s_{i-1},0)\delta(s_i,0)\delta(s_{i+1},0)\\ +K\sum_{\Delta^i}[1-\delta(s_{i-1},0)\delta(s_i,0)\delta(s_{i+1},0)]\\ +K_1\sum_i[1-\delta(s_i,0], \end{aligned} \quad (20)$$

where K_1 is non-classical H-bond. It may be described by zigzag ladder with the triangle interactions (Δ^i) of the classical hydrogen bond (J), the solvent interaction (K) and single site interactions (i) of the non-classical H-bond (K_1).

Using the theory of dynamical systems for two-dimensional mapping, like in the previous section, we have obtained the separating line, which divides the coil (paramagnetic, disordered) phase from helix (modulated, ordered) one (see Fig. 5). Two examples of Arnold tongues for non-classical helix-stabilizing interaction with $Q = 50$ for $\varphi = \frac{5}{6}\pi$ $w = \frac{5}{12}$ and $\varphi = \frac{3}{4}\pi$ $w = \frac{3}{8}$ are shown on Figs. 6 and 7.

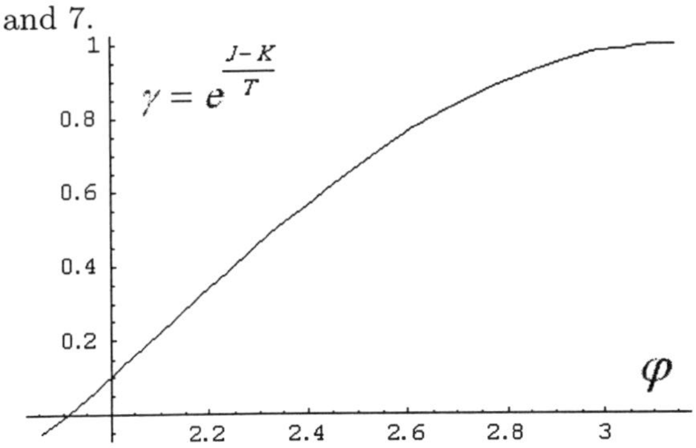

Figure 5. The line separating coil (paramagnetic, disordered) and helix (modulated, ordered) phases.

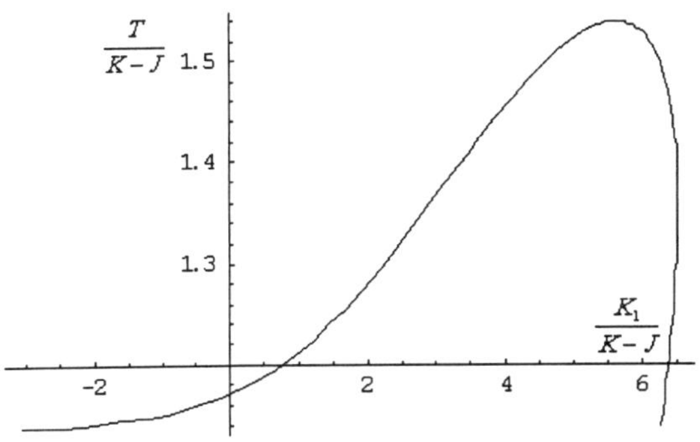

Figure 6. Arnold tongue with winding number $w = \frac{5}{12}, \varphi = \frac{5}{6}\pi$ and $Q = 50$ for non-classical helix-stabilizing interaction.

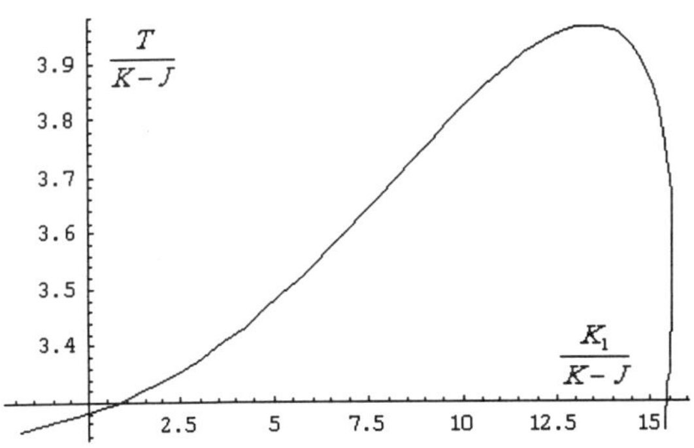

Figure 7. Arnold tongue with winding number $w = \frac{3}{8}, \varphi = \frac{3}{4}\pi$ and $Q = 50$ for non-classical helix-stabilizing interaction.

5 Conclusions

We exhibit the period doubling or Arnold tongue with $w = 1/2$ for anti-ferromagnetic Ising and Potts models on the Bethe (recursive) lattice. We have studied the Yang-Lee complex zeros of the partition function for classical helix-coil transition, using the dynamical system approach of multi-dimensional mapping. We confirm that in microscopic theory there is no real helix-coil phase transition, but only pseudo one. For non-classical helix-stabilizing interaction for proteins and polypeptides we have obtained the real phase transition between coil (paramagnetic, disordered) and helix (modulated, ordered) phases. We have got also two Arnold tongues with different winding numbers in microscopic theory.

Acknowledgments

N.A. thanks for the invitation at XIth Regional Conference on Mathematical Physics and IPM Spring Conference (May 3 - 6, Tehran, Iran). We thank R. Artuso, S. Gevorkyan, A. Grosberg, Sh. Hayryan, C.- K. Hu, N. Ivanov, and M. S. Li for interesting discussion and helpful interactions. This work was partly supported by the Cariplo Foundation The Landau Network-Centro Volta and ANSEF grant.

References

1. C.N. Yang and T.D. Lee, *Phys. Rev.* **87**, 404 (1952); T.D. Lee and C.N. Yang, *Phys. Rev.* **87**, 410 (1952).
2. P.W. Kasteleyn and C.M. Fortuin, *J. Phys. Soc. Japan* **26**, (Suppl.)11 (1969); C. M.Fortuin and P.W. Kasteleyn, *Physica* **57**, 536 (1972); J. Salas and A.D. Sokal, *J. Stat. Phys.* **104**, 609 (2001).
3. S. Coombes and P.C. Bressloff, *Phys. Rev.* E **60**, 2086 (1999); R.S. Mackay and C. Tresser, *Physica* D **19**, 206 (1986).
4. U.H.E. Hansmann and Y. Okamoto, *J. Chem. Phys.* **110**, 1267 (1999); N.A. Alves and U.H.E. Hansmann, *Phys. Rev. Lett.* **84**, 1836 (2000).
5. B.H. Zimm and I.K. Bragg, *J. Chem. Phys.* **31**, 526 (1959); S. Lifson and A. Roig, *J. Chem. Phys.* **34**, 1963 (1961).
6. R. Baxter in *Exactly Solved Models in Statistical Mechanics*, (New York: Academic Press); T.A.

Arakelyan, V.R. Ohanyan, L.N. Ananikyan, N.S. Ananikian and M. Roger, *Phys. Rev.* B **67**, 024424 (2003); N.S. Ananikian , S.K. Dallakian and B. Hu, *Complex Systems* **11**, 213 (1999); A.Z. Akheyan and N.S. Ananikian, *J. Phys.* A **29**, 721 (1996); N.S. Ananikian *et al. Fractals* **5**, 175 (1997).

7. H. Qian and J.A. Schellman, *J. Phys. Chem.* **96**, 3987 (1992).
8. D.J. Jacobs, S. Dallakyan, G.G. Wood and A. Heckathorne, *Phys. Rev.* E **68**, 061109 (2003).
9. G. Salvi and P. Des Los Rios, *Phys. Rev. Lett.* **91**, 258102 (2003); G. Salvi, S. Mölbert and P. Des Los Rios, *Phys. Rev.* E **66**, 061911 (2002); P. Des Los Rios and G. Caldarelli, *Phys. Rev.* E **62**, 8449 (2000); M.I. Marqués, J. M. Borreguero, H.E. Stanley and N.V. Dokholyan, *Phys. Rev. Lett.* **91**, 138103 (2003).
10. C.B. Anfinsen, *Science* **181**, 223 (1973); C.N. Pace and Ch. Tanford, *Biochemistry* **181**, 198 (1968); G.P. Privalov and P.L. Privalov, *Methods Enzymol.* **323**, 31 (2000).
11. N.S. Ananikian, S.A. Hajryan, E.S. Mamasakhlisov, V.F. Morozov. *Biopolymers* **30**, 357 (1990).
12. R.G. Ghulghazaryan, N.S. Ananikian and P.M.A. Sloot, *Phys. Rev.* E **66**, 046110 (2002); A. Alahverdian, N.S. Ananikian, S. Dallakian, *Phys. Rev.* E **57**, 2452 (1998); R.G. Ghulghazaryan and N.S. Ananikian, *J. Phys.* A **36**, 6297 (2003).
13. G.N. Ramachandran, C. Ramakrishnan and V. Sasisekharan, *J. Mol. Biol.* **7**, 95 (1963); J.T. Edsall, P.J. Flory, J.C. Kendrew, A.M. Liquory, G. Nementhy, G.N. Ramachandran and H.A. Scheraga, *J. Mol. Biol.* **15**, 399 (1966).

GENERALIZED INTEGRABLE MULTI-SPECIES REACTION-DIFFUSION PROCESSES

M. ALIMOHAMMADI

Physics Department, University of Tehran
North Karegar Ave., 14395 Tehran, Iran
E-mail: alimohmd@ut.ac.ir

We consider the most general boundary condition for the multi-species asymmetric exclusion processes on a one-dimensional lattice. In this way we may introduce the various interactions in which the number of particles is constant in time, including ones which have been studied yet and the new ones. In these new models, the particles have simultaneous diffusion, the two-particle interactions $A_\alpha A_\beta \rightarrow A_\gamma A_\delta$, and the n-particle extended drop-push interaction. As well as requirement of satisfying of the two-particle S-matrices in spectral Yang-Baxter equation, some constraints are obtained on reaction rates to ensure the consistency of evolution equations. In two-species case, we obtain the explicit solutions of these constraints and equations.

[a]One-dimensional asymmetric simple exclusion processes (ASEP) have been shown to be of physical interest in various problems in the recent years. As a first step, the totally ASEP has been solved exactly in Ref. [1]. In this simple model, each lattice site is occupied by at most one particle and particles hop with rate one to their right-neighboring sites if they are not occupied. The model is completely specified by a master equation and a boundary condition, imposed on probabilities appears in the master equation. The coordinate Bethe Ansatz has been used to show the factorization of the N-particle scattering matrix to the two-particle matrices. By choosing other suitable boundary conditions, without changing the master equation, one may study the more complicated reaction-diffusion processes, even with long-range interaction. See for example Ref. [2] in which the drop-push model has been studied by this method.

If one considers the model with more-than-one species, the situation becomes more complicated. The source of complexity is the above mentioned factorization of N-particle scattering matrix, which in these case restricts the two-particle S-matrices to satisfy some kind of spectral Yang-Baxter (SYB) equation.

Here, we are going to study the most general multi-species model, *i.e.* the most general boundary condition, where *all* the previous known models are the special cases of it. In its general form, the reactions are

$$\begin{aligned} A_\alpha \emptyset &\rightarrow \emptyset A_\alpha \quad \text{with rate } 1, \\ A_\alpha A_\beta &\rightarrow A_\gamma A_\delta \quad \text{with rate } c_{\gamma\delta}^{\alpha\beta}, \\ A_\alpha A_\beta \emptyset &\rightarrow \emptyset A_\gamma A_\delta \quad \text{with rate } b_{\gamma\delta}^{\alpha\beta}, \\ &\vdots \end{aligned} \tag{1}$$

where the dots indicate the other drop-push reactions with n adjacent particles, in which in the meantime the types of the particles can also be changed. We show that the reaction rates must satisfy some specific constraints, in order that we have a set of consistent evolution equations. The two-particle S-matrices of this general multi-species model must also satisfy SYB equation.

Consider a p-species system with particles $A_1, \cdots, A_p$. The basic quantities, that we are interested in, are the probabilities $P_{\alpha_1\cdots\alpha_N}(x_1,\cdots,x_N;t)$ for finding at time t the particle of type α_1 at site x_1, particle of type α_2 at site x_2, etc. We take these functions to define probabilities only in the physical region $x_1 < x_2 < \cdots < x_N$. The most general master equation for an asymmetric exclusion process is

$$\begin{aligned} \frac{\partial}{\partial t} P_{\alpha_1\cdots\alpha_N}(x_1,\cdots,x_N;t) &= P_{\alpha_1\cdots\alpha_N}(x_1-1,\cdots,x_N;t) \\ &+\cdots + P_{\alpha_1\cdots\alpha_N}(x_1,\cdots,x_N-1;t) \\ &-NP_{\alpha_1\cdots\alpha_N}(x_1,\cdots,x_N;t). \end{aligned} \tag{2}$$

This equation describes a collection of N particles drifting to the right with unit rate. The master equation (2) is only valid for

$$x_i < x_{i+1} - 1, \tag{3}$$

since for $x_i = x_{i+1} - 1$, there will be terms with $x_i = x_{i+1}$ on the r.h.s. of (2), which are out of the physical region. One can, however, assume that (2) is correct for all the physical region $x_i < x_{i+1}$, and impose certain boundary conditions for $x_i = x_{i+1}$.

[a]Talk presented at the XIth Regional Conference on Mathematical Physics, May 2004, Tehran, Iran.

These boundary conditions determine the nature of the interactions between particles. Now the question is, what are the possible boundary conditions? It can be shown that the most general boundary condition is (see Ref. [3] for more details)

$$P_{\alpha_1\alpha_2}(x,x) = \sum_\beta b^{\beta_1\beta_2}_{\alpha_1\alpha_2} P_{\beta_1\beta_2}(x-1,x) + \sum_\beta c^{\beta_1\beta_2}_{\alpha_1\alpha_2} P_{\beta_1\beta_2}(x,x+1), \quad (4)$$

where β stands for $(\beta_1\beta_2)$. These b and c matrices introduce interactions to particles. The conservation of probability results in

$$\sum_\alpha (b+c)^{\beta_1\beta_2}_{\alpha_1\alpha_2} = 1. \quad (5)$$

The simple exclusion processes of Ref. [1] is an example of this model with $p=1$ and $b=0$, the drop-push model with equal rate is $p=1$ and $c=0$ case, and with non-equal rate, is an example of one-species case of Refs. [1] and [2].

If we consider the matrix c as a diagonal matrix, it can be shown that we encounter two cases. In the first case, the matrix b must be also diagonal, and thus the model reduces to the ordinary, *i.e.* single-species, drop-push model with variable rate. In the second case, c must be zero matrix, and thus the model becomes the known extended drop-push model. Hence considering c as a diagonal matrix, does not lead to any new model.

If matrix c is non-diagonal, then considering $\dot{P}_{\alpha_1\alpha_2}(x,x+1)$ and using (2), (4) and (5), one can show that the resulting evolution equation describes the following two-particle reactions

$$\begin{aligned} A_\alpha\emptyset &\to \emptyset A_\alpha &&\text{with rate } 1,\\ A_\alpha A_\beta &\to A_\gamma A_\delta &&\text{with rate } c^{\alpha\beta}_{\gamma\delta},\\ A_\alpha A_\beta\emptyset &\to \emptyset A_\gamma A_\delta &&\text{with rate } b^{\alpha\beta}_{\gamma\delta}. \end{aligned} \quad (6)$$

In more-than-two particle reactions, one first need to know $P_{\alpha_1\alpha_2\alpha_3}(x-1,x,x)$. This can be achieved only when the matrices b and c satisfy

$$\sum_{\beta_2} c^{\gamma_1\gamma_2}_{\alpha_1\beta_2} b^{\beta_2\beta_3}_{\alpha_2\alpha_3} = 0. \quad (7)$$

Then, considering $\dot{P}_{\vec{\alpha}}(x,x+1,x+2)$, one can show that besides the reactions (6), the model describes

$$A_{\beta_1}A_{\beta_2}A_{\beta_3}\emptyset \to \emptyset A_{\alpha_1}A_{\alpha_2}A_{\alpha_3} \qquad \text{with rate } b^{\vec{\beta}}_{\vec{\alpha}}, \quad (8)$$

if

$$\sum_\beta b^{\vec{\alpha}}_{\vec{\beta}} := \sum_{\beta\gamma} b^{\alpha_1\alpha_2}_{\beta_1\gamma} b^{\gamma\alpha_3}_{\beta_2\beta_3} = \sum_\beta b^{\alpha_1\alpha_2}_{\beta_1\beta_2}. \quad (9)$$

This is the last constraint that must be satisfied by the elements of matrices b and c. It can be shown that the more-than-three adjacent particles probabilities are consistent with following reactions

$$\begin{aligned} A_\alpha\emptyset &\to \emptyset A_\alpha \quad \text{with rate } 1,\\ A_\alpha A_\beta &\to A_\gamma A_\delta \quad \text{with rate } c^{\alpha\beta}_{\gamma\delta},\\ A_{\alpha_0}\cdots A_{\alpha_n}\emptyset &\to \emptyset A_{\gamma_0}\cdots A_{\gamma_n}\\ &\qquad \text{with rate } (b_{n-1,n}\cdots b_{0,1})^{\alpha_0\cdots\alpha_n}_{\gamma_0\cdots\gamma_n}, \end{aligned} \quad (10)$$

if the constraints (5), (7) and, (9) are satisfied. In (10), we use the following definition

$$b_{k,k+1} = 1\otimes\cdots\otimes 1\otimes \underbrace{b}_{k,k+1} \otimes 1\otimes\cdots\otimes 1. \quad (11)$$

The important point is that we need not any further constraint.

To obtain the probabilities, one may use the Bethe Ansatz. It can be shown that this is possible, only if the matrices b and c are such that the following $p^2\times p^2$ matrix

$$S(z_1,z_2) = -(1-z_2^{-1}b-z_1c)^{-1}(1-z_1^{-1}b-z_2c), \quad (12)$$

satisfies the spectral Yang-Baxter equation

$$\begin{aligned} &S_{12}(z_2,z_3)S_{23}(z_1,z_3)S_{12}(z_1,z_2) =\\ &\qquad S_{23}(z_1,z_2)S_{12}(z_1,z_3)S_{23}(z_2,z_3). \end{aligned} \quad (13)$$

In $p=2$, it can be shown that there are two sets of solution for the constraints (5), (7) and (9). If one considers the simplest one and determines the parameters such that Eq. (13) is also satisfied, one finds that the following reactions are the only integrable model in this set:

$$\begin{aligned} A\emptyset &\to \emptyset A,\\ B\emptyset &\to \emptyset B,\\ AB &\to BA,\\ BA\emptyset &\to \emptyset AB,\\ BB\emptyset &\to \emptyset BB,\\ BAA\emptyset &\to \emptyset AAB,\\ BAB\emptyset &\to \emptyset ABB,\\ BBA\emptyset &\to \emptyset BAB,\\ BBB\emptyset &\to \emptyset BBB,\\ &\vdots \end{aligned} \quad (14)$$

The dots indicates the more-than-three particle drop-push reactions which are specified by (10) and (11), and all reactions occur with the rate one. We denote $A \equiv A_1$ and $B \equiv A_2$.

References

1. G.M. Schütz, *J. Stat. Phys.* **88**, 427 (1997).
2. M. Alimohammadi, V. Karimipour and M. Khorrami, *Phys. Rev.* E **57**, 6370 (1998).
3. M. Alimohammadi, arXiv: cond-mat/0403672.

TWO-BAND GINZBURG-LANDAU THEORY AND ITS APPLICATION TO RECENTLY DISCOVERED SUPERCONDUCTORS

I. N. ASKERZADE

Institute of Physics, Azerbaijan National Academy of Sciences
Baku-Az1143, Azerbaijan
and
Department of Physics, Faculty of Sciences, Ankara University
06100, Tandogan, Ankara, Turkey
E-mail: Iman.Askerzade@science.ankara.edu.tr

Temperature dependence of the upper critical field $H_{c2}(T)$, lower critical field $H_{c1}(T)$, thermodynamic magnetic field $H_{cm}(T)$ and pair breaking critical current density $j_c(T)$ are studied in the vicinity of T_c by using a two-band Ginzburg-Landau (G-L) theory. The results are shown to be in good agreement with experimental data for the superconducting magnesium diboride, MgB_2, and non-magnetic borocarbides $LuNi_2B_2C$, YNi_2B_2C. In addition, two-band G-L theory was applied for the calculation of specific heat jump, which is smaller than in single-band G-L theory. Peculiarities of Little-Parks effect in two-band G-L theory are also studied. It is shown that quantization of the magnetic flux and relation between surface magnetic field $H_{c3}(T)$ and upper critical field $H_{c2}(T)$ is the same as in single band G-L theory. Generalization of two-band Ginzburg-Landau theory to the layered case is discussed.

1 Introduction

[a]The recently discovered superconductor[1], MgB_2, has attracted attention for both experimental and theorethical works due to the fact that it holds the highest superconducting transition temperature of about $T_c = 40K$ for a binary compound of a relatively simple crystal structure. Calculations of the band structure and the phonon spectrum predict a double energy gap[2,3], a larger gap attributed to two-dimensional p_{x-y} orbitals and smaller gap attributed to three-dimensional p_z bonding and anti-bonding orbitals. Two-band characteristic of the superconducting state in MgB_2 is clearly evident in the recently performed tunnel measurements[4,5] and specific heat measurement[6]. Another class of two-band superconductors are the non-magnetic borocarbides[7] Lu $(Y)Ni_2B_2C$.

Magnetic phase diagram for a bulk samples MgB_2 and non-magnetic borocarbides Lu $(Y)Ni_2B_2C$ has also been of interest to researchers. In contrast to conventional superconductors, the upper critical field for a bulk samples of MgB_2 and borocarbides Lu $(Y)Ni_2B_2C$ has a positive curvature near T_c. To understand the nature of the unusual behavior at a microscopic level, a two-band Eliashberg model of superconductivity was first proposed by Shulga *et al.*[7] for $LuNi_2B_2C$ and YNi_2B_2C and recently[8], for the MgB_2. Two-band G-L model for a bulk MgB_2 was successfully applied to fit the experimental results of the temperature dependence of upper and lower critical fields for MgB_2 and non-magnetic borocarbides[9,10,11].

In this paper, we apply two-band G-L theory to determine the temperature dependence of superconducting state parameters of MgB_2 and non-magnetic borocarbides Lu $(Y)Ni_2B_2C$. We show that the presence of two-order parameter in the theory gives a good approximation of experimental data.

2 Basic Equations

In the presence of two order parameters (OP) in a superconductor, G-L functional free energy can be written as[9,10,11]

$$F[\Psi_1,\Psi_2] = \int d^3r(F_1 + F_{12} + F_2 + H^2/8\pi), \quad (1)$$

with

$$F_i = \frac{\hbar^2}{4m_i}|(\nabla - \frac{2\pi iA}{\Phi_0})\Psi_i|^2 + \alpha_i(T)\Psi_i^2 + \frac{\beta_i}{2}\Psi_i^4, \quad (2)$$

and

$$F_{12} = \varepsilon(\Psi_1\Psi_2^* + c.c.) + \varepsilon_1(\nabla + \frac{2\pi iA}{\Phi_0})\Psi_1^*(\nabla - \frac{2\pi iA}{\Phi_0})\Psi_2 + c.c.). \quad (3)$$

Here, m_i denotes the effective mass of the carriers belonging to band i (i=1;2). F_i is the free energy of separate bands. The coefficient α is given

[a]Talk presented at the XIth Regional Conference on Mathematical Physics, May 2004, Tehran, Iran.

as $\alpha_i = \gamma_i(T - T_{ci})$, which depends on temperature linearly, γ is the proportionality constant, while the coefficient β is independent of temperature. $\vec{H}$ is the external magnetic field and $\vec{H} = curlA$. The quantities ε and ε_1 describe inter-band interaction of two order parameters and their gradients, respectively.

Minimization of the free energy functional with respect to the order parameters yields G-L equations for two-band superconductors in one dimension, $A = (0, Hx, 0)$

$$-\frac{\hbar^2}{4m_1}(\frac{d^2}{dx^2} - \frac{x^2}{l_s^4})\Psi_1 + \alpha_1(T)\Psi_1 + \varepsilon\Psi_2$$
$$+\varepsilon_1(\frac{d^2}{dx^2} - \frac{x^2}{l_s^4})\Psi_2 + \beta_1\Psi_1^3 = 0, \quad (4)$$

$$-\frac{\hbar^2}{4m_2}(\frac{d^2}{dx^2} - \frac{x^2}{l_s^4})\Psi_2 + \alpha_2(T)\Psi_2 + \varepsilon\Psi_1$$
$$+\varepsilon_1(\frac{d^2}{dx^2} - \frac{x^2}{l_s^4})\Psi_1 + \beta_2\Psi_2^3 = 0, \quad (5)$$

where $l_s^2 = \frac{\hbar c}{2eH}$ is the so-called magnetic length.

3 Results

The two-band G-L equations (4a) and (4b) was applied to determine the temperature dependence of SC state parameters in non-magnetic borocarbides MgB_2, $LuNi_2B_2C$ and YNi_2B_2C: a) upper critical field $H_{c2}(T)$, b) lower critical field $H_{c1}(T)$, c) thermodynamic magnetic field $H_{cm}(T)$, d) pair-breaking current density $j_c(T)$. The choice of the $T_{c1} = 20K$,$T_{c1} = 10K$, $\varepsilon^2 = 3/8, \varepsilon_1 = 0.0976$ corresponds to MgB_2. According microscopical calculations ratio of masses in different bands is equal $x = 3$. In the case of non-magnetic borocarbides $LuNi_2B_2C$ and YNi_2B_2C, we have following set of parameters: $T_{c1} = 9.8K$, $T_{c1} = 2.3K$, $\varepsilon^2 = 0.33, \varepsilon_1 = 0.12$ and $T_{c1} = 10K$,$T_{c1} = 1.825K$, $\varepsilon^2 = 0.33, \varepsilon_1 = 0.132$, respectively. Ratio of masses for non-magnetic borocarbides is equal $x = 5$.[7]

We conclude that the two-band G-L theory explains the reduced magnitude of the specific heat jump and the small slope of the thermodynamic magnetic field in MgB_2. It is shown that, the relation between upper critical field and so-called surface critical field is the same as in the case of single-band superconductors. Temperature dependence of surface critical field of two-band superconductors must give positive curvature. Quantization of magnetic flux in the case of two-band SC remains the same as in single-band SC. However, periodicity of Little-Parks oscillations of T_c in two-band superconductors is absent. Briefly, is also discussed the possibility of generalization TB GL theory to the case of layered anisotropy.

References

1. J. Nagamatsu, N. Nakagava, T. Muranaka, Y. Zenitani and J. Akimitsu, *Nature* **410**, 63 (2001).
2. J. Kortus, I.I. Mazin, K.D. Belashchenko, V.P. Antropov and L.L. Boyer, *Phys. Rev. Lett.* **87**, 4656 (2001).
3. A. Liu, I.I. Mazin and J. Kortus, *Phys. Rev. Lett.* **87**, 087005 (2001).
4. X.K. Chen, M.J. Konstantinovich, J.C. Irwin, D.D. Lawriea and J.P. Frank, *Phys. Rev. Lett.* **87**, 157002 (2001).
5. H. Giublio, D. Roditchev, W. Sacks, R. Lamy, D.X. Thanh, J. Kleins, S. Miraglia, D. Fruchart, J. Markus and P. Monod, *Phys. Rev. Lett.* **87**, 177008 (2001).
6. F. Bouquet, R.A. Fisher, N.E. Phillips, D.G. Hinks and J.D. Jorgensen *Phys. Rev. Lett.* **87**, 047001 (2001).
7. S.V. Shulga, S.-L. Drechsler, K.-H. Müller, G. Fuchs, K. Winzer, M. Heinecke and K. Krug *Phys. Rev. Lett.* **80**, 1730 (1998).
8. S.V. Shulga, S.-L. Drechsler, H. Echrig, H. Rosner and W. Pickett, arXiv: cond-mat/0103154.
9. I.N. Askerzade, N. Guclu and A. Gencer, *Supercond. Sci. Techn.* **15**, L13 (2002).
10. I.N. Askerzade, N. Guclu, A. Gencer and A. Kiliç, *Supercond. Sci. Techn.* **15**, L17 (2002).
11. I.N. Askerzade, *Physica* **C397**, 99 (2003).

AB-INITIO AND HUBBARD-SHAM MODEL CALCULATIONS OF BAND STRUCTURE OF *GeSe*

G. S. ORUDZHEV, Z. A. JAHANGIRLI

Azerbaijan Technical University
Gusein Javid Avenue, 25, Az 1073, Baku, Azerbaijan
E-mail: horucov@physics.ab.az

D. A. GUSEINOVA, F. M. HASHIMZADE

Institute of Physics, ANAS
Gusein Javid Avenue, 33, Az 1143, Baku, Azerbaijan

The band structure of the *GeSe* crystals has been calculated by the Density Functional Theory (DFT) in the Local Density Approximation and using the Hubbard-Sham model screening. The nonlocal pseudo-potentials constructed by procedure offered G.B. Bachelet *et al.* For the exchange-correlation energy and potentials we use the Ceperley-Alder results parametrized by Perdew and Zunger. The results received within the Hubbard-Sham model screening with selected parameters of the charge distribution around each particular ion coincide with the results received from the Density Functional Theory.

[a]Due to a number of interesting physical properties associated with the strong anisotropy of crystal structure, the layered semiconductor compounds *GeSe* attracts increased attention. The presence of weak interlayer bonds causes interest to carrying out intensive researches on photoelectric and optical properties of this crystal.

The semiconductor compound *GeSe* crystallizes in orthorhombic lattice with space group of symmetry $D_{2h}^{16}(P_{cmn})$. The crystal axis c is perpendicular to layers, the axes a and b lay on the plane of a layer. The crystals can easily cleavage perpendicularly axes c. The elementary cell of *GeSe* contains four formula units. The structure consists of two layers, each of which consisting of two goffered planes from atoms cations and anions. According to Ref. [1], bonds in these crystals are formed of three hybridized c-functions of each atom with the small contribution of s-functions. The partial participation of s-functions weakens the c-functions in three primary directions and amplifies them in the others three directions. This leads to a trivalent homopolar bond which causes a deviation of the structure from the cubic structure of *NaCl*.

The band structure of the *GeSe* crystals has been calculated by the Density Functional Theory in the Local Density Approximation. In this calculation, the Ceperley-Alder's results parametrized by Perdew and Zunger[2] were used for the exchange-correlation energy. The nonlocal pseudo-potentials were calculated from the first principles by the procedure offered by G.B. Bachelet *et al.*[3]. In calculating the charge distribution of valence electrons, the integration on Brillouin zone is replaced by summation over special points using the Monhorst-Pack's scheme. About 2300 plane waves were used to decompose the wave function. Thus the maximal kinetic energy of plane waves taken into account was 20 Ry.

In Ref. [4] the Brillouin zone of simple orthorhombic lattice is presented. The coordinates x, y and z coincide with crystal axes b, c and a, respectively.

The calculation of band structure was carried out along high symmetry lines in Brillouin zone.

The calculated energy band structure of *GeSe* is presented in Fig. 1. The characteristic feature of the energy band structure of *GeSe* is the existence of three groups of valence bands.

There are two pronounced maxima in the valence bands. The absolute maximum of the valence bands is located on the symmetry line $V(0,0,k)$ and corresponds to the irreducible representation V_1, and other maximum at the center of the Brillouin zone corresponds to the irreducible representation Γ_6. Three basic minima of conduction bands are

[a]Talk presented by Z.A.J. at the XIth Regional Conference on Mathematical Physics, May 2004, Tehran, Iran.

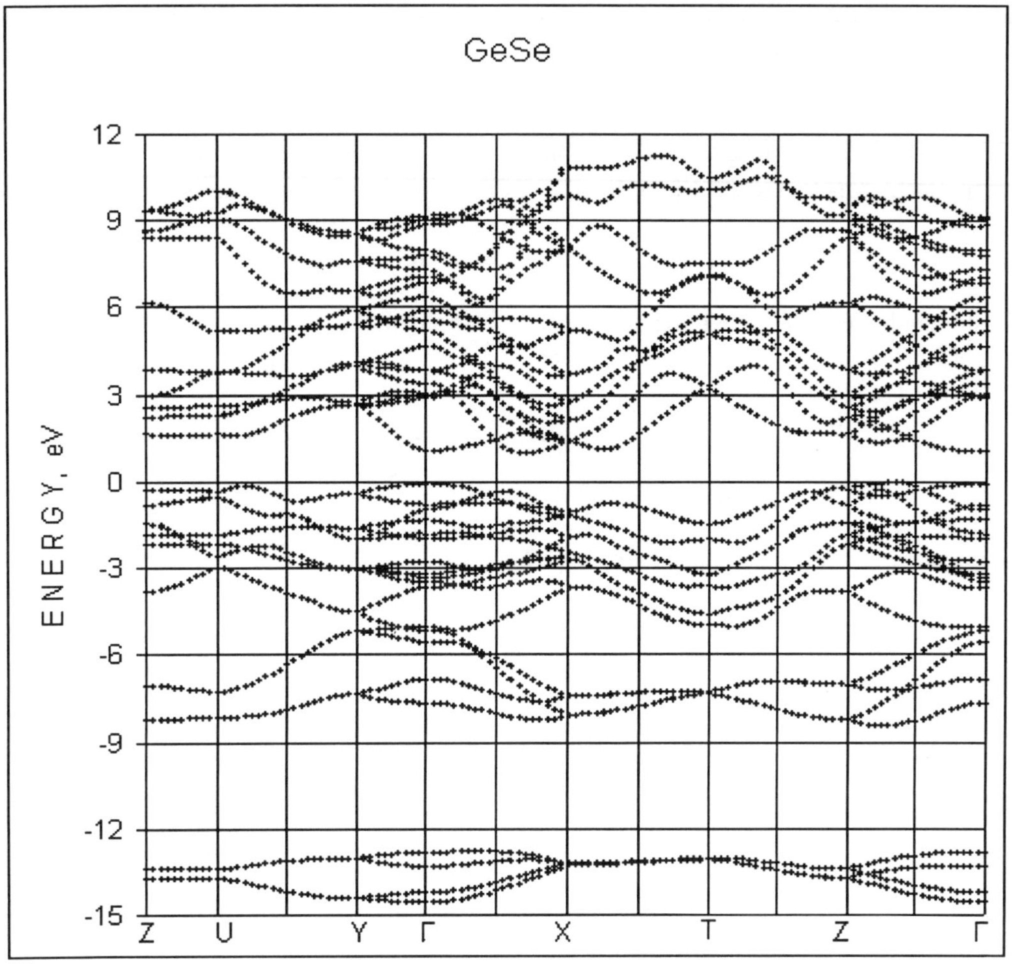

located on the lines L and D.

The lowest indirect band gap $E_{gi} = 1.02\ eV$ corresponds to the transition $\Gamma_6 - V_1$. This result is in good agreement with experimental one $1.1\ eV$ (see Refs. [5], and [6]).

The band structure shows strong anisotropy along the symmetry lines D, L and V. The dispersion of valence bands along symmetry lines F and A' is rather small in comparison with dispersion of bands along other lines. Because of strong anisotropy of crystal structure, along the F and D symmetry lines, which are perpendicular to planes of layers, the dispersion is not enough, since the interaction between layers is too small.

According to the photoelectron spectra of $GeSe$,[7,8,9] three peaks were observed in the density of valence band states. The lowest group consisting of four bands, is separated from others by an energy gap about $5eV$ and almost does not influence the semiconductor properties of the crystal. The group-theoretical analysis shows that, the lowest valence bands located about $-(13-14)\ eV$ originates from the s-states of Se. The following group of valence bands is located at an energy level about $-7\ eV$ and originates predominantly from the s-states of Ge. The large group of twelve bands situated on the top of the valence bands originates predominantly from the c-states of anions and cations.

The band structure calculations are carried out also by taking into account the screening and exchange-correlation effects within the framework of dielectrical formalism theory on the Hubbard-Sham model[10]. In the lowest order of the perturbation theory, the screening charge depends linearly on the potential of bare ions and each Fourier components is screened independently. We have shown, that if the ion charge of each atom is screened individually and thus instead of an average number of valence electrons on atom, the valency of the given ion is taken into account, the results of model calculations will be in better agreement to the results arising from the DFT method.

The use of the model of screening saves enormous machine time in calculating the optical functions.

References

1. F.M. Gashimzade and V.E. Kharchiev, *Fiz. Tverd. Tela (Russian Journal of Solid State Physics)* **4**, 434 (1962).
2. J. Perdew and A. Zunger, *Phys. Rev.* B **23**, 5048 (1981).
3. G.B. Bachelet, D.R. Hamann and M. Schlüter, *Phys. Rev.* B **26**, 4199 (1982).
4. G. Valiukonis, F.M. Gashimzade, D.A. Guseinova, G. Krivaite, A.M. Kulibekov, G.S. Orudzhev and A. Sileika, *Phys. Stat. Sol.* (b) **117**, 81 (1983).
5. F. Lukes, *Czech. J. Phys.* B **18**, 784 (1968).
6. A.M. El-Korashy, A.P. Lambros, A. Thanailakis and N.A. Economou, *Solid State Commun.* **19**, 759 (1976).
7. P.C. Kemeny, J. Azoulay, M. Cardona and L. Ley, *Nuovo Cimento* 39 **B**, 709 (1977).
8. R.B. Shalvoy, G.B. Fisher and P.J. Stiles, *Phys. Rev.* B **15**, 2021 (1977).
9. A. Kosakov, H. Neumann and G. Leonhardt, *J. Electron Spectroscopy related Phenomena* **12**, 181 (1977).
10. V. Heine, M.L. Cohen and D. Weire in *Pseudopotential Theory*, (Moscow, Mir, 1973).

INVERSE PHOTO EMISSION SPECTROSCOPY

A. A. HOSSEINI* and P.T. ANDREWS
*Physics Department, Faculty of Science, Mazandaran University
P.O. Box 47415-416, Babolsar, Iran
E-mail: hos-a-p1@umz.ac.ir

In this work we intended to design and construct a spectrometer called Inverse Photo Emission Spectrometer (I.P.E.S.), which could be used to investigate the density of unoccupied state near the Fermi level (F.L.) in crystalline and non-crystalline materials.

1 Introduction

[a]Until 1980 great attention has been focused on the distribution and density measurement of the occupied electronic state in solids. Our understanding of these states was at an advanced stage theoretically and experimentally. The unoccupied states, however, have not been so much studied before 1980. Although some works have been done to study unoccupied state, but none match the power and range for filled state of the angle resolved PE experiments.

If the energy is above, the vacuum level bands can be examined by LEED. Conservation of $k_{\|}$ holds enabling a specific selection of k_z bands to be made by changing the angles of incident.

Band gaps are detected as peaks in the reflectivity *vs.* energy plots but it is not possible to follow the dispersion of individual band. Techniques used to study unoccupied state till 1980 and the disadvantages of these techniques can be summarized as:

1. *Secondary Electron Emission:* Some information is available here as (LEED). Because of a factor of $1 - R(E)$ in the emission probability (R is the total LEED reflection), the technique is limited to low energy unoccupied states.

2. *P.E.:* It also investigates the unoccupied states because these provide the means of escape from crystal. P.E. was probably the best technique for examining empty states, but was restricted to those states which can escape from the crystal and for interpretation assumes a knowledge of $K(E)$ dispersion in the filled band being excited.

3. *Bremsstrahlung Isochromate Spectroscopy:* Some worker measured the X-rays emitted when an electron in KeV range is decelerated to near the F.L. But high k values of the incident electron means that thermal and other non k-conserving processes dominated the emission and only a total density of states is seen rather than detailed band-by-band mapping, which is possible at very much lower energy.

4. *X-ray:* Near edge absorption of fine structure could be used to measure the variation in absorption cross section as an electron is excited from a core level to the conduction band. With this technique the lowest conduction bands are accessible, but because core levels are localized, no dispersion of the unoccupied state can be obtained.

5. *Appearance Potential Spectroscopy (APS):* It measures the probability of excitation of a core electron as a function of incident electron energy giving information about the lowest energy unoccupied states. States immediately above the F.L. can be studied but the information obtained is in a highly convoluted form and no k resolution could be done.

In contrast to above techniques, I.P.E. can provide detailed information because of k-conservation. Electron in crystal emits bremsstrahlung radiation which is preferred to be called I.P.E. to stress the very close relationship to the $p > E$ processes. The theory of I.P.E. was introduced by J. Pendry[1] from SRC Daresburg Lan in Warrington UK in 1981. He pointed out that it should be possible to derive the distribution in E and k of the unoccupied band from I.P.E. data and that it also can provide, the spin distribution. No other technique applicable to unoccupied bands can give anything but the density of

[a]Talk presented by A.A.H. at the XIth Regional Conference on Mathematical Physics, May 2004, Tehran, Iran.

states in E and over for this purpose I.P.E.S. has far higher resolution.

Some experimentalist called the technique VUV Isochromatic Spectroscopy and used a simple apparatus to obtain the density of states in E for polycrystalline Ta and compare the results with those of X-ray Isochromatic Spectroscopy. The results show the potential of the technique and confirm the accuracy of the calculation made by Pendry. We, as a research group at Oliver Lodge Lab physics Dept. of University of Liverpool, design and constructed an apparatus based on I.P.E. in late 1980 which was considerably more complex than that of those in Germany and could be used for measuring unoccupied band distribution in k as well as E over a wide range. When the apparatus was working satisfactorily, we could go on and add a source of spin polarized electrons, so that the minority and majority spin bands in ferromagnetic material could be distinguished.

There is still a good deal to understand about 3d transition metals and the ferromagnetic Fe, Co and Ni. Our aim from the construction of I.P.E.S. was to obtain $E-k$ curves in the case of Ni, which was the most widely studied but not about the unoccupied states. I.P.E.S. is a very useful apparatus for studying the transition metals with few d-electrons, because most of their bands are unoccupied. In I.P.E. the most energetic photons are studied. Those due to primary processes in which k is conserved $\bar{e}(k,E) \to \bar{e}(k-q, E-\hbar\omega) + \hbar\omega(q,\hbar\omega)$ in a manner analogous to the photo emission experiments such as

$$\hbar\omega(q,\hbar\omega) + \bar{e}(k',E') \to \bar{e}(k'+q, E'+\hbar\omega).$$

2 Possibilities for I.P.E. Experiments

1) The experiment most likely to succeed in maximizing the photon yield for ease of detection requires

(i) deep penetration by the incident electron, so that it has a large overlap with the crystal,

(ii) a large density of unoccupied state above the F.L., such as occurs in partially filled d band

(iii) a strong lattice potential to increase the electron-photon interaction.

In this respect, the W is an excellent candidate. A metal with longest m.f.p. for electron for $E = 0$ (111) and (100) surfaces are satisfactory but not the (110). In W the unfilled section of the d band stretches from around 5eV below to just above the vacuum level affording many transitions in the UV and VIS to a 1-2eV incident electron.

2) SPI.P.E.: Easily obtained spin polarized electron beams puts SPI.P.E. high on the list of priorities. Due to Hubbard model there is a symmetry between filled and empty states in the d band. There is further interesting to see how this symmetry holds. In the case of Ni, it is possible to argue that the proper study of the magnetism ties in the hole states. They are in a sense the active ingredients of the magnetism. The complimentary information brought by I.P.E. gives a full picture of magnetism in the d band.

3) The Fermi surface test: Obviously F. surface is the limit of both the hole and electron excitation spectra. What is not obvious is that P.E. and I.P.E. accurately reflects these spectra. There are many higher order terms having to do with introduction in the filled state that could complicate the spectra. The question is whether P.E. and I.P.E. measure the same F.S.

4) Conduction bands in semiconductors: Optical experiments provide a lot of information about direct gap, but indirect gaps are something fundamental. The dispersion of bands near the edge crucially affects the behavior of exitons, inversion-layers, impurity levels, polaron frequency, etc. P.E. provides dispersion at the valence band maximum and I.P.E. at the conduction band minimum. So, the incident electron energy must be sufficiently energetic so that its path length is limited by electronic excitations. Otherwise, it collides with phonons and diffuses around the crystal acquiring a random k vector. Little information is obtained from I.P.E. or for that matter from P.E. under these circumstances. But usually there is no problem in satisfying the electronic excitation condition.

5) Lifetimes of excited states: Much of the theory of materials is based on the assumption of an elementary excitation spectrum. The lifetimes of these excitations set limits on the validity of this picture, and provide critical tests of theories. Many of these have been measured for hole states and there is some knowledge through LEED studies of lifetime when $E > 0$. The lifetime appears as a broadening of the inter band and surface state excitations.

6) Surface states: Empty as well as filled surface states exist and are particularly important on clean semiconductor surfaces. These unsaturated states can lead to instabilities, such as the reconstruction of the silicon (111) surface the driving force being saturation of the bonds. The band is split by this process. Currently only half the picture can be seen with P.E. and I.P.E. provides the other half.

7) Adsorbate studies: When a molecule bonds to a surface, there may be some considerable rearrangement of the orbitals. P.E. is an invaluable technique for identifying the orbitals often with the aid of selection rules, and making some suggestion as to the molecule-surface bonding. Sometimes unfilled states can play a key role. Carbon monoxide is a molecule which has all its orbitals filled. Therefore, when it bonds to a surface sharing some of its occupied orbitals, there must also be a back flow into the unfilled $2\pi^*$.

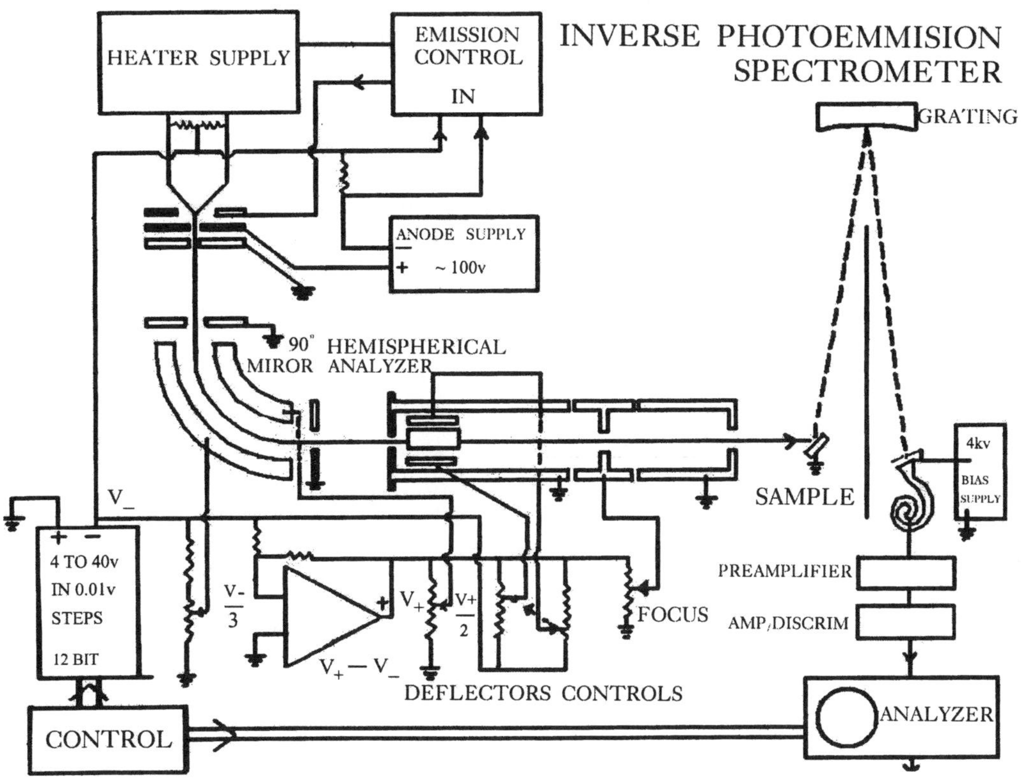

Figure 1.

3 The Apparatus

The apparatus for I.P.E. contains three essential parts; the electron source, the crystal under study and the photon detector.

a) The electron source must provide a beam of electron with well defined energy and momentum (with resolution less than 0.1eV) and directional spread of not more than ±3. The energy of the electron, E, is dictated by the photons being countered ($\hbar\omega$) and the work function of the crystal φ ($E = \hbar\omega - \varphi$). During the measurements the energy of the electron is scanned over a range of a few eV, while the energy of the detected photon is kept fixed. Source of electron meeting these requirements are fairly common (thermo-ionic emitters). But in the case of spin polarized electron, we used hemispherical mirror analyzer giving a 90° deviation, so that it could be adapted to analyze polarized electrons from a Caesiated GaAllium Arsenide photo-emitter and deliver a beam of electron with transverse polarization to the target.

b) The crystal surface under study must be cleaned and maintained in that state during measurements. So all the preparation of the sample and the measurements take place under the U.H.V. condition ($p < 10^{-6} torr$). To meet this condition, we design a clean U.H.V. system using torbomolecular pump backed with roots pump for primary evacuation ($p \simeq 10^{-6} torr$) and using cryogenic pumps to reduce pressure to $10^{-8} torr$ and followed by getter pump, sputter-ion pump and Titanium sublimation pump (TSP) to get a vacuum better than $10^{-10} torr$. We used the Ar ion gun for cleaning the surface and a LEED-Auger system for determining the state of surface of the sample after cleaning. We also used Q.M.S. to analyze the residual gas in the system and leak the detection device. The crystalline sample was mounted on a goniometer, so that the direction of the electron beam altered relative to the crystal axis could be adjusted. All parts of vacuum system were bankable up to $20^o c$ for out gassing.

c) Photon detector: The photon detector system should have an energy resolution of 0.1eV and accept photon in a large solid angle and have a high absolute detection efficiency if the count rate is to be sufficient for measurements to be made in the relatively short time during which the surface remaining uncontaminated. Choice of photon energy for doing the measurements was about 20eV, which implies electron energy of 15-20eV and which is nearly five times larger than typical transition metals work function ϕ. Simplicity and the potentially large solid angle accepted by the Geiger counter make it attractive, but the electron energy at threshold is only 4eV, which is comparable with crystal work function so that possibilities for scanning $k_{\|}$ are limited. For this reason we used grating monochromatic and electron multipliers. The grating monochromator must have large solid angle of acceptance and absolute efficiency together with a large source size so the electron density on the crystal can be kept to a figure which is attainable with electron energy of 15-20eV. Most metals are poor reflectors near $\hbar\omega = 20ev$ except for Osmium which has a reflectivity of 35% at $600A^o$. So, one might expect to achieve an absolute efficiency of 25% at the blaze wavelength. The efficiency of channel-electrons used as photon detectors is about 10%. So, it is possible to detect 2.5% of photons incident on grating. Os coated grating with focal ratio of 3 accepts about 0.1 stradian. With such grating, the predicted count rate is about 1000 per second with a $100\mu A$ electron beam. The range of energy, which it would be useful to scan, is about 5eV build up over hour. For this reason, we used Pt coated grating rather than Os. In order to adjust the photon detection system, we include a capillary discharge UV source in the system and to use light from this reflected from a wire mounted in the specimen holder to confirm the performance of the monochromator before making I.P.E. measurements. During ray tracing calculation on various grating arrangement is performed and we work out that it was possible to attain a resolution of 0.1 to 20eV with a source 0.3mm wide. Typical arrangement and circuit for the apparatus (I.P.E.S.) are shown schematically in Figs (1) and (2).

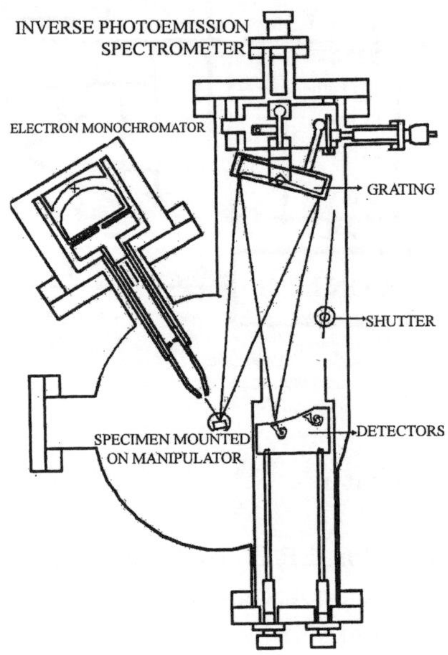

Figure 2.

References

1. J. Pendry *J. Phy.* Solid State **C**, 14 (1981).

STUDYING OF POROUS POLY-SILICON IN PRESENCE OF ETHANOL BY SCANNING TUNNELING SPECTROSCOPY

A. IRAJI ZAD and F. RAHIMI
Department of Physics, Sharif University of Technology
P.O. Box 11365-9161, Tehran, Iran
Email: iraji@sharif.ir

Porous silicon (PS) has a large surface area and the ability to adsorb gas molecules in the environment. This phenomenon changes the surface properties like the electrical properties of porous layers. In this research, we produced porous poly-silicon on the basis of poly-silicon wafer. The topography of the surface was investigated by Scanning Electron Microscopy (SEM), and showed that the structure of the film is micron size islands and the red photoluminescence of the sample proved the existence of nano-pores in addition to the micron size islands. Scanning Tunneling Spectroscopy (STS) revealed that the electrical surface properties of porous layer changes in the presence of ethanol.

[a]In the recent years, porous silicon (PS) has received considerable attention due to its interesting properties. One of these properties is the large surface to volume ratio that gives PS the ability to react with gases and senses them readily[1,2,3]. In these researches the probe for sensing gas is the resistivity of samples which easily changes in the presence of gases. Some investigation has been done so far to study the reaction of gas with the porous surface by Fourier Transform Infrared spectroscopy (FT-IR)[4,5], which revealed that after the introduction of the gas molecules, new bonds were formed on the surface. These new bonds can introduce new surface states and new electrical surface properties[6]. To the best of our knowledge, there are not any observations on the change of electrical surface properties of porous silicon by Scanning Tunneling Spectroscopy.

In this article, we produced porous layer on poly-silicon wafer and observed the topography and photoluminescence of the surface by SEM and UV irradiation respectively. Scanning Tunneling Spectroscopy (STS) measurements were used to probe the local electrical properties of the surface in the presence of ethanol. Porous layers were formed by conventional electrochemical etching[7] of p-type poly-silicon wafer with $0.4-2$ Ωcm resistivity and 330 ± 40 μm thickness. The HF concentrations, current density and the time of etching were 13% 32 mA/cm^2 and 75 min., respectively. SEM image from the cross section of the sample which is broken from a grain boundary is shown in Fig. 1a. As illustrated, micron scale porosity with about 30 μm pores is obtained. While SEM results show micron-scale morphology on the surface of the samples, we expect that the high surface area of the layers originates from the nano-scale pores inside the micron-scale features. The nano-scale pores cannot be observed by SEM, however the red luminescence observed from this sample, when illuminated by UV light, confirms the existence of this structure (Fig. 1b).

Visible luminescence is a feature of nano-porous silicon[8]. Since the resistivity of our sample was changed in the presence of ethanol[3], we observed the change in electrical surface properties in the presence of ethanol by STS measurements.

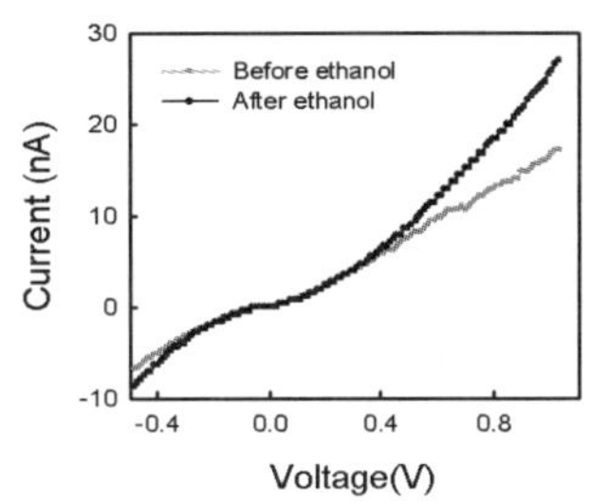

Figure 2. Typical STS results (I-V curves) from the porous layer.

Scanning Tunneling Spectroscopy was done in the ambient atmosphere using a Pt/Ir tip (Nanosurf Easyscan instrument). We kept the tip of a Scanning Tunneling Microscope at a certain distance from the top of the porous layer and measured the current at different tip-sample voltages. Fig. 2 shows the change of the I-V curve as a result of gas exposure. This shows that the tunneling current at a fix volt-

[a]Talk presented by A.I.Z. at the XIth Regional Conference on Mathematical Physics, May 2004, Tehran, Iran.

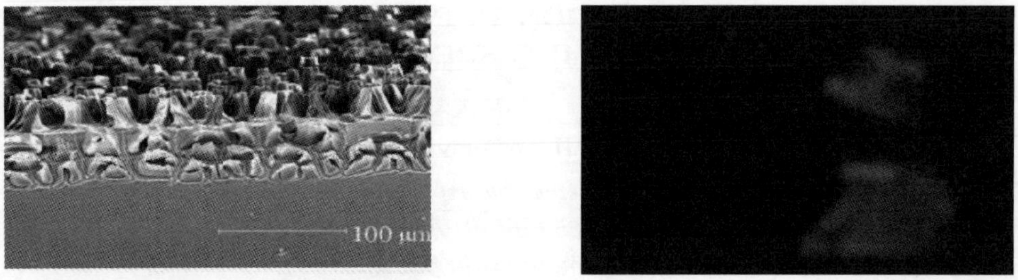

(a) (b)

Figure 1. a) SEM image of a cross-section and b) Red luminescence under UV light of the porous poly-silicon sample.

age was increased, after the introduction of gas. To ensure that the changes observed in the I-V curves are related to the porous layer and not to the adsorption of gases on the tip, we measured the I-V curves when the tip was positioned above an unetched part of the sample. In this case no change in the I-V curve was observed as ethanol was introduced. This implies that the changes seen in Fig. 2 are related to the porous layer.

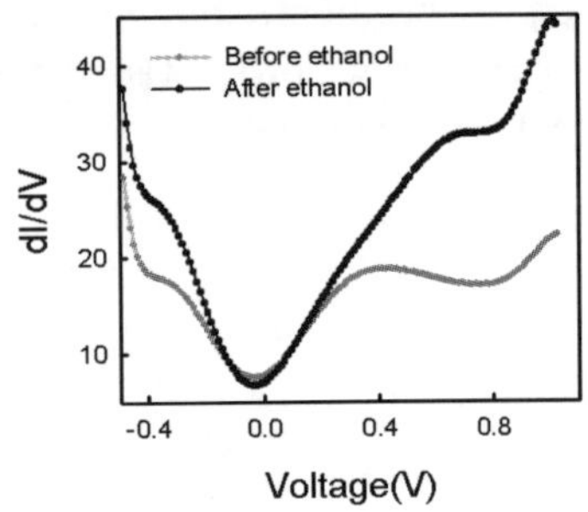

Figure 3. First derivative (dI/dV).

In order to observe the local density of states (LDOS), we should extract (dI/dV)/(I/V).[9] By extracting the (dI/dV)/(I/V) graphs before and after gas exposure, we could perceive that the change in surface states arises from the ethanol exposure (Figs. 3 and 4). Fig. 4 shows these changes very well. The singularity in Fig. 4 around the zero voltage is arisen from the singularity in (I/V) part.

In summary, the ethanol adsorption on the porous poly-silicon sample, changes the surface states which are easily observed by STS. This phenomenon can change the resistivity of the sample.

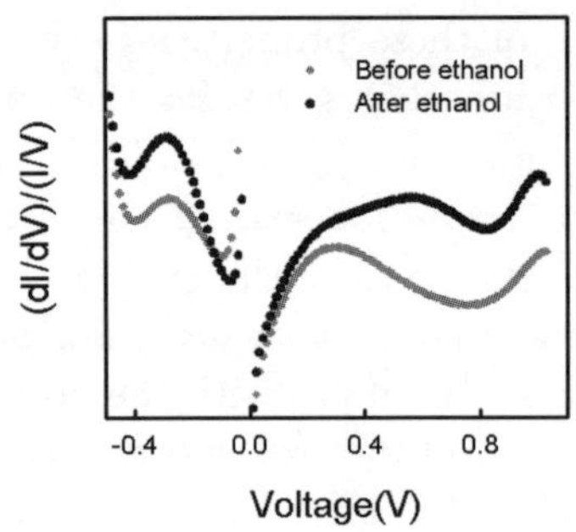

Figure 4. Normalized derivative (dI/dV) extracted from Figure 2.

Acknowledgments

The Sharif University of Technology Research Department supported this work.

References

1. C. Baratto, G. Faglia, E. Comini, G. Sberveglieri, A. Taroni, V. La Ferrara, L. Quercia and G. Di Francia, *Sens. Actuators, B, Chem.* **77**, 62 (2001).
2. L. Pancheri, C.J. Oton, Z. Gaburro, G. Soncini and L. Pavesi, *Sens. Actuators, B, Chem.* **97**, 45 (2004).
3. A. Iraji Zad, F. Rahimi, M. Chavoshi and M.M. Ahadian, *Sens. Actuators, B, Chem.* **100**, 341 (2004).

4. J.A. Glass, E.A. Wovchko and J.T. Yates, *Surf. Sci.* **338**, 125 (1995).
5. L. Boarino, C. Baratto, F. Geobaldo, G. Amato, E. Comini, A.M. Rossi, G. Faglia, G. Lérondel and G. Sberveglieri, *Materials Science and Engineering* B **69-70**, 210 (2000).
6. A. Many, Y. Goldstein and N.B. Grover in *Semiconductor Surfaces*, (North-Holland Publishing Company, Amsterdam, 1965).
7. L. Canham in *Properties of Porous Silicon*, (INSPEC, London, 1997).
8. J.-C. Vial and J. Derrien in *Porous Silicon Science and Technology*, (Springer-Verlag and Les Editions de Physique, Berlin, 1995).
9. N. Li, M. Zinke-Allmang and H. Iwasaki, *Surf. Sci.* **554**, 253 (2004).

PHASE TRANSITION AND SHOCK FORMATION IN REACTION-DIFFUSION SYSTEMS: NUMERICAL APPROACH

F. H. JAFARPOUR
Physics Department, Bu-Ali Sina University, Hamadan, Iran
and
Institute for Studies in Theoretical Physics and Mathematics (IPM), School of Physics
P.O. Box 19395-5531, Tehran, Iran
E-mail: farhad@sadaf.basu.ac.ir

We study a one-dimensional branching-coalescing model on a chain of length L with reflecting boundaries. The phase transitions of this model is investigated in a canonical ensemble by using the Yang-Lee description of the non-equilibrium phase transitions. Numerical study of the canonical partition function zeros reveals two second-order phase transitions in the system. Both transition points are determined by the density of the particles on the chain. In some regions the density profile of the particles has a shock structure.

[a]One-dimensional driven lattice gases are models of particles which diffuse, merge and separate with certain probabilities on a lattice with open, periodic or reflecting boundaries. In the case of open boundaries the particles are allowed to enter or leave the system from both ends or only one end of the chain. In the case of reflecting boundaries or periodic boundaries, however, the number of particles will be a conserved quantity provided that no other reactions other than the diffusion of particles take place. In the stationary state, these models exhibit a variety of interesting properties such as non-equilibrium phase transitions and spontaneous symmetry breaking which cannot be found in equilibrium models (see Ref. [1] and references therein). In the present paper we study the phase transitions in a one-dimensional branching-coalescing model with reflecting boundaries in which the particles diffuse, coagulate and decoagulate on a lattice of length L. The reaction rules are specifically as follows

$$\begin{array}{ll} \emptyset + A \to A + \emptyset & \text{with rate } q, \\ A + \emptyset \to \emptyset + A & \text{with rate } q^{-1}, \\ A + A \to A + \emptyset & \text{with rate } q, \\ A + A \to \emptyset + A & \text{with rate } q^{-1}, \\ \emptyset + A \to A + A & \text{with rate } \Delta q, \\ A + \emptyset \to A + A & \text{with rate } \Delta q^{-1}, \end{array} \tag{1}$$

where A and $\emptyset$ stand for the presence of a particle and a hole, respectively. It is assumed that there is no injection or extraction of particles from the boundaries. We will also assume that the number of particles on the chain is a conserved quantity. This model has already been studied in grand canonical ensemble using the Empty Interval Method (EIM) in Ref. [2]. In this formalism the physical quantities such as the density of particles are calculated from the probabilities to find empty intervals of arbitrary length. Later, this model was studied using so-called the Matrix Product Formalism (MPF)[3]. According to this formalism, the stationary probability distribution function of the system is written in terms of the products of non-commuting operators E and D and the vectors $|V\rangle$ and $\langle W|$ as

$$P(\tau_1, \cdots, \tau_L) = \frac{1}{Z_L}\langle W|\prod_{i=1}^{L}(\tau_i D + (1-\tau_i)E)|V\rangle. \tag{2}$$

Each site of the lattice is occupied by a particle ($\tau_i = 1$) or is empty ($\tau_i = 0$). The factor Z_L in (2) is a normalization factor. The operators D and E stand for the presence of particles and holes respectively and besides the vectors $|V\rangle$ and $\langle W|$ should satisfy the following quadratic algebra[3]

$$\begin{aligned} &[E, \bar{E}] = 0, \\ &\bar{E}D - E\bar{D} = q(1+\Delta)ED - \frac{1}{q}DE - \frac{1}{q}D^2, \\ &\bar{D}E - D\bar{E} = -qED + \frac{1+\Delta}{q}DE - qD^2, \\ &\bar{D}D - D\bar{D} = -q\Delta ED - \frac{\Delta}{q}DE + (q + \frac{1}{q})D^2, \\ &\langle W|\bar{E} = \langle W|\bar{D} = 0, \bar{E}|V\rangle = \bar{D}|V\rangle = 0. \end{aligned} \tag{3}$$

[a]Talk presented at the XIth Regional Conference on Mathematical Physics, May 2004, Tehran, Iran.

The operators $\bar{D}$ and $\bar{E}$ are auxiliary operators and do not enter in the calculation of (2). Having a representation for the quadratic algebra (3), one can easily compute the steady state weights of any configuration of the system using (2). It has been shown that (3) has a four-dimensional representation[3].

A natural question that might arise is whether or not we can see the shocks in our branching-coalescing model defined by (1) with the reflecting boundaries. To answer this question, we will study the model with reflecting boundaries in a canonical ensemble where the number of particles on the chain is equal to M, so that the density of particles $\rho = \frac{M}{L}$ remains constant. We will then investigate the phase transitions and the density profile of particles on the chain.

Recently, it has been shown that the classical Yang-Lee theory[4] can be applied to the one-dimensional out-of-equilibrium systems in order to study the possible phase transitions of these models[5,6,7,8]. According to this theory in the thermodynamic limit, the zeros of the canonical or grand canonical partition function, as a function of an intensive variable of the system, might approach the real positive axis of that parameter at one or more points. Depending on how these zeros approach the real positive axis the system might have one or more phase transition of different orders. If the zeros intersect the real positive axis at a critical point at an angle $\frac{\pi}{2n}$, then n will be the order of phase transition at that point[5].

Let us define the canonical partition of our model as

$$Z_{L,M} = \langle W|\text{Coefficient}[(E + xD)^L, M]|V\rangle, \quad (4)$$

in which Coefficient$[Expr, n]$ gives the coefficient of x^n in the polynomial $Expr$. Using the matrix representations D, E, $\langle W|$ and $|V\rangle$ given in Ref. [3], we have been able to calculate the canonical partition zeros numerically. One can use MATHEMATICA to calculate $\langle W|(E + xD)^L|V\rangle$ for arbitrary q and Δ and finite L in which x is a free parameter. The result will be a polynomial of x. The coefficient x^M in this polynomial gives the canonical partition function of the system. In Fig. 1 we have plotted the numerical estimates for the zeros of $Z_{L,M}$ obtained from (4) on the complex-q plane for $L = 80$, $M = 42$. The canonical partition function (4) has $4(L - M)$ zeros in the complex-q plane. We have found that for large L and M, the locations of these zeros are not sensitive to the value of Δ. We have also calculated the numerical estimates for the roots of (4) as a function of Δ for fixed values of q. It turns out that (4), as a function of Δ, does not have any positive root; therefore, we expect that the phase transition points do not depend on Δ.

As can be seen in Fig. 1 the zeros lie on two different curves and accumulate towards two different points on the positive real-q axis. By extrapolating the real part of the nearest roots to the positive real-q axis for large L and M, we have found that the transition points are $q_c = \frac{1}{\sqrt{1-\rho}}$ $(1 < q_c < \infty)$ and $q'_c = \sqrt{1-\rho}$ $(0 < q'_c < 1)$. As $\rho \to 0$ the two curves lie on each other and we will find only one transition point at $q_c = q'_c = 1$. It appears also that the zeros on both curves approach the real-q axis at an angle $\frac{\pi}{4}$ (the smaller angle). This predicts two second-order phase transitions at q_c and q'_c. The reason that the system has two phase transitions can easily be understood. The parameter q determines the asymmetry of the system and for any q the system is invariant under the following transformations

$$\begin{aligned} q &\longrightarrow q^{-1}, \\ i &\longrightarrow L - i + 1. \end{aligned} \quad (5)$$

Therefore, one can expect to distinguish two critical points which are related according to the symmetry of the system.

Let us now study the density profile of particles on the chain $\rho(i)$ in each phase. The density of particles at site i is defined as

$$\rho(i) = \frac{\sum_{\mathcal{C}} \tau_i P(\tau_1, \cdots, \tau_L)}{\sum_{\mathcal{C}} P(\tau_1, \cdots, \tau_L)}, \quad (6)$$

in which $\mathcal{C}$ is any configuration of the system with fixed number of particles M and $P(\tau_1, \cdots, \tau_L)$ is given by (2). One can use the matrix representation of the quadratic algebra 3 to calculate (6) using MATHEMATICA. In Fig. 2 we have plotted (6) for two different values of q with $L = 60$ and $M = 36$. For this choice of the parameters the transition points are $q_c = 1.581$ and $q'_c = 0.633$. The density of particle has two general behaviors for $q > 1$. For $q > q_c$ and in the thermodynamic limit $(L \to \infty, M \to \infty, \rho = \frac{M}{L})$ the density profile of particles is a shock in the bulk of the chain; while in the close vicinity of the left boundary, it increases exponentially. The density of particles in the hight-density region of the shock is equal to $\rho_{High} = 1 - q^{-2}$. This region is separated by a rather sharp interface from the low-density region in which the density of particles is equal to $\rho_{Low} = 0$.

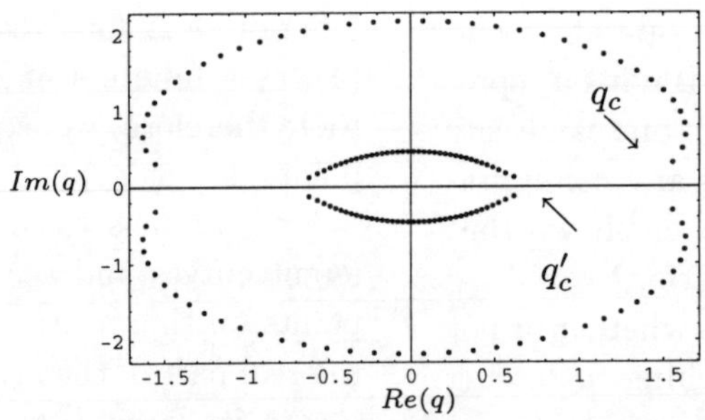

Figure 1. Plot of the numerical estimates for the canonical partition function zeros obtained from (4) for $L = 80$ and $M = 42$.

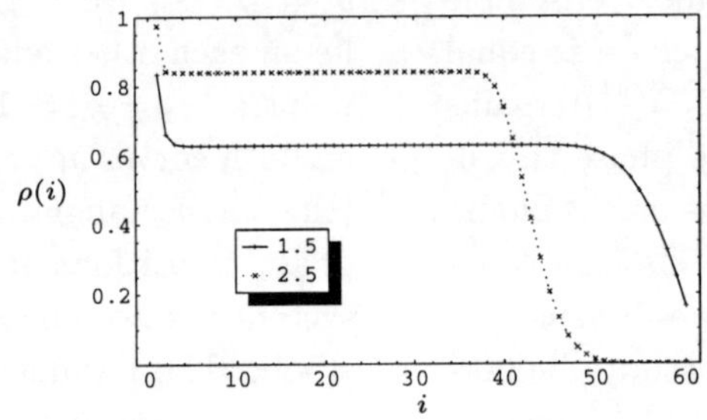

Figure 2. Plot of the density profile of the particles (6) on a chain of length $L = 60$ for $M = 36$ and two values of q ($q > 1$) above and below the critical point $q_c = 1.581$.

The low-density region is extended over $(1-\frac{\rho}{1-q^{-2}})L$ sites. This can be seen in Fig. 2 for $q = 2.5$; however, the reason that the shock interface is not sharp is that our calculations are not in real thermodynamic limit. One should expect that the shock front becomes sharper and sharper as the length of the system L and also the number of particles on the chain M increase. For $1 < q < q_c$ the density of particles in the bulk of the chain is constant equal to ρ, it drops near the right boundary exponentially and increases exponentially in the close vicinity of the left boundary. The exponential behavior of the density profile of particles near the boundaries in this phase is due to the finiteness of the representation of the algebra (3). It is known that if the associated quadratic algebra of the model has finite dimensional representations, the density-density correlation functions cannot have algebraic behaviors[3]. At $q = 1$ one finds $\rho(i) = \rho$. The density profile of particles in the region $q < 1$ is related to that of $q > 1$ through (5) that is

$$\rho(i,q) = \rho(L-i+1, q^{-1}). \qquad (7)$$

In this paper we have studied a branching-coalescing model in which particles hop, coagulate and decoagulate on one-dimensional lattice of length L. By working in a canonical ensemble we have restricted ourself to the case where the total number of particles on the chain is constant. The Yang-Lee theory predicts that the model has two second-order phase transitions. Both phase transition points are determined by the density of particles on the system ρ. The study of the mean particle concentration at each site of the chain for $q > 1$ shows that the density profile of the particles has a shock-like structure in the region $q > q_c = \frac{1}{\sqrt{1-\rho}}$. The exception is near the left boundary where the density of particles increases exponentially. This is the first time that shocks are seen in one-dimensional reaction-diffusion models with reflecting boundaries. In the region $1 < q < q_c = \frac{1}{\sqrt{1-\rho}}$ the density profile of the particles is constant in the bulk of the chain; however, near the left (right) boundary it increases (decreases) exponentially. Our numerical investigations also show that the width of the shock scales as $L^{-\nu}$ with $\nu = \frac{1}{2}$. In the thermodynamic limit $L \to \infty$, one finds a very sharp shock interface. Since the system is invariant under the transformation (5), the density profile of the particles for $q < 1$ can be obtained from (7).

References

1. G.M. Schütz in *Phase Transitions and Critical Phenomena* vol. 19, ed C. Domb and J.

Lebowitz (Academic Press, New York, 1999).
2. H. Hinrichsen, K. Krebs and I. Peschel, *Z. Phys.* B **100**, 105 (1996).
3. H. Hinrichsen, K. Krebs and I. Peschel, *J. Phys. A* **29**, 2643 (1996).
4. C.N. Yang and T.D. Lee, *Phys. Rev.* **87**, 404 (1952); *Phys. Rev.* **87**, 410 (1952).
5. R.A. Blythe and M.R. Evans, *Brazilian Journal of Physics* **33**, 464 (2003).
6. P.F. Arndt, *Phys. Rev. Lett.* **84**, 814 (2000).
7. F.H. Jafarpour, *J. Stat. Phys.* **113**, 269 (2003).
8. F.H. Jafarpour, *J. Phys.* A **36**, 7497 (2003).

CHARGE AND MAGNETIZATION PLATEAUX IN STRONGLY CORRELATED SYSTEMS

A. LANGARI
Institute for Advanced Studies in Basic Sciences
Zanjan 45195-159, Iran
and
Institute for Studies in Theoretical Physics and Mathematics (IPM), School of Physics
P.O. Box 19395-5531, Tehran, Iran
E-mail: langari@iasbs.ac.ir

The opening of energy gap in the charge and spin degrees of freedom is reported on the ladder geometry. *I*) Our recent works on ferrimagnetic (S_1, S_2) two legs ladder is reviewed. We study the emergence of magnetization plateau by exact diagonalization using Lanczos method. We also explain the reason for different magnetization plateaux in the phase diagram of bond-alternation ladder. *II*) We study the charge density plateau in the two and three legs $t-J$ ladder, where a charge gap can be opened as a function of the chemical potential in different regime of rung and leg couplings.

1 Introduction

[a]Charge and magnetization plateaux are the result of energy gap in the spectrum of charge and spin degrees of freedom, respectively. The study of magnetization process in one-dimensional quantum spin systems has been spurred since the discovery of quantized plateaux in the magnetization curve[1,2,3]. This behavior is a sort of magnetic analogy of the quantum Hall effect for charge degrees of freedom. Likewise, one-dimensional (S_1, S_2) systems have shown to display unusual quantum properties[4]. Similarly a two leg ferrimagnetic ladder responds differently to a magnetic field comparing with the homogeneous spin ladders[5]. We will discuss the properties of a two leg bond alternation $(S_1 = 1, S_2 = 1/2)$ ferrimagnetic model in the presence of magnetic field in next sections.

The charge density plateau in $\rho(\mu)$, where ρ is charge density and μ is the chemical potential, looks similar to the magnetization plateaux in the magnetization curve $m(h)$ found in spin ladders. A metal–insulator transition is accompanied by the opening of a gap, which appears as a plateau in the charge density $\rho(\mu)$. It is signaled by discontinuous changes in the slope of ground state energy per site as a function of ρ. This emerges immediately in the "local rung approximation"[6] or "bond operator theory"[7]. In the following we go beyond this approximations by means of a systematic perturbation theory in the leg hopping coupling constant.

2 Magnetization Plateaux in Ferrimagnetic Ladders

We have considered a two leg ferrimagnetic ladder — a bipartite lattice — composed of two different spins $(S_1 = 1, S_2 = 1/2)$ where each type of spin sits on a sublattice (see Fig. 1 of Ref. [8]). The Hamiltonian of this model is defined by

$$
\begin{aligned}
H = J\sum_{\alpha=1}^{2}\sum_{n=0}^{\frac{L}{2}-1} & \big([1+\gamma^{(\alpha)}(i)]\ \mathbf{S}_1^{(\alpha)}(i)\cdot\mathbf{S}_2^{(\alpha)}(i+1) \\
& + [1+\gamma^{(\alpha)}(i+1)]\ \mathbf{S}_2^{(\alpha)}(i+1)\cdot\mathbf{S}_1^{(\alpha)}(i+2)\big) \\
& + J'\sum_{\alpha\neq\beta}^{1,2}\sum_{n=1}^{L}\mathbf{S}_1^{(\alpha)}(n)\cdot\mathbf{S}_2^{(\beta)}(n) - hS^z_{total}, \qquad (1)
\end{aligned}
$$

where $\mathbf{S}_1^{(\alpha)}(n)$ denotes the quantum spin-1 at site n in the leg $\alpha = 1, 2$ of the ladder, similarly $\mathbf{S}_2^{(\alpha)}(n)$ for the spin-1/2 and the number of sites is $N = 2 \times L$, L is the number of rungs. Alternatively, we may use m, the normalized value of the magnetization with respect to the saturation value as $m = M/M_{sat}$.

[a]Talk presented at the XIth Regional Conference on Mathematical Physics, May 2004, Tehran, Iran.
This presentation is a collaboration with the following contributors:
M. Abolfath (*Institute for Microstructural Sciences, National Research Council, Ottawa, Canada*)
A. Fledderjohann and K.-H. Mütter (*Physics Department, University of Wuppertal, 42097 Wuppertal, Germany*)
M. A. Martín-Delgado (*Departamento de Física Teórica I, Universidad Complutense, 28040-Madrid, Spain*)

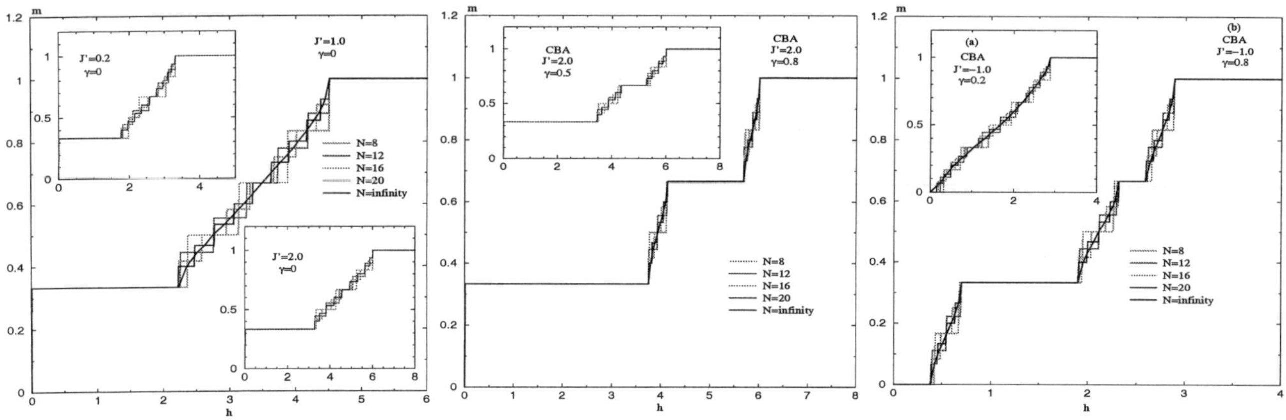

(a) (b) (c)

Figure 1. Magnetization plateaux of two leg ferrimagnetic ($S_1 = 1, S_2 = 1/2$) ladder for different rung coupling (J'). (a) No dimerization is assumed ($\gamma = 0$), it also represents the case of SBA configuration ($J' > 0$) where the only plateau exists at $m = 1/3$. (b) The case of CBA and $J' > 0$ where the 2nd plateau at $m = 2/3$ appears. (c) CBA and $J' < 0$ with three non-trivial plateaux $m = 0, 1/3, 2/3$ for $\gamma > \gamma_c$. The inset shows the gapless phase in $\gamma < \gamma_c$ with no plateau.

The magnetic field is h, J and J' are coupling constants along the legs and the rungs, respectively, and $\gamma^{(\alpha)}(n)$ is the dimerization patterns. We consider two different dimerization patterns : i) CBA (Columnar Bond Alternation), for which $\gamma^{(\alpha)}(n) = (-1)^{n+1}\gamma$, and ii) SBA (Staggered Bond Alternation), where $\gamma^{(\alpha)}(n) = (-1)^{\alpha+1+n}\gamma$. We use periodic boundary conditions along the legs of the ladder. The phase diagram of this system shows a rich structure[9] in the absence of magnetic field. Its low energy spectrum consists of two bands, where an anti-ferromagnetic band is separated from the lower ferromagnetic one by an energy gap[10]. Adding a magnetic field causes level crossings at different field strength (h) which lead to the magnetization process ($m(h)$). An energy gap in the spin excitation spectrum appears as a plateau in $m(h)$. We have studied numerically the formation of magnetization plateaux by Lanczos method. For inter-chain coupling $J' > 0$ we found a normalized plateau at $m = 1/3$ starting at zero field and the trivial one at $m = 1$, Fig. 1a.

The magnetization of SBA case is similar to what presented in Fig. 1a. This can be explained by the effective XXZ Hamiltonian in a longitudinal filed[4,8]. The CBA configuration shows an extra plateau at $m = 2/3$ when $J' > 0$, Fig. 1b. If we switch the rung coupling to ferromagnetic interaction, $J' < 0$, we will observe two different phases for CBA configuration. For $\gamma < \gamma_c$ the model is gapless for all range of magnetic field and results to no plateau, while for $\gamma > \gamma_c$ the model is gapful where the first plateau appears at $m = 0$. There are also two other plateaux at $m = 1/3$ and $2/3$ which can be explained by the effective Hamiltonian of dimerized $S = 3/2$ anti-ferromagnetic Heisenberg chain[8], Fig. 1c.

3 Charge Density Plateaux in t-J Ladder

We have considered the $t - J$ model on a two and three legs ladder where $t(t')$ is the hopping parameter and $J(J')$ is the exchange coupling on rungs (legs), (see Fig. 1 and Fig. 2 in Ref. [11]). The Hamiltonian of the ladder is $H = H_{leg}(t', J') + H_{rung}(t, J)$. Where each term is the usual $t-J$ Hamiltonian composed of a hopping term and an exchange interaction. In the limit of vanishing leg couplings $t' = 0$, the system is composed of decoupled rungs, so the ground state is the direct product of rung states. For $\rho = \frac{Q}{N} < 1/2$, it is the product of rung ground states with charge $q = 0, 1$, if $\alpha = J/t < 2$ and it is the direct product of $q = 1, 2$ rung ground states for $\rho > 1/2$. The chemical potential $\mu = \frac{d}{d\rho}(E/N)$ is

$$\mu = \begin{cases} -1 & \text{for } \rho \leq \frac{1}{2}, \\ 1 - \alpha & \text{for } \rho \geq \frac{1}{2}, \end{cases} \tag{2}$$

where E is the ground state energy and N is the total number of sites. The discontinuity at $\rho = 1/2$ is the first indication of charge density plateau at $\rho = 1/2$. Then we have considered a first order perturbation theory in t' by obtaining the effective Hamiltonian of

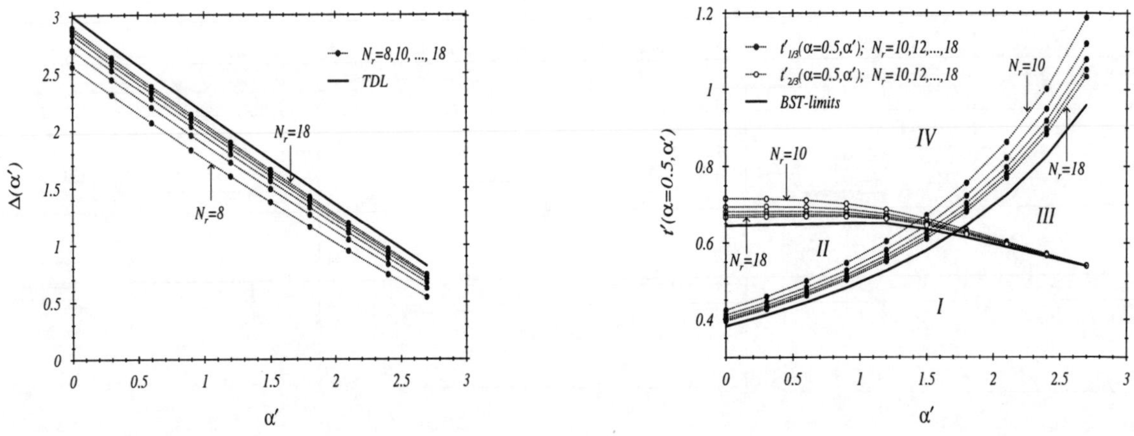

(a) (b)

Figure 2. Phase boundary in the parameter space to specify the existence of a specified plateau. Lanczos results for different number of rungs (N_r) and an extrapolation to the thermodynamic limit (TDL, solid line). (a) Two legs ladder. If the value of $\frac{2t-J}{tt'}$ falls below the solid line, then no plateau at $\rho = 1/2$ appears while in the upper part of phase diagram a plateau appears (b) Three legs ladder (see the explanation in the text).

ladder which is a $t-J$ model on a chain with renormalized parameters and additional diagonal term[11]. We have implemented the numerical Lanczos method to study the effective Hamiltonian. We found two different regimes in the parameter space which is presented in Fig. 2a. We have plotted $\Delta(\alpha')$ versus $\alpha' = J'/t'$ for finite system sizes and an extrapolation to thermodynamic limit. The gapped phase with a non-vanishing plateau at $\rho = 1/2$ is characterized by $\Delta(\alpha') < \frac{2t-J}{tt'}$. We then generalized our approach to three leg $t-J$ ladder. We have found the boundaries in parameter space which separate different phases with specified plateaux. In Fig. 2b, the region (I) contains two plateaux at $m = 1/3, 2/3$, passing the boundary to region (II), we observe only a plateau at $m = 1/3$ while in region (III) only the $m = 2/3$ plateau appears and finally in region (IV) no plateau exists.

We have improved our perturbative calculation by using four sites clusters, where it is possible to observe the plateaux at $\rho = 1/4$ and $3/4$. Moreover we have found a systematic way to trace the spin-charge phase separation of $t-J$ model using finite clusters[12].

Acknowledgments

The author would like to thank the fruitful collaboration with M. Abolfath, A. Fledderjohann, K.-H. Mütter and M.A. Martín-Delgado who are the contributors to this set of works.

References

1. K. Hida, *J. Phys. Soc. Jpn.* **63**, 2359 (1994).
2. M. Oshikawa *et al.*, *Phys. Rev. Lett.* **78**, 1984 (1997).
3. K. Totsuka, *Phys. Lett.* A **228**, 103 (1997).
4. M. Abolfath *et al.*, *Phys. Rev.* B **63**, 144414 (2001) and references therein.
5. R.M. Wiessner *et al.*, *Eur. Phys. J.* B **15**, 475 (2000) and references therein.
6. J. Riera *et al.*, *Eur. Phys. J.* B **7**, 53 (1999).
7. K. Park *et al.*, *Phys. Rev.* B **64**, 184510 (2001).
8. A. Langari *et al.*, *Phys. Rev.* B **62**, 11725 (2000).
9. A. Langari *et al.*, *Phys. Rev.* B **61**, 343 (2001).
10. A. Langari *et al.*, *Phys. Rev.* B **63**, 54432 (2001).
11. A. Fledderjohann *et al.*, *Eur. Phys. J.* B **36**, 193 (2003).
12. A. Fledderjohann, A. Langari and K.-H. Mütter, in prepration.

MAGNETIZATION PLATEAUX IN THE ISING LIMIT OF THE MULTIPLE-SPIN EXCHANGE MODEL ON PLAQUETTE CHAIN

V. R. OHANYAN and N. S. ANANIKIAN

Department of Theoretical Physics, Yerevan Physics Institute
Alikhanian Brothers 2, 375036 Yerevan, Armenia
E-mail: vohanyan@www.physdep.r.am

We consider the Ising spin system, which emerges from the corresponding Multiple-Spin Exchange (MSE) Hamiltonian, on the special one-dimensional lattice, diamond-plaquette chain. Using the technique of transfer-matrix we obtain the exact expression for free energy of the system with the aid of which we obtain the magnetization function. Analyzing magnetization curves for various values of temperature and coupling constants, we find the magnetization plateaux at 1/3 and 2/3 of the full moment. The corresponding microscopic spin configurations are unknown because of high frustration.

[a]The Heisenberg model is widely recognized as a lattice model for magnetism of materials. However, this model is by no means universal, because it is based on several assumptions. One of these assumptions suggests the pair character of exchange interactions. This means that only the exchange processes of no more than two particles (nearest-neighbour or next-nearest-neighbour and so on) are taken into consideration

$$\mathcal{H}_{Heis} = 2J \sum_{\langle i,j \rangle} P_{ij}. \tag{1}$$

Here P_{ij} are the pair exchange operators, which implement the transposition of two spin states in i-th and j-th sites of the lattice

$$P_{ij} |\xi_i\rangle \otimes |\xi_j\rangle = |\xi_j\rangle \otimes |\xi_i\rangle. \tag{2}$$

For the $SU(2)$ spins and $s = 1/2$ the expression for P_{ij} is

$$P_{ij} = \frac{1}{2} \left(1 + \boldsymbol{\sigma}_i \cdot \boldsymbol{\sigma}_j\right), \tag{3}$$

where $\boldsymbol{\sigma}_i$ are the Pauli matrices. The generalization of this picture has been known since 60-s[1] and was called the multiple-spin exchange (MSE) model. This model describes the magnetism of the system of almost localized fermions by means of the concept of many particle permutation. The general form of MSE Hamiltonian is

$$\mathcal{H}_{MSE} = -\sum_n J_n \left(-1\right)^p P_n. \tag{4}$$

Here P_n denotes the $n-$particle cyclic permutation operator, J_n ($J_n < 0$) is corresponding exchange energy and p is the parity of the permutation, which indicates how many pair transpositions are contained in the given cyclic permutation. Many peculiar magnetic and thermodynamical properties of ^{3}He adsorbed on graphite surface can be understood only within the framework of MSE model[2].

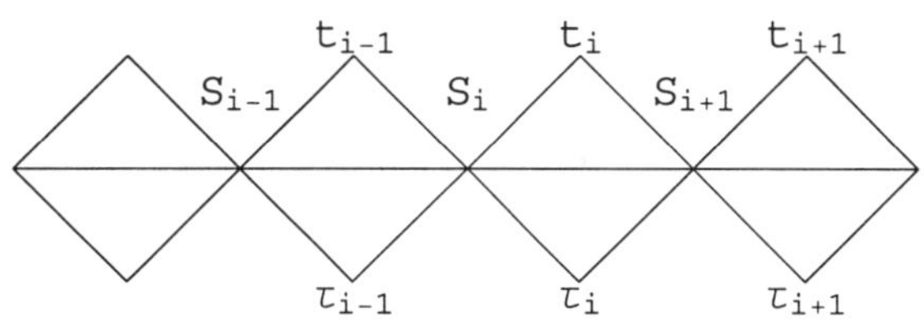

Figure 1. The diamond plaquette spin chain.

Recently the significant role of MSE interaction was revealed in low-dimensional cuprate compounds[3], which initiated the interest in the two-leg spin ladders with four-spin cyclic interaction[4]. The MSE model by itself exhibits rich phase structure even at classical level[5]. Another interesting feature of MSE model is the possibility of complex magnetic behavior, including such a phenomenon as magnetization plateau, which was established in the MSE model on triangular lattice[6]. The study of the magnetization plateaux[7] is the one of the main directions of present-day investigations of macroscopic non-trivial quantum effects in condensed matter physics, which have a number of fundamental and applied values. Despite the purely quantum origin of this effect it was shown recently that magnetization plateaux can appear in the Ising spin systems as well[8,9,10], exhibiting in some cases fully qualitative correspondence with its Heisenberg counterpart[9]. The latter fact is very important because it can serve for more

[a]Talk presented by V.O. at the XIth Regional Conference on Mathematical Physics, May 2004, Tehran, Iran.

profound understanding of magnetization plateaux physics, can provide it with new methods and can initiate a search of novel magnetic materials with huge axial anisotropy.

We consider a one-dimensional lattice consisting of corner shared diamond plaquettes (Fig. 1). The diamond plaquette is a square plaquette of 4 spins with nearest neighbor interaction and additional bound, connecting two opposite spins. If we consider the Hamiltonian of MSE model and restrict ourselves to the two-, three-, and four-spin exchanges, which is the simplest case, we arrive at he following Hamiltonian

$$\mathcal{H} = J\sum_{(i,j)} \boldsymbol{\sigma}_i \cdot \boldsymbol{\sigma}_j + K\sum_p h_p - \mu H\sum_i {\sigma_i}^z, \quad (5)$$

$$\begin{aligned} h_p = &\sum_{1\leq i\leq j\leq 4} \boldsymbol{\sigma}_i \cdot \boldsymbol{\sigma}_j + (\boldsymbol{\sigma}_1 \cdot \boldsymbol{\sigma}_2)(\boldsymbol{\sigma}_3 \cdot \boldsymbol{\sigma}_4) + \\ &+ (\boldsymbol{\sigma}_1 \cdot \boldsymbol{\sigma}_4)(\boldsymbol{\sigma}_2 \cdot \boldsymbol{\sigma}_3) - (\boldsymbol{\sigma}_1 \cdot \boldsymbol{\sigma}_3)(\boldsymbol{\sigma}_2 \cdot \boldsymbol{\sigma}_4), \end{aligned} \quad (6)$$

where $J = J_3 - J_2/2$ and $K = -J_4/4$. This model has been successfully applied for describing the properties of solid ^{3}He films. In our case we omit the off-diagonal part of the Hamiltonian (6), which is equivalent to the formal replacement of all spin operators by simple Ising variables that take values ± 1. Putting the system described by the Hamiltonian 6 onto the lattice depicted in Fig. 1, we get the following model

$$\begin{aligned} -\beta\mathcal{H} = &\sum_i \alpha_1 s_i s_{i+1} + \alpha_2 (s_i + s_{i+1})(t_i + \tau_i) + \\ &+ \alpha_3 (t_i\tau_i + s_i s_{i+1} t_i \tau_i) + h(s_i + t_i + \tau_i), \end{aligned} \quad (7)$$

where

$$\begin{aligned} \alpha_1 &= \beta(J_3 - J_2/2 - J_4/4), \quad (8) \\ \alpha_2 &= \beta(J_3/2 - J_2/2 - J_4/4), \\ \alpha_3 &= -\beta J_4/4. \end{aligned}$$

Here the Ising spins placed in the corners of diamond plaquette chain are denoted by t_i and τ_i, and spins placed in the middle line by s_i. The system allows the exact calculation of the partition function, which can be represented as a trace of the N-th power of the corresponding transfer-matrix[11]

$$Z = \sum_{(s,t,\tau)} e^{-\beta\mathcal{H}} = Sp\,\mathbf{T}^N. \quad (9)$$

The components of transfer-matrix

$$\mathbf{T} = \begin{pmatrix} T_{++}, & T_{+-} \\ T_{-+}, & T_{--} \end{pmatrix}, \quad (10)$$

are

$$\begin{aligned} T_{++} &= 2e^{\alpha_1+h}\left(e^{2\alpha_3}\cosh(4\alpha_2+2h) + e^{2\alpha_3}\right), \\ T_{+-} &= T_{-+} = 2e^{-\alpha_1}\left(\cosh(2h)+1\right), \\ T_{--} &= 2e^{\alpha_1-h}\left(e^{2\alpha_3}\cosh(4\alpha_2-2h) + e^{2\alpha_3}\right). \end{aligned} \quad (11)$$

As usual, to calculate the partition function, we should obtain the eigenvalues of the transfer matrix. To be precise we need only the maximal eigenvalue, because only this one survives in the thermodynamical limit $N \to \infty$. So, $Z_N = \lambda^N_{max}$. The maximal eigenvalue is

$$\begin{aligned} \lambda = &A\cosh(h) + B\cosh(3h) + \\ &+ \sqrt{\sum_{k=o}^{3} C_k \cosh(2kh)}, \end{aligned} \quad (12)$$

where the coefficients are

$$\begin{aligned} A &= e^{\alpha_1-4\alpha_2+2\alpha_3} + 2e^{\alpha_1-2\alpha_3}, \\ B &= e^{\alpha_1+4\alpha_2+2\alpha_3}, \\ C_0 &= e^{2\alpha_1}(6e^{-4\alpha_1} + 2e^{-4\alpha_2} - 2e^{-4\alpha_3} \\ &\quad -e^{4\alpha_3}\cosh(8\alpha_2)), \\ C_1 &= e^{2\alpha_1+4\alpha_3}\left(1 + \frac{1}{2}e^{-8\alpha_2} + 2e^{-8\alpha_3}\right) \\ &\quad -4e^{2\alpha 1}\cosh(4\alpha_2) + 8e^{-2\alpha_1}, \\ C_2 &= e^{2\alpha_2}\left(2e^{-4\alpha_1} + 2e^{4\alpha_2} - e^{4\alpha_3}\right), \\ C_3 &= \frac{1}{2}2e^{2\alpha_1+8\alpha_2+4\alpha_3}. \end{aligned} \quad (13)$$

Having the partition function, we can obtain the free energy of the system per one spin in the thermodynamical limit

$$f = -\frac{1}{\beta}\lim_{N\to\infty}\frac{\log\lambda^N}{3N} = -\frac{1}{3\beta}\log\lambda. \quad (14)$$

Then, using the conventional thermodynamical relations $m = -\left(\frac{\partial f}{\partial H}\right)$, we obtain the magnetization per spin as a explicit function

$$m = \frac{A\sinh(h) + 3B\sinh(3h) + \frac{\sqrt{\sum_{k=1}^{3} kC_k\sinh(2kh)}}{\sqrt{\sum_{k=o}^{3} C_k\cosh(2kh)}}}{3\left(A\cosh(h) + B\cosh(3h) + \sqrt{\sum_{k=o}^{3} C_k\cosh(2kh)}\right)}. \quad (15)$$

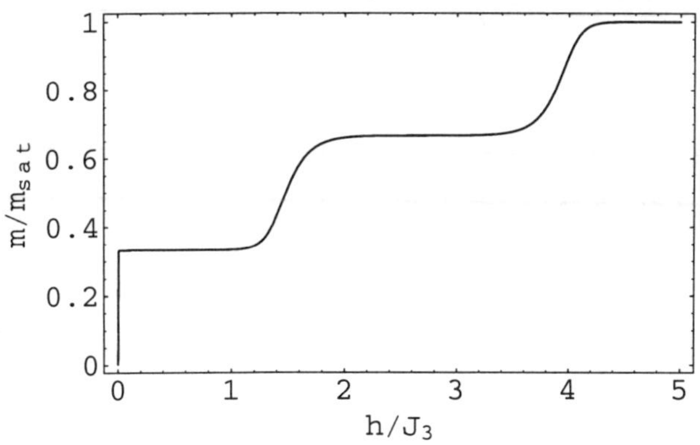

Figure 2. The magnetization curve for Ising plaquette chain at $T/J_3 = 0.17$, $J_2/J_3 = 1.5$ and $J_4/J_3 = 1.7$.

Having this function, we can draw the plots of the magnetization processes for all finite temperatures and arbitrary values of coupling constants. First of all, due to its geometry, the system is highly frustrated in case of anti-ferromagnetic effective coupling constants. This means that even at $T = 0$ the ground state of the system is disordered, because the arrangement of spins on lattice precludes the satisfying of all interaction simultaneously. However, it is possible to choose such MSE coupling constants at which the frustration will be partially or entirely removed in the Ising limit. Among the variety of magnetization curves obtained for different sets of J_2, J_3 and J_4, the most remarkable is the one with two magnetization plateaux at $m = 1/3$ and $m = 2/3$ in the units of full moment, Fig. 2. At that region of coupling constants the system is highly frustrated and it is not so easy to determinate the microscopic spin configurations corresponding to these plateaux. Apparently, they are some complex periodic structures with spatial period being at lest 3 for 1/3-plateau and 6 for 2/3-plateau. The investigation of the entire MSE model on diamond plaquette chain may drastically change the picture obtained by us. On the one hand, the appearance of other plateaux is possible, on the other hand, the plateaux pertained to the Ising system might not survive in quantum case. The typical example is the simple $S = 1/2$ Heisenberg chain which is gapless, whereas the corresponding Ising system exhibits the plateau at $m = 0$. It is noteworthy that apparently among the overdoped $RCuO_{2+x}$ (R=Y, La, *etc.*) compounds[12] the ones are possible whose magnetic lattice are analogous to that considered here.

References

1. D.J. Thouless, *Proc. Phys. Soc. London* **86**, 893 (1965).
2. G. Misguich, B. Bernu, C. Lhuillier and C. Waldtmann, *Phys. Rev. Lett.* **81**, 1098 (1998); G. Misguich, C. Lhuillier, B. Bernu and C. Waldtmann, *Phys. Rev.* B **60**, 1064 (1999).
3. R. Coldea *et al.* *Phys. Rev. Lett.* **86**, 5377 (2001); M. Matsuda, K. Katsumata, R.S. Eccleston, S. Brehmer and H.-J. Mikeska *Phys. Rev.* B **62**, 8903 (2000); M. Windt *et al.* *Phys. Rev. Lett.* **87**, 127002 (2001); K.P. Schmidt, C. Knetter and G.S. Uhrig, *Europhys. Lett.* **56**, 877 (2001).
4. M. Müller, T. Vekua and H.-J. Mikeska, *Phys. Rev.* B **66**, 134423 (2002); V. Gritsev, B. Normand and D. Baeriswyl, *Phys. Rev.* B **69**, 094431 (2004) and references therein.
5. K. Kubo and T. Momoi, *Z. Phys.* B **103**, 485 (1997).
6. T. Momoi, H. Sakamoto and K. Kubo, *Phys. Rev.* B **59**, 9491 (1999).
7. D.C. Cabra, M.D. Grynberg, A. Honecker and P. Pujol in *Condensed Matter Theories*, vol. 16. eds. S. Hernández and J. W. Clark (Nova Science Publishers, New York, 2000); A. Honecker, J. Schulenburg and J. Richter, *J. Phys.: Condens. Matter* **16**, S749 (2004) and references therein.
8. J. Strečka and M. Jaščur, *J. Phys.: Condens. Matter* **15**, 4519 (2003).
9. V.R. Ohanyan and N.S. Ananikian, *Phys. Lett.* A **307**, 76 (2003).
10. T.R. Arakelyan, V.R. Ohanyan, L.N. Ananikyan, N.S. Ananikian and M. Roger, *Phys. Rev.* B **67**, 024424 (2003).
11. R. Baxter in *Exactly Solved Models in Statistical Mechanics* (Academic Press, 1982).
12. R.J. Cava *et al.*, *J. Solid State Chem.* **104**, 437 (1993).

EXACTLY-SOLVABLE PROBLEMS FOR TWO-DIMENSIONAL EXCITONS

D. G. W. PARFITT and M. E. PORTNOI
School of Physics, University of Exeter
Stocker Road, Exeter EX4 4QL, United Kingdom
E-mail: d.g.w.parfitt@exeter.ac.uk

Several problems in mathematical physics relating to excitons in two dimensions are considered. First, a fascinating numerical result from a theoretical treatment of screened excitons stimulates a re-evaluation of the two-dimensional hydrogen atom. This yields a new integral relation in terms of special functions, and fresh insights into the dynamical symmetry of the system are also obtained. The second problem relates to excitons in a quantizing magnetic field in the fractional quantum Hall regime. An exciton against the background of an incompressible quantum liquid is modeled as a few-particle neutral composite. A complete set of exciton basis functions is derived and classified, and some exact results are obtained for this complex few-particle problem.

1 Two-Dimensional Hydrogen Atom Revisited

[a]The optical properties of semiconductor nanostructures are often governed by excitons and their screening. The simplest way to model an exciton is via the Thomas-Fermi approximation. This results in the so-called Stern-Howard potential[1], which has the following form in momentum space

$$V_s(q) = -\frac{4\pi}{(q+q_s)}. \tag{1}$$

Note that we use excitonic Rydberg units throughout this Section, so that all lengths are measures in terms of exciton Bohr radius and energies in units of excitonic Rydberg. Numerical calculations[2] have shown that with increasing screening, bound states in this potential disappear at integer values of the inverse screening wave number $1/q_s$. The most compact way to formulate this numerical observation is in terms of a homogeneous integral equation

$$q^2\Phi(\mathbf{q}) = \frac{\lambda}{\pi}\int \frac{\Phi(\mathbf{q}')}{|\mathbf{q}-\mathbf{q}'|+1}\, d^2q', \tag{2}$$

where $\lambda = 1/q_s$ is an eigenvalue. It follows[2] that $\Phi(q)$ is square integrable when

$$\lambda = \frac{(2|m|+\nu)(2|m|+\nu+1)}{2}, \tag{3}$$

where m is the internal angular momentum of the exciton, and $\nu = 0, 1, 2, \ldots$ is the number of non-zero nodes in the radial wave function. It should be emphasized that this result is of a numerical nature and has not been proven analytically. In an attempt to find such a proof we address the problem of an unscreened 2D exciton, which has a well-known solution in real space.

The unscreened exciton problem is analogous to the familiar two-dimensional hydrogen atom, the only difference being that the proton is replaced by a valence hole. In momentum space, the Schrödinger equation for the relative motion of the 2D unscreened exciton is given by an integral equation:

$$(q^2+q_0^2)\,\Phi(\mathbf{q}) = \frac{1}{\pi}\int \frac{\Phi(\mathbf{q}')\,d\mathbf{q}'}{|\mathbf{q}-\mathbf{q}'|}, \tag{4}$$

where q_0 is related to the energy eigenvalue via $q_0^2 = -E$. These energy eigenvalues are given by

$$E = -\frac{1}{(n+1/2)^2}, \quad n = 0, 1, 2, \ldots, \tag{5}$$

where n is the principal quantum number. Note that the energy is degenerate in the azimuthal quantum number m: this "accidental" degeneracy is due to the presence of a conserved quantity known as the *Runge-Lenz vector* $\hat{\mathbf{A}}$, which has two components in the plane of relative motion and is defined by

$$\hat{\mathbf{A}} = (\hat{\mathbf{q}}\times\hat{\mathbf{L}}_z - \hat{\mathbf{L}}_z\times\hat{\mathbf{q}}) - \frac{2}{\rho}\boldsymbol{\rho}, \tag{6}$$

where $\hat{\mathbf{L}}_z$ is an operator corresponding to the projection of the angular momentum perpendicular to the plane of motion, and $\boldsymbol{\rho}$ is the radius vector.

Equation (4) can be solved by projecting the 2D momentum space onto the surface of a three-dimensional (3D) unit sphere, in parallel with Fock's well-known approach to the 3D hydrogenic problem[3]. Each point on a unit sphere is completely defined by

[a]Talk presented by D.G.W.P. at the XIth Regional Conference on Mathematical Physics, May 2004, Tehran, Iran.

two polar angles, θ and ϕ, and the Cartesian coordinates of a point on the sphere are given by

$$u_x = \sin\theta\cos\phi = \frac{2q_0 q_x}{q^2 + {q_0}^2}, \tag{7}$$

$$u_y = \sin\theta\sin\phi = \frac{2q_0 q_y}{q^2 + {q_0}^2}, \tag{8}$$

$$u_z = \cos\theta = \frac{q^2 - {q_0}^2}{q^2 + {q_0}^2}. \tag{9}$$

An element of surface area on the unit sphere is

$$d\Omega = \sin\theta\, d\theta\, d\phi = \left(\frac{2q_0}{q^2 + {q_0}^2}\right)^2 d\mathbf{q}, \tag{10}$$

and the distance between two points transforms as

$$|\mathbf{u} - \mathbf{u}'| = \frac{2q_0}{(q^2 + {q_0}^2)^{1/2}(q'^2 + {q_0}^2)^{1/2}}|\mathbf{q} - \mathbf{q}'|. \tag{11}$$

If the wave function on the sphere is written as

$$\chi(\mathbf{u}) = \frac{1}{\sqrt{q_0}}\left(\frac{q^2 + {q_0}^2}{2q_0}\right)^{3/2}\Phi(\mathbf{q}), \tag{12}$$

then (4) reduces to the simple form:

$$\chi(\mathbf{u}) = \frac{1}{2\pi q_0}\int \frac{\chi(\mathbf{u}')\, d\Omega'}{|\mathbf{u} - \mathbf{u}'|}. \tag{13}$$

The eigenfunctions are then expanded in terms of spherical harmonics, and inverting the transformation in (12) yields for the 2D momentum-space eigenfunctions

$$\Phi_{nm}(\mathbf{q}) = (-i)^{|m|}\sqrt{2\pi\frac{(n-|m|)!}{(n+|m|)!}}\left(\frac{2q_0}{q^2 + q_0^2}\right)^{3/2} \times P_n^{|m|}\left(\frac{q^2 - q_0^2}{q^2 + q_0^2}\right)e^{im\phi}, \tag{14}$$

where $P_n^m(z)$ is an associated Legendre function.

Fourier transforming the above functions into real space and comparing them with the standard real-space eigenfunctions leads to a previously un-tabulated integral relation in terms of special functions[4]

$$\int_0^\infty P_n^m\left(\frac{1-y}{1+y}\right)\frac{J_m(x\sqrt{y})}{(1+y)^{3/2}}\,dy = \frac{(-1)^n(2x)^m e^{-x}}{n + 1/2}L_{n-m}^{2m}(2x), \tag{15}$$

where $n, m = 0, 1, 2, \ldots;\ m \le n$, $J_m(z)$ is a Bessel function and $L_n^m(z)$ is an associated Laguerre polynomial.

It can also be shown[4] that the two components of the Runge-Lenz vector in real space, together with the operator $\hat{\mathbf{L}}_z$, correspond to the generators of infinitesimal rotations in 3D momentum space.

2 Anyon Exciton: Exact Solutions

An anyon exciton is a composite neutral particle consisting of a valence hole and several fractionally-charged quasi-electrons. These quasi-electrons are the elementary excitations of an incompressible quantum liquid (IQL), which underlies the fractional quantum Hall effect. They are *anyons*, *i.e.* particles obeying fractional statistics, which means that their wave function acquires a complex phase-factor upon particle interchange. The anyon exciton model[5] was introduced a decade ago to explain intrinsic photoluminescence (PL) experiments in the fractional quantum Hall regime. It was later developed[6] to model IQLs with filling factors $\nu = 1/3$ and $2/3$, and is valid at large separation between the hole and the two-dimensional electron gas, so that the Coulomb field of the hole does not destroy the IQL. Recent developments in experimental techniques[7] have allowed the effective electron-hole separation (in units of magnetic length) to be changed while keeping the filling factor constant, and thus direct verification of the AEM is now possible.

We generalize the model to an exciton consisting of a valence hole and N anyons with charge $-e/N$ and statistical factor α. The hole and anyons reside in different layers, separated by a distance of h magnetic lengths, and are subject to a magnetic field $\mathbf{H} = H\hat{\mathbf{z}}$ perpendicular to their planes of confinement. We assume that the hole and quasi-electrons are in their corresponding lowest Landau levels, and as the exciton is neutral we can assign it an in-plane momentum $\mathbf{k}$.

The Hamiltonian for $N+1$ non-interacting particles in a quantizing magnetic field is

$$\hat{H}_0 = \frac{1}{2m_h}\left(\frac{1}{i}\nabla_h - [\hat{\mathbf{z}} \times \mathbf{r}_h]\right)^2 + \sum_{j=1}^{N}\frac{1}{2m_a}\left(\frac{1}{i}\nabla_j + \frac{1}{N}[\hat{\mathbf{z}} \times \mathbf{r}_j]\right)^2, \tag{16}$$

where $\mathbf{r}_h$ and $\mathbf{r}_j$ are the position vectors of the hole and anyons, respectively, and we have chosen the symmetric gauge $\mathbf{A} = [\mathbf{H} \times \mathbf{r}]/2$. Note that all distances are scaled with the magnetic length $l_H = (c\hbar/eH)^{1/2}$, and as usual, e, $\hbar$, c, and the dielectric constant are assumed equal to unity.

It can be shown[8] that the wave function satisfying this Hamiltonian can be written in the most general form as

$$\Psi = \exp\{i\mathbf{k}\cdot\mathbf{R} + i\hat{\mathbf{z}}\cdot[\mathbf{R}\times\boldsymbol{\rho}]/2 - (\boldsymbol{\rho}-\mathbf{d})^2/4\} \times P_L(\bar{\zeta}_i)\prod_{j<l}(\bar{\zeta}_j - \bar{\zeta}_l)^\alpha \prod_p e^{-|\zeta_p|^2/4N}, \quad (17)$$

where $\mathbf{d} = \mathbf{k}\times\hat{\mathbf{z}}$ is the exciton dipole moment, and we have introduced the following new coordinates

$$\mathbf{R} = \frac{1}{2}\left(\mathbf{r}_h + \frac{1}{N}\sum_{j=1}^{N}\mathbf{r}_j\right), \quad (18)$$

$$\boldsymbol{\rho} = \mathbf{r}_h - \frac{1}{N}\sum_{j=1}^{N}\mathbf{r}_j, \quad (19)$$

$$\boldsymbol{\xi}_j = \mathbf{r}_j - \frac{1}{N}\sum_{l=1}^{N}\mathbf{r}_l, \quad (20)$$

together with the set of complex coordinates $\zeta_j = \xi_{xj} + i\xi_{yj}$. Note also the following constraint on these coordinates

$$\sum_{j=1}^{N}\boldsymbol{\xi}_j = \sum_{j=1}^{N}\zeta_j = 0. \quad (21)$$

In (17), P_L is a symmetric polynomial of degree L in the variables $\bar{\zeta}_i$. Note that for $\mathbf{k} = 0$ the problem has rotational symmetry about the z-axis, and the degree of the symmetric polynomial L is related to the exciton angular momentum $[L_z = -L - N(N-1)\alpha/2]$. Thus, the problem of classifying all states of a $(N+1)$-particle anyon exciton can be reduced to finding all possible symmetric polynomials P_L for a particular value of L, taking into account constraint (21). This constraint means that there is no first-order polynomial ($L = 1$) for any N.

All linearly-independent symmetric basis polynomials of a particular degree may be enumerated by considering the possible products of elementary symmetric polynomials[9], as the total degree L is the sum of the degrees of the constituent polynomials. All such products may be enumerated by using the following result from the theory of partitions[10].

The number of ways of partitioning a number L into partitions of size $1, 2, \ldots, M$ is given by the coefficient of x^L in the expansion of

$$\frac{1}{1-x}\cdot\frac{1}{1-x^2}\cdots\frac{1}{1-x^M} = \prod_{k=1}^{M}\frac{1}{1-x^k}. \quad (22)$$

The number of ways of partitioning increases rapidly with L and does not follow any pattern. Furthermore, the different products of elementary symmetric polynomials for a particular value of L must be determined by hand. Note, however, that the presence of constraint (21) significantly reduces the number of possible symmetric polynomials by removing the first factor in the product (22).

We now introduce anyon-anyon and anyon-hole Coulomb interactions as follows:

$$V_{aa} = \frac{1}{N^2}\sum_{j<l}\frac{1}{|\boldsymbol{\xi}_j - \boldsymbol{\xi}_l|}, \quad (23)$$

$$V_{ah} = -\frac{1}{N}\sum_{j=1}^{N}\frac{1}{|\boldsymbol{\xi}_j - \boldsymbol{\rho} + h\hat{\mathbf{z}}|}. \quad (24)$$

Solving the Schrödinger equation for the few-particle system in a boson approximation ($\alpha = 0$) allows us to derive some exact results for the $(N+1)$-particle anyon exciton problem. For example, the binding energy for a $(N+1)$-particle exciton with $\mathbf{k} = 0$ and $L = 0$ is given by

$$E_b = \sqrt{\frac{\pi}{2N}}e^{h^2/2N}\operatorname{erfc}\left(\frac{h}{\sqrt{2N}}\right) - \frac{(N-1)\sqrt{\pi}}{4N\sqrt{N}}, \quad (25)$$

where $\operatorname{erfc}(x)$ is the complementary error function and the energy is measured in units of $e^2/(\epsilon l_H)$, where ϵ is the dielectric constant.

References

1. F. Stern and W.E. Howard, *Phys. Rev.* **163**, 816 (1967).
2. M.E. Portnoi and I. Galbraith, *Solid State Commun.* **103**, 325 (1997).
3. V.A. Fock, *Z. Phys.* **98**, 145 (1935).
4. D.G.W. Parfitt and M.E. Portnoi, *J. Math. Phys.* **43**, 4681 (2002).
5. E.I. Rashba and M.E. Portnoi, *Phys. Rev. Lett.* **70**, 3315 (1993).
6. M.E. Portnoi and E.I. Rashba, *Phys. Rev.* B **54**, 13791 (1996).
7. G. Yusa, H. Shtrikman and I. Bar-Joseph, *Phys. Rev. Lett.* **87**, 216402 (2001).
8. D.G.W. Parfitt and M.E. Portnoi, *Phys. Rev.* B **68**, 035306 (2003).
9. G. Birkhoff and S. MacLane in *A Survey of Modern Algebra*, 4th ed. (Macmillan, New York, 1977).
10. A. Slomson in *An Introduction to Combinatorics*, (Chapman and Hall, London, 1991).

CHAPTER 3: HIGH ENERGY PHYSICS; PHENOMENOLOGY

HADRONIC STRUCTURE FUNCTIONS FROM THE UNIVERSAL AND THE BASIC STRUCTURES

F. ARASH
Physics Department, Tafresh University, Tafresh, Iran
E-mail: farash@cic.aut.ac.ir

It is shown that there is a basic structure common to all hadrons, which is generated perturbatively in QCD. Basically, it is a valence quark with its own cloud of quarks and gluons, a quasi-particle that we will call it a *valon*. In the valon representation, structure functions of nucleon and pion are calculated and is shown that there is an excellent agreement between the data and the model results in a wide range of kinematics. Calculation of the polarized structure functions also shows that there is a sizeable orbital angular momentum contribution to the spin of a valon coming from the partonic cloud.

1 Introduction

[a]The parton distributions in proton have been studied extensively in recent years both theoretically and experimentally over a wide range of Q^2 and x. Some data are also available for pion structure function from ZEUS collaboration and from the Drell-Yan experiments[1]. The global analysis of parton distribution functions (PDF's) relies on the fitting some 1300 data points with an input PDF function in the Next-to-Leading Order (NLO). It is not clear, however, that if one moves away from proton, where the data points are ample, and wants to study the structure of other hadrons, like pion, with lesser amount of data points, if the insight gained from the analysis of proton data will be useful. Therefore, it is desirable to present a model, which provides a simple parameterization of parton distributions for a basic structure which is common to all hadrons. In other words, the structure of a valon is independent of the hosting hadron and once this structure is determined, it can be used to obtain insights into other hadronic structures. The valon plays a role in scattering problems as the constituent quark does in bound-state problems. The notion of valon was introduced in early 80's by R.C. Hwa[2], and since then, it is shown that it can be applied to a number of hadronic processes, successfully.

The valon structure originates from QCD processes, and its dressing is recently calculated in the NLO in QCD[3], where it is also applied to proton structure function, $F_2^p(x,Q^2)$ in a wide range of kinematics, namely: $1 < Q^2 < 5000$ and $10^{-4} < x < 1$. An excellent agreement with the experimental data is obtained and also it compares very nicely with the various global fit results. Due to the lack of space here, the interested reader should see Ref. [3].

The purpose of this paper is to show that the same universal structure yields accurate results for pion structure functions and also elaborates on the polarized structure function of proton.

2 Formalism

Within the framework of NLO calculations the structure of a valon is calculated and a Q^2 dependent parameterization is obtained, see Ref. [3]. It suffices here to give the results in its parameterized form as follows

$$zq^{valon}_{valence}(z,Q^2) = az^b(1-z)^c, \tag{1}$$

$$zq^{valon}_{sea}(z,Q^2) = \alpha z^\beta (1-z)^\gamma [1+\eta z + \xi z^{0.5}]. \tag{2}$$

The parameters a, b, c, α, β, γ, η, and ξ are functions of Q^2 and are given in the appendix of Ref. [3]. Gluon distribution in a valon has an identical form as in (2) but with different parameter values.

The structure of any hadron can be written as

$$F_2^h(x,Q^2) = \sum_{valon} \int_x^1 \frac{dy}{y} G_{\frac{CQ}{h}}(y) F_2^{valon}(\frac{x}{y},Q^2), \tag{3}$$

where the summation runs over the number of valons in a particular hadron. F_2^{valon} is the valon structure function whose components are given in Eqs. (1) and (2). $G_{\frac{valon}{h}}(y)$ is the probability of finding a valon with the momentum fraction y of the hadron h. Their explicit forms can be found in Refs. [2] and [3] for proton. In Figs. 9 of Ref. [3], we have plotted the $F_2^p(x,Q^2)$ for the range of $Q^2 = 1 - 2000\ GeV^2$.

[a]Talk presented at the XIth Regional Conference on Mathematical Physics, May 2004, Tehran, Iran.

The predictions of the model agree rather well with the experimental data.

ZEUS collaboration has also measured the pion structure function at small x region[2]. The data are normalized in two different ways, namely, using pion flux and additive quark model. The two method of normalization differ by a factor of two. I have calculated F_2^π based on valon structure in Refs. [4] and [5] and indeed it yields interesting results. As an example, in Fig. 1, the model results at two values of Q^2 are shown. For the entire range of Q^2 see Ref. [4].

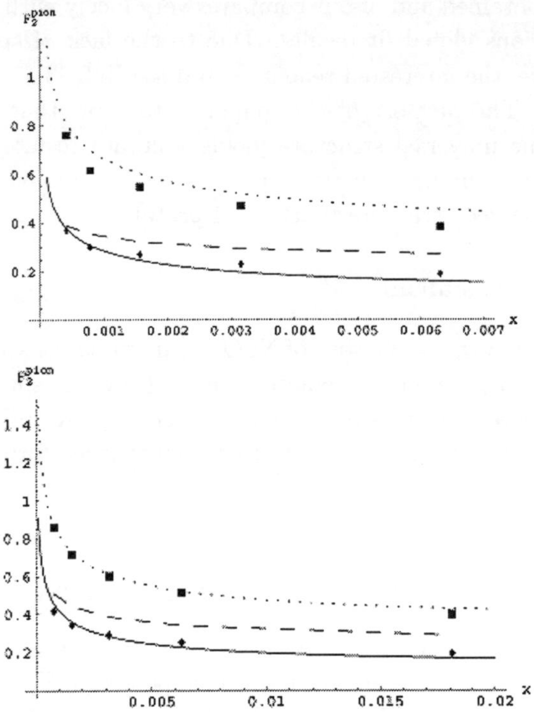

Figure 1. Pion structure function at $Q^2 = 7GeV^2$ (Left) and at $Q^2 = 15GeV^2$ (Right). The diamonds and squares are pion flux and additive quark model normalization of the data [1], respectively. The solid line represents the calculated results from the valon model. The dashed line is the result from SMRS and the dotted line corresponds to GRS determination.

An interesting observation is made that there is a simple relationship between F_2^p and the pion flux normalization of F_2^π, namely

$$F_2^{\pi(EF)}(x,Q^2) \approx kF_2^p(x,Q^2), \tag{4}$$

with $k = 0.361$. We have calculated the r.h.s. of (7) in our model and compared it with the effective flux normalization, $F_2^{\pi(EF)}(x,Q^2)$, in the l.h.s. The comparison is presented in Fig. 2. As one can see from the figures, the relationship holds rather well at all Q^2 values. Since our model produces very good fit to the proton structure function data in a wide range kinematics, we have also investigated the relation

$$F_2^\pi(x,Q^2) \approx \frac{2}{3}F_2^p(\frac{2}{3}x,Q^2), \tag{5}$$

which is based on color-dipole BFKL-Regee expansion and corresponds to the ZEUS's additive quark model normalization. Both sides of (5) is calculated in our model. The results are shown in Fig. 2. It is worth to note that the following relationship also holds very well between ZEUS's pion flux normalization data and F_2^p

$$F_2^{\pi(EF)}(x,Q^2) = \frac{1}{3}F_2^p(\frac{2}{3}x,Q^2). \tag{6}$$

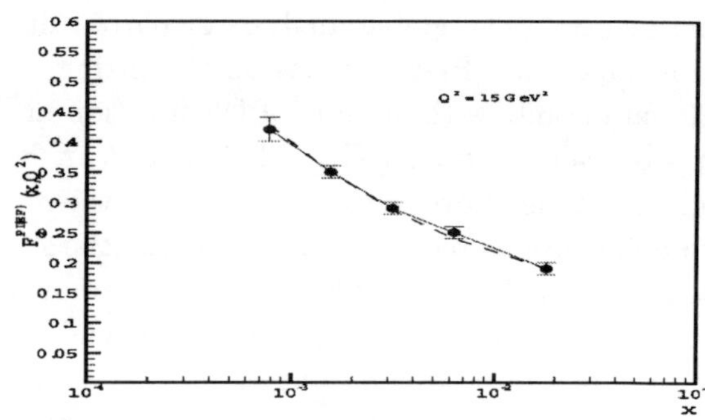

Figure 2. The effective flux normalization of F_2^π data as compared with the scaled F_2^p at $Q^2 = 15GeV^2$ of (5) and (6). The data points are from Ref. [1]. The solid line is $0.371F_2^p(x,Q^2)$ pertinent to (4) and the dashed line is that of (6), both are calculated for proton directly from the model.

A detailed calculations and results for pion, kaon and proton can be found in Refs. [3,4,5].

3 Polarized Structure of Nucleon

The case of polarized structure functions requires two extra ingredients, namely, finding the polarized valon distribution in the polarized hadron and the polarized parton distribution in a polarized valon. The first is achieved by defining a function $\delta F_j(y)$ via

$$\Delta G_j(y) = \delta F_j(y)G_j(y), \tag{7}$$

where the subscript j stands for a particular valon flavor. An additional constraint is imposed on $\delta F_j(y)$ such that the polarization of a valon in a proton to be fixed by the $SU(6)$ model, namely

$$P_U = \int_0^1 dy \delta F_U(y) G_U(y) = 2/3,$$

for U-type valon and similarly $P_D = -1/3$ for the D-type valon. The second task is carried out in the same way as for the unpolarized case: the moments of polarized structure function are evaluated and then using inverse Mellin transformation technique, the polarized parton distributions in a valon are obtained. This procedure provides a good agreement with the experimental data for g_1^p, g_1^n, and g_1^d. But it fails to give a spin 1/2 for the valon, and consequently, it does not account for the spin of nucleon. This means that there are additional contributions besides the partonic polarization. One way to overcome this shortcoming is to look for the orbital angular momentum of partons in a valon. That is, in the sum rule

$$S_z^U = \frac{1}{2}(S_z^v + S_z^{sea})^U + (S_z^{gluon})^U + L_z^U = \frac{1}{2}, \quad (8)$$

for a U-type valon, gluon polarization grows rapidly with increasing Q^2, where as the first term in the above sum rule remains largely constant. Therefore, the last term must compensate for the gluon helicity growth. In fact, valonic structure is very similar to the anisotropic superconductivity, where the *cooper* pairs need to rotate collectively around the core valence quark in opposite direction to that of the quark itself, giving rise to a negative back flow. Thus, there must be a large orbital angular momentum contribution to fulfill the above sum rule. In Fig. 3 this contribution along with the gluon polarization is shown.

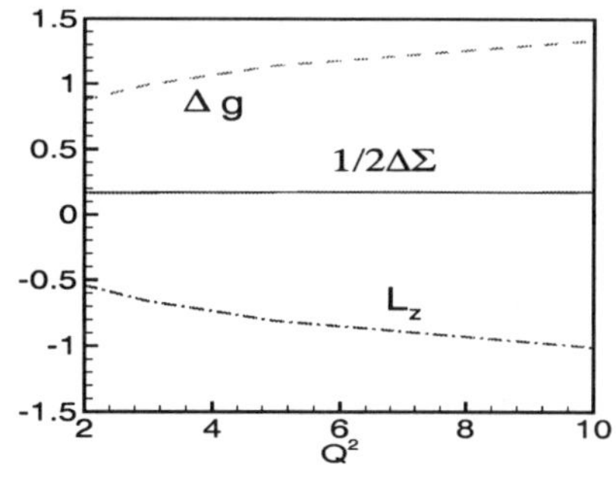

Figure 3. Various contributions to the spin of a U-type valon. The orbital angular momentum contribution is large and negative and almost cancels that of the gluon.

References

1. ZEUS Collaboration, S. Chrkanov, *et al.*, *Nucl. Phys.* B **637**, 3 (2002); J.S. Conway *et al.*, *Phys. Rev.* D **39**, 92 (1989).
2. R.C. Hwa and M.S. Zahir, *Phys. Rev.* D **23**, 2539 (1981); R.C. Hwa and C.B. Yang, *Phys. Rev.* C **66**, 025204.(2002).
3. F. Arash and A.N. Khorramian, *Phys. Rev.* C **67**, 045201 (2003).
4. F. Arash, *Phys. Lett.* B **557**, 38 (2003).
5. F. Arash, *Phys. Rev.* D **69**, 054024 (2004).

CONFINEMENT AND FAT-CENTER-VORTICES MODEL

S. DELDAR
Department of Physics, University of Tehran
North Karegar Ave., 14395 Tehran, Iran
E-mail: sdeldar@khayam.ut.ac.ir

In this paper, I review shortly potentials obtained for $SU(2)$, $SU(3)$ and $SU(4)$ static sources from fat-center-vortices model. Results confirm the confinement of quarks in all three gauge groups. Proportionality of string tensions with flux tube counting is better than Casimir scaling especially for $SU(4)$.

1 Introduction

[a]Today there is no doubt about the confinement of quarks which claims that quarks are not free in the nature and all we see are in color singlet states. In other words, the potential between quarks is linear and increases with distance

$$V(r) = -\frac{A}{r} + Kr + c \tag{1}$$

where the first term is Coulombic and K is the coefficient of the linear part of the potential and is called string tension. Many numerical calculations in $SU(2)$, $SU(3)$ and $SU(4)$ for fundamental and higher representations[1,2] confirm the confinement of quarks in QCD. Also there are some phenomenological models which explain the confinement in terms of gauge field configurations included instantons, merons, abelian monoploes and center vortices. QCD vacuum is assumed to be constructed from these gauge field configurations. In this paper, I review very shortly fat center vortices model and take a look at the results obtained using this model for $SU(2)$[3], $SU(3)$[4] and $SU(4)$[5]. I discuss the "Casimir scaling" and "flux tube counting", the two possible candidates for explaining the behavior of string tension of quarks in higher representations at intermediate distances.

2 Fat-Center Vortices Model

The center vortex theory was introduced by t'Hooft[6] in the late 1970's. The theory is able to explain quark confinement by describing the interaction between a center vortex – which is indeed a topological field configuration – and a Wilson loop. Potential between two quarks can be measured by studying the Wilson loop at large T

$$W(R,t) \simeq \exp^{-V(R)T}. \tag{2}$$

$W(R,T)$ is the Wilson loop as a function of R, the spatial separation of quarks, and the propagation time T, and $V(R)$ is the gauge field energy associated with the static quark-antiquark source. Briefly, the vortex theory states that the area law falloff for large T of Wilson loop is due to fluctuations in the number of center vortices linking the loop.

Potential between adjoint quarks is zero at large distances. This is because adjoint quarks are screened by gluons at this distance. Therefore center vortices do not have any effect on the Wilson loop of higher representations. But before the onset of screening, there exists a region where the potential between two sources is linear. Fat center vortices introduced by Greensite *et al.*[3] has been able to predict this linearity of the potential for all representations. Based on this model the vortex is thick enough such that its core somewhere overlaps the perimeter of the Wilson loop. More details about this improved vortex theory can be found in Ref. [3]. Here, I only introduce the Wilson loop obtained from this model and discuss results for $SU(2)$, $SU(3)$ and $SU(4)$ quarks.

The average Wilson loop is given by

$$< W(C) >= \prod_x \{1 - \sum_{n=1}^{N-1} f_n (1 - Re\mathcal{G}_r[\vec{\alpha}^n_C(x)])\}, \tag{3}$$

where x is the location of the center of the vortex and A is the area of the loop C and $\mathcal{G}_r$ is

$$\mathcal{G}_r[\vec{\alpha}] = \frac{1}{d_r} Tr \exp[i\vec{\alpha}.\vec{H}], \tag{4}$$

with d_r the dimension of the representation and $\{H_i, i = 1, 2, ..., N-1\}$ the generators of the group.

[a]Talk presented at the XIth Regional Conference on Mathematical Physics, May 2004, Tehran, Iran.

f is the probability that any given unit is pierced by a vortex and depends on what fraction of the vortex core is enclosed by the Wilson loop. It also depends on the shape of the loop and the position of the center of the vortex in the plane of loop C relative to the perimeter. $\alpha_R(x)$, the flux distribution of the vortex can be chosen in such a way that a good physical behavior for the potential is achieved. Examples for $SU(3)$ gauge group can be found in Ref. [4]. The physical flux distribution leads to confinement of sources at intermediate distances and screening at large distances for zero n-laity representations.

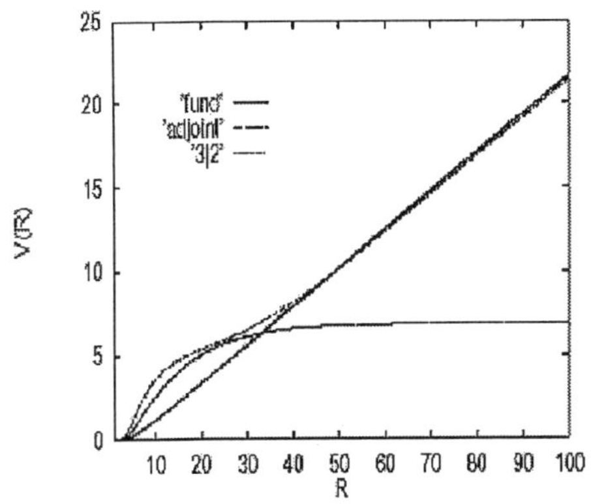

Figure 1. Interquark potential V(R) induced by center vortices for quarks in the $j = \frac{1}{2}$, 1, $\frac{3}{2}$ representations.

3 Results and Discussion

Figure 1 shows the potential versus distance for $SU(2)$ quarks in the fundamental and representations with $j = 1$ and $j = \frac{3}{2}$.[3] As seen from the plot, at large distances adjoint quarks are screened and potential for quarks in the $j = \frac{3}{2}$ representation gets the same slop as that of the fundamental representation. This is because as the distance between quarks increases, a pair of quarks in the adjoint representation (gluon-antigluon) pops out of the vacuum and makes a pair of gluon-antigluon with initial adjoint quark-antiquark (gluon-antigluon) such that the initial sources are not able to see each other (screened) and the force between them is zero. For quarks in representation $\frac{3}{2}$ the gluon-antigluon pair which is created from the vacuum, interacts with the initial sources and the result is a pair of quark-antiquark in the fundamental representation. That is why the slop of the potential changes to the slop of that of the fundamental representation. This behavior is seen for $SU(3)$[4] and $SU(4)$[5] sources as well. Figure 2 shows the potential versus R for representations 6, 8, 10, 15-symmetric, 15-antisymmetric and 27 in $SU(3)$. At large distances, the force between quarks of zero-triality representations 8, 10 and 27, is zero and for non zero triality representations, 6, 15a and 15s, potentials get parallel to that of the fundamental representation.

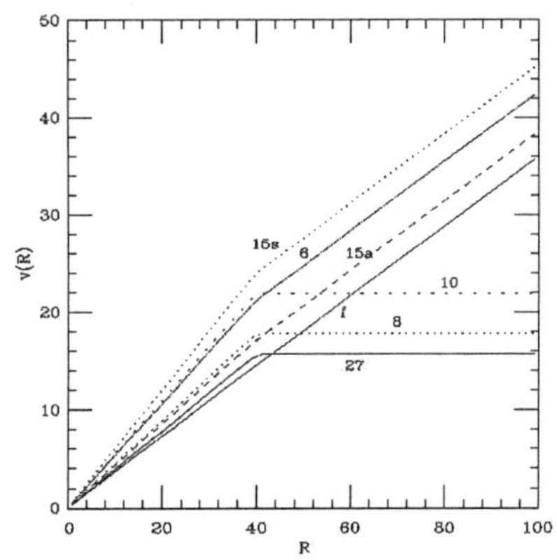

Figure 2. Potentials between static sources of $SU(3)$. Zero triality representations 8, 27 and 10 are screened at large distances and others become parallel to that of the fundamental representation. At intermediate distances $R \in [3, 5]$, the quark potentials are linear for all representations.

Also for $SU(4)$, Fig. 3, quarks in representations 15 and 35 are screened and non-zero 4-ality representations, 6, 10 and 20, become parallel to either representation 6 or 10. More details are available in Ref. [5]. On the other hand at intermediate distances for all sources in all representations for all three gauge groups, there exists a linear potential. The string tension which is the coefficient of the linear term of the potential has an interesting property. Numerical calculations[2] show that the string tension is proportional to the quadratic operator of the representation which is called "Casimir scaling"[7]. On the other hand, it seems that the string tension is also proportional to the number of fundamental tubes embedded into the higher representation. This idea is the so called "flux tube counting"[8]. Our recent results especially for static sources in $SU(4)$ agree better with flux tube counting than Casimir scaling.

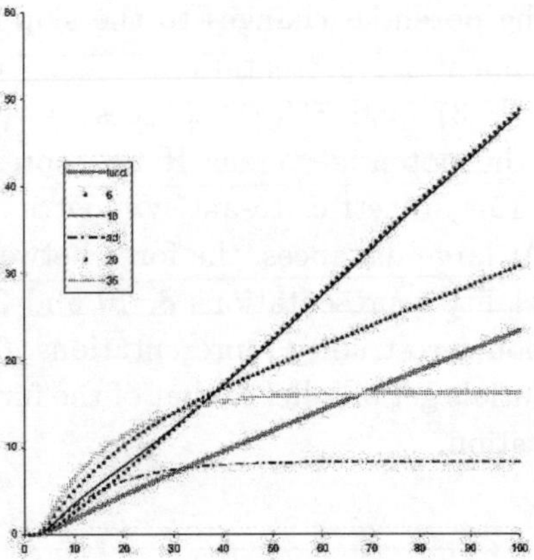

Figure 3. Potentials of static $SU(4)$ sources. At large distances, representations 15(adj) and 35 are screened; representation 20 gets the same slope as fundamental representation and representation 10 is paralleled to representation 6. String tensions at intermediate distances are qualitatively in agreement with the number of fundamental flux tubes.

Tables 1 and 2 show the results in $SU(3)$ and $SU(4)$, respectively. In each table, the first row indicates the representation. In the second row the number of original quarks and anti-quarks, (n,m), anticipated in each representation is shown. The ratio of Casimir scaling of each representation to the fundamental representation and the ratio of string tensions are given in the third and fourth row, respectively. The last row of each table indicates the number of fundamental fluxes, which exists in each representation. The agreement between the ratio of string tensions with both Casimir scaling and flux tube counting is qualitative but as seen from tables this agreement is better with flux tube counting than Casimir scaling, especially for $SU(4)$.

Acknowledgments

I would like to thank Tehran University Research Council for support of this work.

Table 1. This table shows Casimir numbers ratios, number of flux tubes and string tensions ratios for $SU(3)$ gauge group. Comparing to Casimir scaling, a better agreement between string tensions and flux tube counting is observed, especially for higher representations.

Repn.	$3(fund.)$	8	6	$15a$	10	27	$15s$
(n,m)	$(1,0)$	$(1,1)$	$(2,0)$	$(2,1)$	$(3,0)$	$(2,2)$	$(4,0)$
c_r/c_f	1	2.25	2.5	4	4.5	6.	7
fund. fluxes	1	2	2	3	3	4	4
k_r/k_f	1	2.02	2.21	3.1	3.4	3.8	5.6

Table 2. The same as Table 1 but for quarks in $SU(4)$.

Repn.	$4(fund.)$	6	$15(adj.)$	10	20	35
(n,m)	$(1,0)$	$(2,0)$	$(1,1)$	$(2,0)$	$(3,0)$	$(4,0)$
c_r/c_f	1	1.33	2.13	2.4	4.2	6.4
fund. fluxes	1	2	2	2	3	4
k_r/k_f	1	1.51	1.56	1.76	2.31	2.66

References

1. C. Bernard, *Phys. Lett.* B **108**, 431 (1982); *ibid. Nucl. Phys.* B **219**, 341 (1983); J. Ambjorn, P. Olesen and C. Peterson, *Nucl. Phys.* B **240**, 189 (1984); C. Michael, *Nucl. Phys.* B **259**, 58 (1985); G. Poulis and H. Trottier, *Phys. Lett.* B **400**, 358 (1997); S. Ohta and M. Wingate, *Nucl. Phys.* B **83**, 381 (2000).
2. G.S. Bali *Phys. Rev.* D **62**, 114503 (2000); S. Deldar *Phys. Rev.* D **62**, 034509 (2000).
3. M. Faber, J. Greensite and S. Olejnk, *Phys. Rev.* D **57**, 2603 (1998).
4. S. Deldar, *JHEP* **01**, 013 (2001).
5. S. Deldar, S Rafibaksh, hep-ph/0411184.
6. G. 't Hooft, *Nucl. Phys.* B **153**, 141 (1979).
7. J. Ambjorn, P. Olesen, C. Peterson, *Nucl. Phys.* B **240**, 533 (1984).
8. G.S. Bali, *Phys. Rept.* **343**, 1 (2001); A. Armoni and M. Shifman, *Nucl. Phys.* B **67**, 671 (2003).

$SU_L(4) \times U(1)$ MODEL FOR ELECTROWEAK UNIFICATION

FAYYAZUDDIN

National Center for Physics, Quaid-i-Azam University
Islamabad 45320, Pakistan
E-mail: ncp@comsats.net.pk

After some general remarks about unification, the electroweak unification group $SU_L(4) \times U_X(1)$ is discussed. Some salient features of this group are summarized.

1 Introduction

[a]The standard model of the particle physics is based on the group

$$G_1 \equiv SU_c(3) \times SU_L(2) \times U_Y(1).$$

The fermion content of G_1 for the each generation in their left-handed chirality state is

$$\begin{aligned} \begin{pmatrix} u \\ d \end{pmatrix}_{1/3} &: q(3,2,1/3), \\ \bar{u} &: (3,1,-4/3), \\ \bar{d} &: (\bar{3},1,2/3), \\ \begin{pmatrix} \nu_e \\ e^- \end{pmatrix}_{-1/2} &: \ell(1,2,-1/2), \\ e^+ &: (1,1,1). \end{aligned}$$

Thus, there are 15 two-component fermion states per generation. Unification based on the standard model is not true unification because i) there are three independent coupling constants ii) no charge quantization because of $U(1)$ factor.

To overcome these difficulties, one route is through grand unification (GUT) of electroweak and strong quark-gluon forces. The basic hypothesis is that there exists a group

$$G \supset G_1 \equiv SU_c(3) \times SU_L(2) \times U_Y(1).$$

This has been done for the group such as $SU(5)$, $O(10)$ or Pati-Salam group $SU_c(4) \times SU_L(2) \times SU_R(2)$. All these groups predict $\sin^2\theta_W = 3/8$ at unification scale; This unification scale is of order 10^{16} GeV. The reason for the huge desert is the large disparity between the measured value of $\sin^2\theta_W = 0.231$ from $\sin^2\theta_W = 3/8$ at the unification scale.

The other route is through extension of electroweak group $SU_L(2) \times U_Y(1)$ to some simple group such as $SU(3)$ or $SU(4)$. The electroweak unification models are based on these groups give $\sin^2\theta_W = 1/4$ close to the measured value of $\sin^2\theta_W$. Thus one would expect the unification scale to be in TeV range.

Unification model based on the group $SU_L(4)$ was considered by us in 1984[1].

It not only accommodates leptons nicely; but also a right-handed Majorana neutrino. The see-saw mechanism for generation of masses of the standard model neutrinos is a natural feature of this model. The charge operator in this model is given by

$$Q = \frac{1}{2}\left[\lambda_3 + \frac{1}{\sqrt{3}}\left(-\lambda_8 + \sqrt{2}\lambda_{15}\right) + \frac{1}{\sqrt{3}}\left(2\lambda_8 + \sqrt{2}\lambda_{15}\right)\right], \quad (1)$$

$$= \frac{1}{2}\left[\tau_{3L} + \tau_{3R} + Y_1\right]. \quad (2)$$

The leptons can be assigned to the fundamental representation of this group (first generation) as follows

$$F_e = \begin{pmatrix} e^+ \\ \nu_e^c \\ -N_e \\ e^- \end{pmatrix}_R : 4_R. \quad (3)$$

The vector bosons belong to the adjoint representation of $SU_L(4)$; Thus, there are 15 vector bosons: $(W_{12}, W_{21}) : \sqrt{2}W_L^{\mp}$, $(W_{34}, W_{43}) : \sqrt{2}W_R^{\pm}$, $(W_{13}, W_{31}) : \sqrt{2}X_1^{\mp}$, $(W_{23}, W_{32}) : \sqrt{2}\left(X_2^0, \bar{X}_2^0\right)$, $(W_{14}, W_{41}) : \sqrt{2}\left(Y_1^{--}, Y_1^{++}\right)$, $(W_{24}, W_{42}) : \sqrt{2}\left(Y_2^-, Y_2^+\right)$, and three neutral vector bosons, various linear combinations of these can be identified with photon A_μ, neutral vector boson Z_μ and an extra neutral vector boson Z'_μ. We note

$$\frac{1}{e^2} = \frac{2}{g^2} + \frac{1}{g'^2}, \quad (4)$$

$$\frac{1}{g'^2} = \frac{1}{g^2} + \frac{1}{g_1^2}. \quad (5)$$

[a]Talk presented at the $\mathrm{XI}^{\mathrm{th}}XI^{th}$ Regional Conference on Mathematical Physics, May 2004, Tehran, Iran.

In the symmetry limit $g_1 = g/\sqrt{2}$ and hence at the unification scale $g_1 = g/\sqrt{3}$, and $\sin^2\theta_W = 1/4$.

We note that representation (3) is not anomaly free. An anomaly free model can be constructed by putting Hahn and Nambu quarks with integral gauge charges in the fundamental representation $SU_L(4)$ as follows (for first generation)

$$F_d^r = \begin{pmatrix} d'^c \\ u^c \\ -u \\ d' \end{pmatrix}_R^r : 4_R \Longrightarrow 4_L^*,$$
$$F_d^{g,b} = \begin{pmatrix} u \\ -d \\ d^c \\ u^c \end{pmatrix}_L^{g,b} : 4_L. \tag{6}$$

It is clear that representations (3) and (6) together cancel the anomalies. The gauge symmetry $SU_L(4)$ is broken as follows

$$\begin{aligned} SU_L(4) \underset{m_{X,Y}}{\longrightarrow} & \; SU_L(2)\times SU_R(2)\times U_1(1), \\ \underset{m_R}{\longrightarrow} & \; SU_L(2)\times U_1(1) \underset{m_L}{\longrightarrow} U_{em}(1). \end{aligned} \tag{7}$$

As we have emphasized earlier this model works well for leptons. For quarks, the following comments are in order. The electromagnetic current can be split into two parts *viz.*, the color singlet (containing the fractionally charged quarks) and a color octet. The currents coupled to X and Y bosons do not contain any color singlets. Thus with the singlet part of the electromagnet current, it effectively reduces to the standard model of electroweak interactions[1], since known hadrons are color singlets. However, this model becomes more appealing if we could formulate this model in higher dimensions and gauge symmetry is broken by orbifold compactification to four dimensions so that X and Y bosons and non-color singlet part of electromagnetic current is relegated to the bulk and we have effectively fractional quarks on the brane.

2 Symmetry Group $SU_L(4)\times U_X(1)$

In order to accommodate the quarks with fractional charges, it is necessary to extend the group to $SU_L(4)\times U_X(1)$. For anomaly cancelation, the leptons and quarks must be assigned as follows

$$Y_X = 0,$$
$$4_R: \begin{pmatrix} e^+ \\ -\nu_e^c \\ N_e' \\ e^- \end{pmatrix}_R, \begin{pmatrix} \mu^+ \\ -\nu_\mu^c \\ N_\mu' \\ \mu^- \end{pmatrix}_R, \begin{pmatrix} \tau^+ \\ -\nu_\tau^c \\ N_\tau' \\ \tau^- \end{pmatrix}_R \Rightarrow 4_L^*,$$
$$Y_X = -2/3,$$
$$4_L: \begin{pmatrix} u^a \\ d'^a \\ D^a \\ H_d^a \end{pmatrix}_L, \begin{pmatrix} c^a \\ s'^a \\ S^a \\ H_s^a \end{pmatrix}_L,$$
$$\begin{aligned} I_R:\ & (u^a)_R,\ (d'^a)_R\ (D^a)_R\ (H_d^a)_R \\ Y_X =\ & \ 4/3 \quad -2/3 \quad -2/3 \quad -8/3\ , \\ & (c^a)_R,\ (s'^a)_R\ (S^a)_R\ (H_s^a)_R \\ & \ 4/3 \quad -2/3 \quad -2/3 \quad -8/3, \end{aligned} \tag{8}$$

$$4_R: \begin{pmatrix} b'^a \\ t^a \\ U^a \\ T^a \end{pmatrix}_R^c, \qquad Y_X = -4/3$$
$$\begin{aligned} I_L:\ & (b'^a)_L^c,\ (t'^a)_L^c,\ (U^a)_L^c\ (T_d^a)_L^c \\ Y_X =\ & +2/3 \quad -4/3 \quad -4/3 \quad -10/3. \end{aligned} \tag{9}$$

Here $a = 1, 2, 3$ is the color index. Thus we have six extra quarks (exotic) with charges

$$\begin{aligned} U, D, S &: (2/3, -1/3, -1/3), \\ H_d, H_s, T &: (-4/3, -4/3, 5/3). \end{aligned} \tag{10}$$

In the symmetry limit of $SU_L(4)$, $g_1 = g/\sqrt{2}$; thus we get

$$\frac{1}{g'^2} = \frac{3}{g^2} + \frac{1}{g_X^2}, \tag{11}$$
$$\frac{g^2}{g_X^2} = \frac{1 - 3\tan^2\theta_W(m_X)}{\tan^2\theta_W(m_X)},$$
$$\frac{g_X^2}{g^2} = \frac{\sin^2\theta_W(m_X)}{1 - 4\sin^2\theta_W(m_X)}. \tag{12}$$

Thus $\sin^2\theta_W(m_X) < 1/4$, implying the unification scale to be in TeV range. For $\sin^2\theta_W(m_X) = 1/4$, $g_X \to \infty$ and V_μ will decouple at m_X.

Here, it is relevant to give an estimate of the unification scale. We first consider the unification group $SU_L(4)$ and its breaking direct to the group $SU_L(2)\times U_Y(1)$. A straight forward application of renormalization group equations give[1]

$$\alpha^{-1}(m_Z)\left[1 - 4\sin^2\theta_W\right] = 2\left(-C_1^2\beta_2 + \beta_1\right)\ln\frac{m_X}{m_Z}, \tag{13}$$

where

$$\beta_2 = \frac{1}{4\pi}\left[-\frac{22}{3} + \frac{4}{3}\frac{n_f}{2}\right],$$

$$\beta_1 = \frac{1}{4\pi}\left[\frac{4}{3}\frac{n_f}{2}C_1^2\right],$$

$$C_1^2 = 2.$$

Using $\sin^2\theta_W(m_Z) = 0.2311$, $\alpha^{-1}(m_Z) = 128$, we get unification mass scale

$$m_X = 631 m_Z \simeq 5.8 TeV. \quad (14)$$

For the group $\mathrm{SU}_L(4)\times\mathrm{U}_X(1)$, we get

$$\frac{\alpha_X^{-1}(m_Z)}{\alpha_2^{-1}(m_Z)} = \frac{1-4\sin^2\theta_W}{\sin^2\theta_W},$$
$$-2\alpha_2(m_Z)\left(-C_1^2\beta_2+\beta_1\right)\ln\frac{m_X}{m_Z}.$$

This relation reduces to (13) for $\alpha_X^{-1} = 0$. Thus from (14), we conclude that upper bound for the unification mass scale is 5.8 TeV, for which $\alpha_X^{-1} = 0$. The unification scale of order 900 GeV, would correspond to $\alpha_X^{-1}(m_Z)/\alpha_2^{-1}(m_Z) \simeq 0.126$. The lower bound on unification mass scale would be determined by the experimental limit on lepton number violating processes. Salient features of this model are summarized below:

1. The lepton mass term is

$$m_\ell = m_\ell \bar{\ell}_L \ell_R + m'_\nu \bar{\nu}_{\ell L} U_{\ell\ell'} N_{\ell R} - m_N N_{\ell R}^T C N_{\ell R} + h.c.,$$

   where the matrix $U_{\ell\ell'}$ relates the weak eigenstates N'_ℓ with mass eigenstates N_ℓ. Thus the see-saw mechanism is a natural consequence of this model. The CKM type matrix U can accommodate many features of neutrino oscillations.

2. The lepton number violation occurs at a TeV scale. These processes are mediated by vector bosons X, Y. Some interesting lepton number violating processes that can occur at tree level, through the exchange of Y_2 and Y_1 bosons are:

$$\mu^- \to \bar{\nu}_\mu + e^- + \nu_e, \quad \Delta L_\mu = -2, \quad \Delta L_e = 2,$$
$$\mu^+ e^- \to \mu^- e^+,$$
$$\bar{\nu}_\mu e^- \to \bar{\nu}_e \mu^-.$$

   The experimental limit on the decay $\mu^- \to \bar{\nu}_\mu + e^- + \nu_e$

$$R = \frac{\Gamma(\mu^- \to \bar{\nu}_\mu + e^- + \nu_e)}{\Gamma(\mu^- \to all)} < 1.2\times 10^{-2};$$

   Since

$$R \propto \left(\frac{m_W}{m_Y}\right)^4,$$

   the above limit implies $m_Y > 3m_W$. This implies a lower limit for the unification scale *viz.*, $m_X > 3m_W$. Thus the upper limit for the unification scale (5.8 TeV) is much above the present experimental limit.

3. In the quark sector we have six extra heavy quarks; three of them *viz.*, (U, D, S) have charges $(2/3, -1/3, -1/3)$ and are mirrors of (u, d, s) quarks. However, these quarks would decay to light quarks by β-decay with the emission of right-handed Majorana neutrino at tree level through the exchange of X-bosons. A typical process is depicted below $(\Delta L_e = 2) D \to u + e^- + N_e$. The superheavy quarks (U, D, S) can form bound states with the light quarks u, d, s, c and b. Thus this model predict a replica of existing hadrons at a scale of few hundred GeV to TeV. The superheavy quarks H_d, H_s and T, have exotic charges $(-4/3, -4/3, 5/3)$. If these quarks have masses greater than Y bosons, they would decay quickly to light quarks by a process of the form $H_d \to u + Y_1^{--}$. Even if they have masses below Y-bosons, they are expected to decay quickly to light leptons by a typical process: (neutrinoless β-decay; $\Delta L_e = 2$) $H_d \to u + l^- + l^-$. Thus H_d, H_s and T may not live long enough to form bound states with light quarks.

In the end, it is interesting to note that any viable extension of the standard model $(SU(3)_c \times SU_L(2) \times U_Y(1))$ is not possible without extending the existing elementary constituents of matter (*viz.* six leptons and six quarks). In this sense, the standard model is unique if there are no additional constituents of matter.

Acknowledgments

The author wishes to thank IPM for the hospitality and Higher Education Commission of Pakistan for travel grant.

References

1. Fayyazuddin and Riazuddin, *Phys. Rev. D* **30**, 1041 (1984), arXiv: hep-ph/0403042.

THE ROLE OF HIGHER ORDER CORRECTIONS IN DETERMINING POLARIZED PARTON DENSITIES IN THE NUCLEON

A. N. KHORRAMIAN[*,†], A. MIRJALILI[*,‡] and S. ATASHBAR TEHRANI[*,♭]
[†]*Physics Department, Semnan University, Semnan, Iran*
[‡]*Physics Department, Yazd University, Yazd, Iran*
[♭]*Physics Department, Persian Gulf University 75168, Boushehr, Iran*
and
[*]*Institute for Studies in Theoretical Physics and Mathematics (IPM), School of Physics*
P.O. Box 19395-5531, Tehran, Iran
E-mails: khorramiana@theory.ipm.ac.ir, mirjalili, atashbar@ipm.ir

We study moments of polarized valon distributions in leading and next-to-leading order approximation. By computing the internal structure of polarized valons from inverse Mellin transformation and using their distributions in the nucleon, we will be able to calculate the polarized parton distributions and structure function in the nucleon.

1 Introduction

[a]In constituent quark model, the nucleon is envisaged as a bound state of a valence quark cluster called valon. Hwa[1,2] found evidence for valons in the deep inelastic neutrino scattering data, suggested their existence and applied it to a variety of phenomena. In Ref. [3], the valon model is used to extract new information for parton distribution and hadron structure functions. Here, we study the higher order corrections in moment space which are very important to calculate the polarized parton distributions (PPD's) in the nucleon.

2 Moments Analysis of PPD's

In this section, we will introduce moments of polarized parton distributions in the valon. To calculate the NLO evolutions of the polarized parton distributions for $\delta f(x, Q^2)$, we can have its Mellin moments by

$$\delta f^n(Q^2) = \int_0^1 x^{n-1} \delta f(x, Q^2) dx \, . \quad (1)$$

The Q^2 evolutions are governed by the anomalous dimension

$$\delta d_{NS\pm} = \frac{\alpha_s(Q^2)}{2\pi} \delta d_{qq}^{(0)}(x) + \left(\frac{\alpha_s(Q^2)}{2\pi}\right)^2 \delta d_{NS\pm}^{(1)}(x) \, , \quad (2)$$

whose detailed n-dependence will be specified[4]

$$L(Q^2) \equiv \frac{\alpha_s(Q^2)}{\alpha_s(Q_0^2)} \, . \quad (3)$$

The non-singlet (NS) part evolves according to

$$\delta \mathcal{M}_{NS\pm} = \left(1 - \frac{\alpha_s(Q^2) - \alpha_s(Q_0^2)}{2\pi} \delta d_{NS\pm}^{(1)n} - \frac{2\pi b'}{b} \delta d_{qq}^{(0)n}\right) L^{\delta d_{qq}^{(0)n}} , \quad (4)$$

and the NLO running coupling is given by

$$\alpha_s(Q^2) \cong \frac{1}{b \log \frac{Q^2}{\Lambda^2}} - \frac{b'}{b^3} \frac{\ln\left(\ln \frac{Q^2}{\Lambda^2}\right)}{\left(\ln \frac{Q^2}{\Lambda^2}\right)^2} , \quad (5)$$

where $b = \frac{33-2f}{12\pi}$ and $b' = \frac{306-38f}{48\pi^2}$. In the above equation, we chose $Q_0 = 1\ GeV^2$ as a fixed parameter and Λ is an unknown parameter which can be calculated by fitting to experimental data.

The evolution in the flavor singlet and gluon sector are governed by 2×2 anomalous dimension matrix with the explicit solution given by

$$\begin{pmatrix} \delta \mathcal{M}_S \\ \delta \mathcal{M}_{gQ} \end{pmatrix} = \left(L^{\delta \hat{d}^{(0)n}} + \frac{\alpha_s(Q^2)}{2\pi} \hat{U} L^{\delta \hat{d}^{(0)n}} - \frac{\alpha_s(Q_0^2)}{2\pi} L^{\delta \hat{d}^{(0)n}} \hat{U}\right) \begin{pmatrix} 1 \\ 1 \end{pmatrix} ,$$

where $\delta \mathcal{M}_{gQ}$ is the spin dependent quark-to-gluon evolution function. All associated functions in the above equation have been defined [4]. According to (4) and (6), we can calculate the moments of NS, S and gq as a function of n and for $Q^2 = 3\ GeV^2$, for example. The results of LO and NLO calculations are presented in Fig. 1.

---

[a]Talk presented by A.N.K. at the XIth Regional Conference on Mathematical Physics, May 2004, Tehran, Iran.

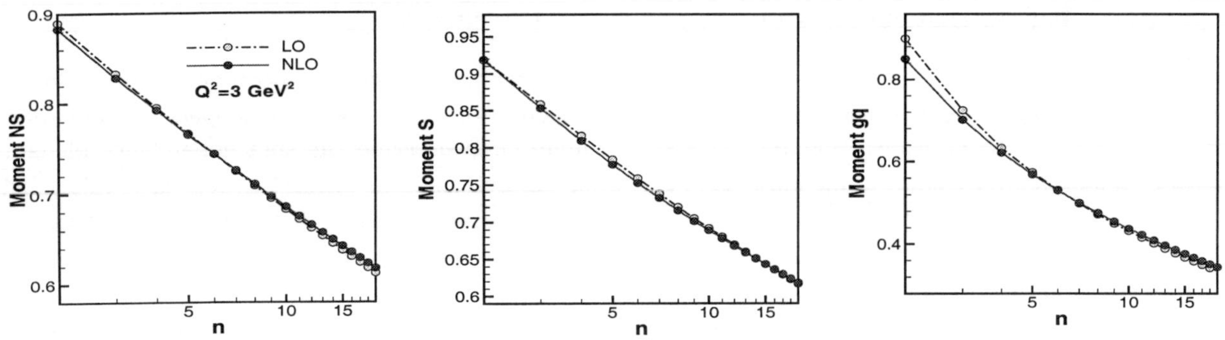

Figure 1. The contribution of moments NS, S and gq from Eqs. (4) and (6) as a function of n and for $Q^2 = 3\ GeV^2$. Solid line indicates NLO and dashed line indicates LO approximation.

3 Proton PDF's and Structure Function in Moments Space

By having the moments of polarized valon distributions[5], the moments of parton distributions in a proton can be strictly determined. The distributions that we shall calculate are δu_v, δd_v, $\delta\Sigma$ and δg. Their moments are denoted respectively by: $\delta\mathcal{M}_{u_v}$, $\delta\mathcal{M}_{d_v}$, $\delta\mathcal{M}_{\Sigma}$ and $\delta\mathcal{M}_g$. Therefore the moments of polarized u and d-valence quark in a proton can be indicated by

$$\delta\mathcal{M}_{u_v}(n,Q^2) = 2\delta\mathcal{M}^{NS}(n,Q^2) \times \delta\mathcal{M}'_{U/p}(n), \quad (6)$$

$$\delta\mathcal{M}_{d_v}(n,Q^2) = \delta\mathcal{M}^{NS}(n,Q^2) \times \delta\mathcal{M}'_{D/p}(n). \quad (7)$$

The moment of polarized singlet distribution (Σ) is given by

$$\delta\mathcal{M}_{\Sigma}(n,Q^2) = \delta\mathcal{M}^{S}(n,Q^2)(2\delta\mathcal{M}''_{U/p} + \delta\mathcal{M}''_{D/p}) \ , \quad (8)$$

where Σ denotes $\sum_{q=u,d,s}(q+\bar{q})$. Consequently we can have the following definitions

$$\delta\mathcal{M}_{\Sigma}(n,Q^2) = \sum_{q=u,d,s} (\delta\mathcal{M}_q(n,Q^2) + \delta\mathcal{M}_{\bar{q}}(n,Q^2)). \quad (9)$$

Thus by having Σ, and all valence quark in moment space, the contribution of $\delta\bar{q}$ can be specified directly.

For the gluon distribution we have

$$\delta\mathcal{M}_g(n,Q^2) = \delta\mathcal{M}_{gq}(n,Q^2)(2\delta\mathcal{M}'_{U/p}(n) + \delta\mathcal{M}'_{D/p}(n)) \ , \quad (10)$$

where $\delta\mathcal{M}_{gq}(n,Q^2)$ is the quark-to-gluon evolution function.

As it has been shown[4] in the NLO contributions to $g_1(x,Q^2)$ we can use directly its moment in following form

$$g_1^n(Q^2) = \frac{1}{2}\sum_q e_q^2\{(1+\frac{\alpha_s}{2\pi}\delta C_q^n)[\delta\mathcal{M}_q + \delta\mathcal{M}_{\bar{q}}] + \frac{\alpha_s}{2\pi}2\delta C_g^n\delta\mathcal{M}_g\} \ , \quad (11)$$

where $\delta q^n(Q^2)$, $\delta\bar{q}^n(Q^2)$ and $\delta g^n(Q^2)$ are moments of polarized parton distributions in a proton. Also δC_q^n, δC_g^n are the n-th moment of spin-dependent Wilson coefficients given by

$$\delta C_q^n = \frac{4}{3}\Big[-S_2(n) + (S_1(n))^2 + \left(\frac{3}{2} - \frac{1}{n(n+1)}\right)S_1(n) + \frac{1}{n^2} + \frac{1}{2n} + \frac{1}{n+1} - \frac{9}{2}\Big], \quad (12)$$

and

$$\delta C_g^n = \frac{1}{2}[-\frac{n-1}{n(n+1)}(S_1(n)+1) - \frac{1}{n^2} + \frac{2}{n(n+1)}]. \quad (13)$$

Using the moments of polarized parton distributions which has been determined by now, we can obtain the moment of polarized proton structure function in NLO by inserting the required distributions function in (12). If we calculate the unknown parameters then the computation of all the moments of polarized parton distributions and structure function, $g_1^n(Q^2)$, will be possible.

4 Polarized Parton Distributions in x-Space

In Ref. [5] the polarized valon distribution was determined in a proton by having the polarized parton structure functions in a valon. Using these results, it is now possible to extract the polarized parton structure in a proton. To obtain the z-dependence of parton distributions from the n-dependent exact analytical solutions in Mellin-moment space, one has to perform a numerical integration in order to invert the Mellin-transformation

$$\delta f(z, Q^2) = \frac{1}{\pi}\int_0^{\infty} dw Im[e^{i\phi} z^{-c-we^{i\phi}} \Delta M(n = c + we^{i\phi}, Q^2)]. \tag{14}$$

Here, the contour of integration, and thus the value of c, has to lie to the right of all singularities of $\Delta M(n = c + we^{i\phi}, Q^2)$ in the complex n-plane, *i.e.*, $c \geq 0$ since according to splitting functions the dominant pole of all δd_{ij}^n is located at $n = 0$. For all practical purposes, one may choose $c \simeq 1, \phi = 135°$ and an upper limit of integration, for any Q^2 , of about $5 + 10/\ln z^{-1}$, instead of ∞, which guarantee stable numerical results.

Using the relation between polarized parton distributions of a proton and the polarized parton distributions in a valon, which is given by the convolution

$$\delta q_{i/p}(x, Q^2) = \sum_j \int_x^1 \delta q_i(\frac{x}{y}, Q^2) \delta G_{j/p}(y) \frac{dy}{y}, \tag{15}$$

the polarized parton distributions and structure function in proton can be obtained.

5 Conclusion

Here we have studied the polarized valon and parton distribution in Mellin-Moment space in NLO approximation to describe the spin dependence of hadron structure function. Using the polarized valon distribution and all parton distributions in a valon, the polarized parton density in a proton is calculable. Our NLO calculation has indicated more sensible improvement comparing to the leading order approximation in QCD.

Acknowledgments

A.N.K. thanks Semnan university for partial financial support to perform this project. We acknowledge the organizers of this conference and IPM for giving us the opportunity to present our research work.

References

1. R.C. Hwa, *Phys. Rev.* D **22**, 759 (1980).
2. R.C. Hwa, *Phys. Rev.* D **22**, 1593 (1980).
3. F. Arash and A.N. Khorramian, *Phys. Rev.* C **67**, 045201 (2003); R.C. Hwa and C.B. Yang, *Phys. Rev.* C **66**, 025204 (2002); *ibid. Phys. Rev.* C **66**, 025205 (2002), arXiv: hep-ph/0303031; H.W. Kua, *et al.*, *Phys. Rev.* D **59**, 074025 (1999); C.B. Yang, *Chin. Phys. Lett.* **20**, 821 (2003).
4. B. Lampe and E. Reya, *Phys. Rept.* **332**, 1 (2000).
5. A.N. Khorramian, A. Mirjalili and S. Atashbar Tehrani in *Proceedings of International Workshop on Quantum Chromodynamics – Theory and Experiment* (Conversano, Bali, Italy, 2003).

INVESTIGATING THE QCD SCALE DEPENDENCE OF TOTAL CROSS SECTION FOR HEAVY QUARK PRODUCTION IN $P\overline{P}$ COLLISIONS

A. MIRJALILI[*,†], A. N. KHORRAMIAN[*,‡] and S. ATASHBAR TEHRANI[*,♭]

[†]*Physics Department, Yazd University, Yazd, Iran*

[‡]*Physics Department, Semnan University, Semnan, Iran*

[♭]*Physics Department, Persian Gulf University 75168, Boushehr, Iran*

and

[*]*Institute for Studies in Theoretical Physics and Mathematics (IPM), School of Physics*

P.O. Box 19395-5531, Tehran, Iran

E-mails: khorramiana@theory.ipm.ac.ir, mirjalili, atashbar@ipm.ir

There is a sizable and systematic discrepancy between experimental data on the $b\bar{b}$ production in $p\bar{p}$, γp and $\gamma\gamma$ collisions and the existing theoretical calculations within perturbative QCD. The uncertainty is coming from renormalization and factorization scale dependence of finite order perturbative calculations of the total cross section of $b\bar{b}$ production in such collisions and will be discussed for $p\bar{p}$ collision in detail. If we employ the approach of "Complete RG-improvement (CORGI)", in which one should perform a resummation to all-orders of renormalization and factorization group-predictable terms at each order of perturbation theory, then the scales dependence will be avoided and the mentioned discrepancy is reduced significantly.

1 Introduction

[a]One of the clean test of perturbative QCD is heavy quark production in hard collisions of hadrons, leptons and photons. Recent data on $b\bar{b}$ production in $p\bar{p}$ collisions at the Tevatron, γp collision in HERA and $\gamma\gamma$ collisions in LEP2[1,2] indicates that they lie systematically by a factor of about $2-4$ above the median of current theoretical calculations. The QCD calculations for the above cases depend on a number of inputs, for instance, α_s, parton distribution functions (PDF) of colliding hadrons or photons, $\cdots$, and finally, the choice of renormalization (RS) and factorization (FS) scales μ and M. We are going to investigate the dependence of existing fixed order (LO and NLO) QCD calculation of the total cross section in $p\bar{p}$ collisions on the choice of the renormalization and factorization scales[3]. Further, we indicate whether these scales will be avoided from our calculation, if we do a resummation on all ultraviolet terms which involve these two scales[4]. Finally, we expect to have a more consistent theoretical result comparing to the available experimental data.

2 Basic Definitions

The basic quantity of perturbative QCD calculations, the renormalized color coupling $\alpha_s(\mu)$, depends on the renormalization scale μ which is governed by

$$\mu\frac{\partial\alpha_s(\mu)}{\partial\ln\mu^2} = \beta(\alpha_s(\mu)) = -\frac{\beta_0}{4\pi}\alpha_s(\mu)^2 - \frac{\beta_1}{16\pi^2}\beta_1\alpha_s(\mu)^3 + \cdots, \quad (1)$$

where for n_f massless quarks $\beta_0 = 11 - (\frac{2}{3})n_f$ and $\beta_1 = 102 - (\frac{38}{3})n_f$.

The solutions of the above equation depend, beside μ, on renormalization scheme (RS). At NLO calculation this RS can be specified, for instance, via the parameter Λ_{RS}, corresponding to the renormalization scale for which $\alpha_s(\mu)$ diverges.

At two loop calculations the coupling μ is then given as the solution of the equation

$$\frac{\beta_0}{4\pi}\ln(\frac{\mu^2}{\Lambda_{RS}^2}) = \frac{1}{\alpha_s(\mu)} + c\ln\frac{c\alpha_s(\mu)}{1+c\alpha_s(\mu)}, \quad (2)$$

where $c = \frac{\beta_1}{4\pi\beta_0}$. For hadrons, the factorization scale dependence of PDF (M) is determined by the system of evolution equations for quark singlet, non-singlet and gluon distribution functions[5]

$$\frac{d\Sigma(M)}{d\ln M^2} = P_{qq}(M)\otimes\Sigma(M) + P_{qG}(M)\otimes G(M), \quad (3)$$

$$\frac{dG(M)}{d\ln M^2} = = P_{Gq}(M)\otimes\Sigma(M) + P_{GG}(M)\otimes G(M), \quad (4)$$

[a]Talk presented by A.M. at the XIth Regional Conference on Mathematical Physics, May 2004, Tehran, Iran.

$$\frac{dq_{NS}(M)}{d\ln M^2} = \\ = P_{NS}(M) \otimes q_{NS}(M) + P_{qG}(M) \otimes G(M), \quad (5)$$

where

$$\Sigma(x, M) = \sum_{i=1}^{N_f} (q_i(x, M) + \bar{q}_i(x, M)),$$

$$q_{NS,i}(x, M) = (q_i(x, M) + \bar{q}_i(x, M)) - \frac{1}{n_f}\Sigma(x, M), \quad (6)$$

for all i. The cross symbol in above relations indicates the convolution

$$f(x) \otimes g(x) = \int_x^1 \frac{dy}{y} f(\frac{x}{y}) g(y). \quad (7)$$

The splitting functions admit expansion in powers of $\alpha_s(M)$

$$P_{ij}(x, M) = \frac{\alpha_s(M)}{2\pi} P_{ij}^{(0)}(x) + \left(\frac{\alpha_s(M)}{2\pi}\right)^2 P_{ij}^{(1)}(x) \\ + \cdots, \quad (8)$$

where $P_{ij}^{(0)}(0)$ are unique and as all higher splitting functions $P_{kl}^{(j)}$, $j > 0$ depend on the choice of the factorization scheme (FS). Conversely, they can be taken to define the FS.

3 RS and FS Dependence of Coefficients

For a single case of a dimensional observable $R(Q)$ with

$$\mathcal{R}(Q) = a + r_1 a^2 + r_2 a^3 + \ldots + r_n a^{n+1} + \cdots, \quad (9)$$

the RS can be labeled by the non-universal coefficients of the β-function and Λ_{RS}.[6]

Self-consistence of perturbation theory that is the derivative of N-th order approximant $R(n)$ with respect to $\ln(\mu)$ is of higher order than the approximant itself. It will yield

$$r_1(\frac{Q}{\mu}, RS) = kb\ln(\frac{\mu}{Q}) + r_1(1, RS), \\ = kb\ln(\frac{\mu}{\Lambda_{RS}}) - \rho(Q), \quad (10)$$

where

$$\rho(Q) = kb\ln(\frac{Q}{\Lambda_{RS}}) .- r_1(1, RS), \quad (11)$$

is a renormalization scale and scheme invariant. For other coefficients, we will arrive first on following partial derivatives for r_2, $r_3, \cdots$

$$\frac{\partial r_2}{\partial r_1} = 2r_1 + c, \; \frac{\partial r_2}{\partial c_2} = -1, \; \frac{\partial r_2}{\partial c_3} = 0, \cdots,$$

$$\frac{\partial r_3}{\partial r_1} = c_2 + 2cr_1 + 3r_2, \; \frac{\partial r_3}{\partial c_2} = -2r_1, \; \frac{\partial r_3}{\partial c_3} = -\frac{1}{2} \; \cdots,$$

$$\vdots$$

After integration, one finds

$$r_2(r_1, c_2) = {r_1}^2 + cr_1 + X_2 - c_2,$$

$$r_3(r_1, c_2, c_3) = {r_1}^3 + \frac{5}{2}c{r_1}^2 \\ + (3X_2 - 2c_2)r_1 + X_3 - \frac{1}{2}c_3, \quad (12)$$

$$\vdots$$

The general structure is as follows

$$r_n(r_1, c_2, \ldots, c_n) = \hat{r}_n(r_1, c_2, \cdots, c_{n-1}) \\ + X_n - c_n/(n-1). \quad (13)$$

Where X_n are Q-independent and RS-invariant and are unknown unless a complete $N^n LO$ calculation has been performed.

4 Complete RG Improvement

From pervious results for r_1 , $r_2, \cdots$, we can reformulated $R(Q)$ as

$$\mathcal{R}(Q) = a + r_1 a^2 + (r_1^2 + cr_1 + X_2 - c_2)a^3 + \cdots. \quad (14)$$

Given a NLO calculation, r_1 is known but $X_2, X_3, \cdots$ are unknown. Thus, the complete subset of known terms in (13) at NLO is

$$a_0 \equiv a + r_1 a^2 + (r_1^2 + cr_1 - c_2)a^3 + \cdots, \quad (15)$$

and it is RS-invariant. Choose $r_1 = 0, c_2 = c_3 = \cdots = c_n = 0$, we obtain $a_0 = a$. At NNLO calculation X_2 is unknown. Further, infinite subset of terms are known and can be resummed to all orders[4]

$$X_2 {a_0}^3 = X_2 a^3 + 3X_2 r_1 a^4 + \cdots. \quad (16)$$

Finally, we will arrive at

$$\mathcal{R}(Q) = a_0 + X_2 {a_0}^3 + X_3 {a_0}^4 + \cdots + X_n {a_0}^{n+1} + \ldots, \quad (17)$$

where $a_0 = a(0, 0, 0, \cdots)$ is the coupling in this scheme and satisfies

$$\frac{1}{a_0} + c\ln\left(\frac{ca_0}{1 + ca_0}\right) = b\ln\left(\frac{Q}{\Lambda_{\mathcal{R}}}\right). \quad (18)$$

In fact, the solution of this transcendental equation can be written in closed form in terms of the Lambert W-function[7], defined implicitly by $W(z)\exp(W(z)) = z$,

$$a_0 = -\frac{1}{c[1+W(z(Q))]},$$
$$z(Q) \equiv -\frac{1}{e}\left(\frac{Q}{\Lambda_{\mathcal{R}}}\right)^{-b/c}. \qquad (19)$$

5 Other Approaches in Investigating Total Cross Section of Heavy Pair Production

In addition to the previous approach, to deal perturbative QCD for what we concern, we can have the following approaches:

Principle of Minimal Sensitivity (PMS): Here, the emphasize is put on the stability of the results with respect to variable μ. In absence of information on higher perturbative terms, the PMS approach is natural as it selects the point μ where the truncated perturbative expansion is most stable and has thus locally the property possessed by all the other result globally.

The effective charge (EC): It is based on the criterion of apparent convergence of perturbation expansions and the renormalization scale μ is chosen in such a way that all higher order contributions vanish *i.e.* demanding $R(n) = R(k)$ for all k.

Complete RG Improvement (CORGI): This approach is explained in more details in the previous section. Due to the properties of this approach where we encounter with RS invariant quantities, we expect to have more consistent result comparing to experimental data, specially for total cross section of heavy quark pair production.

6 Conclusions

Here, the dependence on the factorization scale for total cross section is canceled by that of PDF, provided the splitting function P_{ij} in the evolution equation is taken to all orders. At NLO, *i.e.* taking into account the first two terms in related perturbative series, we get[3]

$$\sigma_{tot}^{NLO}(M,\mu) = \alpha_s^2(\mu)\int\int dxdy \sum_{i=1}^{2n_f} q_i(x,M)$$
$$\times q_i(y,M)[\sigma_{q\bar{q}}^{(2)}(xy) + \alpha_s^2(\mu)\sigma_{q\bar{q}}^{(3)}(xy,M,\mu)]$$
$$+2\,\alpha_s^2(\mu)\int\int dx\, dy\Sigma(x,M)G(y,M)\alpha_s(\mu)\sigma_{qG}^{(3)}(xy,M)$$
$$+\alpha_s^2(\mu)\int\int dx\, dyG(x,M)G(y,M)[\sigma_{GG}^{(2)}(xy) + \alpha_s(\mu)$$
$$\times\sigma_{GG}^{(3)}(xy,M,\mu)].$$

Here, the sum is over n_f quarks and antiquarks and the relation between PDF of protons and antiprotons was taken into account. At this stage, we can employ the CORGI approach, where we need to use it first in moment space to extract parton distributions and then, we have to reformulate in this approach the partonic cross section. Consequently, we expect to get better consistency between our theoretical model and experimental data for cross section of $b\bar{b}$ production, which will be a new task in our future attempts.

Acknowledgements

We would like to thank the "Institute for studies in theoretical Physics and Mathematics (IPM)" to perform such a rich conference and to support and give us the opportunity to present our research works.

References

1. M. Acciarri *et al.* (L3 Collab.), *Phys. Lett.* B **503**, 10 (2001).
2. OPAL Collab., *OPAL Physics Note PN* , 455 (2000).
3. P. Nason, S. Dawson and R.K. Ellis, *Nucl. Phys.* B **303**, 607 (1988).
4. C.J. Maxwell and A. Mirjalili, *Nucl. Phys.* B **577**, 209 (2000).
5. R.K. Ellis, W.J. Stirling and B.R. Webber in *QCD and Collider Physics*, (Cambridge University Press 1996).
6. P.M. Stevenson, *Phys. Rev.* D **23**, 2916 (1981).
7. E. Gardi, G. Grunberg and M. Karliner, *JHEP* **07**, 007 (1998).

EOS OF THE UNIFORM ELECTRON FLUID IN LOCV FRAMEWORK

H. R. MOSHFEGH, M. MODARRES and H. DANESHVAR
Physics Department, University of Tehran
North Karegar Ave., 14395 Tehran, Iran
E-mail: moshfegh@khayam.ut.ac.ir

The equation of state (EOS) of the homogeneous electron is obtained using the lowest order constrained variational (LOCV) method at zero temperature. Using further an appropriate correlation function, an effective two-body interaction is proposed for electrons in the presence of a positive background.

1 Introduction

[a]The uniform electron fluid consists of N electrons embedded in N positively charged ion background. The density of system $\rho = \frac{N}{\Omega}$ (Ω is the volume) is constant in the thermodynamic limit. We work in intermediate coulomb coupling region $1 \leq r_s \leq 10$, where r_s is the Wigner-Seitz radius in units of Bohr radius[1]. There are two points in studying such an ideal system, first to obtain the metal properties and second to test different many-body methods against each other.

Because of the simplicity of the interaction, most of the available many-body methods have been applied to the uniform electron fluid at zero temperature. Thus, this system can be considered as a good testing ground for comparing different many-body techniques. On the other hand, some of the many-body methods have been only designed and applied to the problems which involve the short-range interactions. Therefore, it would be interesting to apply these techniques to the uniform electron fluid, with its long-range coulomb interaction, to find out the accuracy and the performance of these methods.

Originally the lowest order constrained variational (LOCV) method was developed to investigate the properties of symmetric nuclear matter with realistic nucleon-nucleon interactions[2]. For a wide range of models our LOCV calculations agree well with the results of fermion hypernetted chain (FHNC) calculations, where these have been performed and for a number of central potentials, there is in agreement with the essentially exact numerical solutions obtained by Monte-Carlo technique. Our LOCV calculation is a fully self-consistent technique with the state dependent correlation functions. It considers constraint in form of normalization condition. Recently, we applied LOCV formalism to the liquid ^3He and the homogeneous electron fluid at finite temperature[3]. According to the above arguments, in this article we shall attempt to propose the effective two-body interaction between pairs of electrons in homogeneous electron fluid in the presence of a positive background. First, we introduce the Hamiltonian of the system, then we briefly describe the LOCV method and finally we construct an effective two-body interaction.

2 Hamiltonian of Uniform Electron Fluid

The total Hamiltonian of uniform electron fluid with positive background is usually written as[1]

$$H = H_{el} + H_b + H_{el-b}, \tag{1}$$

where H_{el}, H_b and H_{el-b} are the electron, positive-background and electron-positive-background Hamiltonians, respectively. This Hamiltonian can be reduced to[1]

$$H = -\frac{1}{2}e^2 N^2 \Omega^{-1} 4\pi\mu^{-2} + H_{el}, \tag{2}$$

where

$$\begin{aligned} H_{el} &= \sum_{i=1}^{N} \frac{p_i^2}{2m} + \frac{1}{2}e^2 \sum_{i \neq j}^{N} \frac{e^{-\mu|\mathbf{r_i}-\mathbf{r_j}|}}{|\mathbf{r_i}-\mathbf{r_j}|} \\ &= \sum_{i=1}^{N} \frac{p_i^2}{2m} + \frac{1}{2}\sum_{i \neq j}^{N} V(ij). \end{aligned} \tag{3}$$

The exponential factor and the parameter μ are introduced in this Hamiltonian to make the calculations finite. Then, after performing many-body calculations, μ is usually set to zero, if one intends to have only the coulomb interaction.

[a]Talk presented by H.R.M. at the XIth Regional Conference on Mathematical Physics, May 2004, Tehran, Iran.

3 LOCV Method

In the LOCV method, we use an ideal Fermi gas type wave function for the single particle states and variational techniques, to find the wave function of interacting system, *i.e.*,

$$\psi = \mathcal{F}\Phi, \tag{4}$$

where

$$\mathcal{F} = \mathcal{S}\prod_{i>j} f(ij), \tag{5}$$

and $\mathcal{S}$ a symmetrizing operator. In general, the Jastrow correlation functions $f(ij)$ are operators. But, in the case of uniform electron fluid, because of the simplicity of coulomb interaction, we assume them to have the following form

$$f(r) = 1 - exp(-ur) = 1 + C(r), \tag{6}$$

where u is a variational parameter. There are other choices for choosing the correlation functions with further parameters (parameters are calculated variationally). In Ref. [3], we have shown that the above choice has reasonable results at zero and finite temperature.

The electron fluid energy is written as[2,3]

$$E_{in} = T_F + E_{MB}[f]. \tag{7}$$

T_F is simply the Fermi gas kinetic energy. The many-body energy term $E_{MB}[f]$ is calculated by constructing a cluster expansion for the expectation value of our Hamiltonian H_{el}. We keep only the first two terms in a cluster expansion of the energy functional

$$\begin{aligned} E[f] &= \frac{1}{A}\frac{< \Psi|H|\Psi >}{< \Psi|\Psi >} = T_F + E_{MB} \\ &= T_F + E_2 + E_3 + \cdots. \end{aligned} \tag{8}$$

The two-body energy term is defined as

$$E_2 = (2A)^{-1}\sum_{ij} < ij|\mathcal{V}|ij >_a, \tag{9}$$

where

$$\mathcal{V}(12) = -\frac{\hbar^2}{2m}[f(12),[\nabla_{12}^2, f(12]] + f(12)V(12)f(12). \tag{10}$$

Here, the two-body antisymmetrized matrix element $< ij|\mathcal{V}|ij >_a$ are calculated using plane-waves.

In LOCV formalism E_{MB} is approximated by E_2 and one hopes that the normalization constraint makes the cluster expansion to converge very rapidly and bring the many-body effect into E_2 term.

For the electron fluid, because of the choice of correlation functions, the $\mathcal{V}(12)$ is reduced to

$$\mathcal{V}(12) = \frac{\hbar^2}{m}(\nabla C(r))^2 + C(r)(2 + C(r))V(r) + V(r). \tag{11}$$

Now, using plane-waves as the single-particle states, (9) reduces to

$$\begin{aligned} E_2 = \frac{1}{2}\rho[\int V(r)d\mathbf{r} + \int C_F(r)V(r)d\mathbf{r} \\ + \int [(1 + C_F(r))[\frac{\hbar^2}{m}(\nabla C(r))^2 \\ + C(r)(2 + C(r))V(r)]d\mathbf{r}]], \end{aligned} \tag{12}$$

where

$$C_F(r) = -\{9(rk_f cos(rk_f) - sin(rk_f))/(k_f r)^3\}^2. \tag{13}$$

The first integral in E_2 cancels exactly the first term in the total Hamiltonian of (1). While the second term, the Hartree-Fock energy, and the third term can be considered as the correlation energy.

The normalization constraint can also be written as

$$< \psi|\psi >= \rho\int(1 + C_F(r))(1 + C(r))^2 d\mathbf{r}. \tag{14}$$

The above constraint introduces another parameter in our formalism, *i.e.*, the Lagrange multiplier λ. At zero temperature, we minimize the functional $\{E_2 + \lambda < \psi|\psi >\}[f]$ with respect to the parameter u and we choose λ such that the above normalization constraint is satisfied.

4 Results

We find the minimum energy $E_{min} = -1.26$eV occurs at $r_s = 4.83$ with $u = 42.73$. In this case, the normalization condition could be agreeably satisfied ($< \psi|\psi > -1 \cong 2 * 10^{-8}$). This calculation shows a very good agreement with other methods such as perturbation expansion method, variational Monte-Carlo and FHNC method.

At this stage, one can minimize the two-body energy (12) with respect to the variations in the function $f(r)$ (or $C(r)$) but subject to normalization constraint

$$< \psi|\psi >= \rho\int(1 + C_F(r))(1 + C(r))^2 d\mathbf{r} = 1. \tag{15}$$

This leads to the Euler-Lagrange differential equation

$$g''(r)-[a''(r)/a(r)+m\hbar^{-2}(V(r)+\lambda)]g(r)=0, \quad (16)$$

where $g(r)=f(r)a(r)$ and

$$a^2(r)=(2k_f r)^2(1-C_F(r)).$$

The constraint is incorporated by solving the above Euler-Lagrange equation only out to certain distances, until the logarithmic derivative of $C(r)$ matches to $C_F(r)$. Then, we set the $C(r)$ equal to $C_F(r)$. In this way, there is no free parameter in our LOCV formalism, *i.e.*, the healing distance is determined directly by the constraints and the initial conditions. For simple long range coulomb potential (without any effect of the background), the correlation function vanishes because of purely repulsive interaction. The effective potential must include both repulsive and attractive regions, due to electrons and positive fix background ions, respectively.

For constructing such an effective potential, we start with one-parameter trial correlation function and insert it in the Euler-Lagrange (16) by setting $\lambda=0$ and solve the algebraic equation with respect to $V(r)$. By fitting the data for various densities, we find the following form for effective potential:

$$V(r)=14.39[(a+\frac{b}{r_s})/r-(\frac{c}{r})^5]\{1-(\frac{2k_f r}{f_0(r)a(r)})^2\}, \quad (17)$$

where $f_0(r)$ is trial correlation function and the fitting parameters are: $a=2.13$, $b=3.74$ and $c=0.17$.

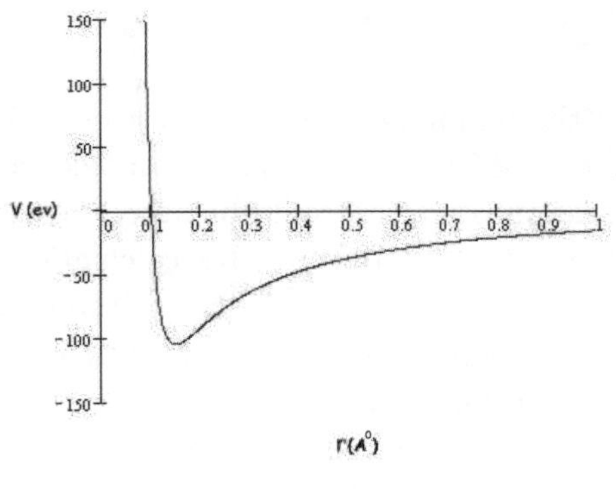

Figure 1.

In Fig. 1, we plot this density dependent effective potential (in eV) for $r_s=5$ versus r (in A^o). This potential contains both electron-electron and electron-background screening effects. It have both attractive and repulsive regions and a minimum which occurs at about $0.15A^o$.

Next, we put this potential in Euler-Lagrange (16) and solve it with respect to the correlation function and find the new Lagrange multiplier and the correlation function. Then, by inserting this new correlation function and the effective potential in (12), we can find the energy of system versus its density.

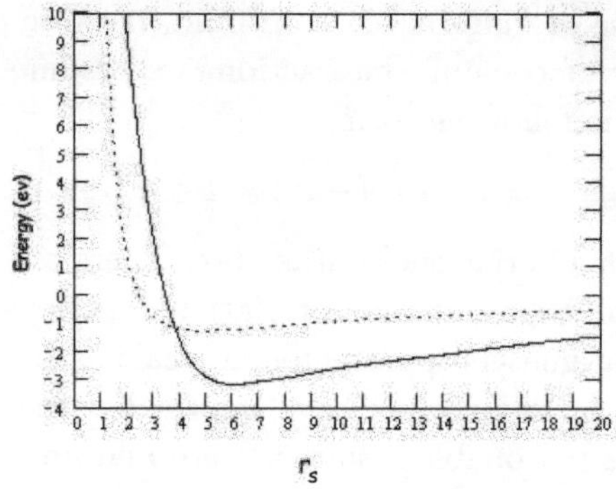

Figure 2.

Fig. 2 shows the EOS of system: Dashed curves shows the EOS calculated from $f_0(r)$ and full curve shows EOS calculated from correlation function that was obtained from the above effective potential. In order to obtain a better results, one can continue this process using a recurring algorithm, till the solution of Euler-Lagrange for correlation function does not vary very much from the previous correlation function.

References

1. A.L. Fetter and J.D. Walecka in *Quantum Theory of Many Body Systems*, (McGraw-Hill, New York, 1971).
2. M. Modarres and J.M. Irvine, *J. Phys.* G **5**, 511 (1979); M. Modarres and H.R. Moshfegh, *Phys. Rev.* C **62**, 4308 (2000).
3. M. Modarres, H.R. Moshfegh and A. Sepahvand *Eur. Phys. J.* B **31**, 159 (2003).

EMISSION ANGLE DEPENDENCE OF FISSION FRAGMENTS SPIN IN $B^{10,11} + Np^{237}$ FUSION-FISSION REACTIONS

M. R. PAHLAVANI

Institute for Studies in Theoretical Physics and Mathematics (IPM), School of Physics
P.O. Box 19395-5531, Tehran, Iran
and
Physics Department, Faculty of Science, Mazandaran University
P.O. Box 47415-416, Babolsar, Iran
E-mail: m.pahlavani@umz.ac.ir

The total average spin of the fragments is measured for $B^{10,11}$ as projectiles and Np^{237} as target at energies around coulomb barrier for fragments emitted in 165°,150°, 105° and 90° relative to beam direction using γ-ray multiplicity measurement method. Theoretical calculations have been performed by using Transition State Model (TSM). Measured fragments spin show a weak angle dependence when compared to the calculated results using TSM. In the framework of TSM the angular distribution of fission fragments and emission angle dependence of fission fragments spin have been studied by distribution of K degree of freedom. To explain the weak measured emission angle dependence of fragments spin, the Gaussian distribution that was considered for K should be considerably modified.

1 Introduction

[a]Study of heavy ion induced fusion-fission reactions at beam energies especially around the coulomb barrier has attracted a great deal of attention in recent years[1]. The study of the spin distribution in fission fragments provides important information on the mechanism of spin generation and the excitation of various spin bearing modes in fission. The total spin of the fragments is largely determined by the excitation of various angular momentum bearing modes, such as wriggling, bending, twisting and tilting[2]. In heavy ion induced fission reaction, where the fissile compound nucleus is populated, a part of the initial angular momentum also gets transferred as the spin of the fission fragments. The total spin of the fragments thus arises from the tilting as well as the other statistical modes. Thus, fission fragments angular distribution depends on the tilting mode variance, K_0^2, assumed to be Gaussian (K is the component of total angular momentum in the direction of fission or symmetry axis).

Among the various angular momentum bearing modes, the tilting mode has been studied well[3], because of its role in determining the angular distribution of fission fragments. The excitation of the tilting mode also determines the emission angle dependence of the fragment spin in the fission process. Recently, a number of experiments are performed to study the spin carried out by fission fragments as a function of emission angle for a few sets of target and projectile[4]. The measured weak emission angle dependence of fragments spin opposed the prediction of statistical transition state model.

In order to study this ambiguity between theory and experiment, we have carried out a number of experiments to measure the total average spin of fragments in various angles relative to beam direction for $B^{10,11}$ on Np^{237} in a few energies around compound nucleus coulomb barrier.

2 Experiments

The experiments was carried out using $B^{10,11}$ ion beams from the 14-Mev BARC-TIFR pelletron accelerator at Mumbai. A thin target of Np^{237} was placed at the center of a cylindrical chamber. Fourteen hexagonal (57cm $\times$ 63 cm) Bismuthed Germanate scintillator crystals (BGO) in a close-packed geometry were positioned in upper and lower sides of cylindrical chamber, were used to detect the γ-rays emitted from fission fragments. The BGO detectors have been chosen because of their high efficiency for detection of γ-rays. Fission fragments were identified and measured by two surface barrier detector with different thickness (thick for E and thin for $\triangle$ E used back to back). By making use of a complicated electronic setup the γ-ray emitted by fragments was de-

[a]Talk presented at the XIth Regional Conference on Mathematical Physics, May 2004, Tehran, Iran.

tected simultaneously with fission fragments in two different angles 90° and 165° (or 105° and 150° in second run) relative to beam direction. Then collected data in list mode type were used to separate the timing spectrum of γ-ray and energy peaks of fragments. A computer code that was constructed on the basis of statistical points of view for multi-detector system, was used to extract the γ-ray multiplicity. The fission fragments average spin could be determined from the γ-multiplicity using following semi-empirical relation[5]

$$\langle S_T \rangle = 2(M_\gamma - \alpha) + \beta M_n. \tag{1}$$

In this expression, α is the total number of statistical γ-rays, M_n is the average neutron emitted by the fragments and β is the average spin removed by each neutron. This equation implies that the statistical γ-rays remove no angular momentum, while the remaining γ-rays each remove two units of spin (because of E-2 transition). The values $\alpha = 4$ and $\beta = 0.5$ are used widely in the literature[6].

3 Theoretical Calculations

In TSM, it is assumed that the saddle point is a transition state between equilibrated compound system that is formed after fusion of projectiles and targets in fusion-fission reactions and scission point. Also for spin zero targets and projectiles, the density of states for K is taken to be a Gaussian centered with the mean equal to zero and having a variance equal to K_0^2.

According to the transition state model, the angular distribution of fission fragments are determined by K, the component of compound nucleus total spin I on symmetry or fission axis. The mean square of the K, $\langle K^2 \rangle$ is given by[7]

$$\langle K^2 \rangle = \frac{\sum_{K=-I,M}^{I} K^2 W_{M=0,K}^{I}(\theta) \exp(-K^2/2K_0^2)}{\sum_{K=-I}^{I} W_{M=0,K}^{I}(\theta) \exp(-K^2/2K_0^2)}. \tag{2}$$

The above expression after substitution for the angular yields, $W_{M=0,K}^{I}(\theta)$, and can be written in an analytical form as

$$\langle K^2 \rangle = 0.5(I+0.5)^2 \sin^2(\theta) \left[1 - \frac{I_1(\beta)^2}{I_0(\beta)^2}\right], \tag{3}$$

where $\beta = \sin(\theta)/\sqrt{2}$ and $\alpha =$ (I + 0.5)/($2K_0^2$), I_0 and I_1 are the zeroth order and first order modified Bessel functions, respectively. Using this expression, one can show that the fragments spin is strongly related to parameter α. The Gaussian standard deviation of K distribution, K_0^2, can be obtained from measured fission fragments anisotropy data that are given by the approximated expression,

$$A = 1 + \frac{\langle I^2 \rangle}{4K_0^2}. \tag{4}$$

The anisotropy is defined as the ratio of the fission probability at 180° to 90° and can be obtained experimentally. The average square of compound nucleus spin, $\langle I^2 \rangle$ can be calculated using couple channel analysis. Then, knowing $\langle I^2 \rangle$ and A, one can obtain K_0^2. Another way to calculate K_0^2 is to use the relation

$$K_0^2 = \frac{J_{eff} T}{\hbar^2}. \tag{5}$$

In this expression J_{eff} and T are the effective moments of inertia and temperature of compound nucleus.

4 Results and Discussion

The experimentally observed values for fission fragments average spins in various angle were analyzed within the framework of the transition state model mentioned above. The result have been shown in Fig. 1 and Fig. 2.

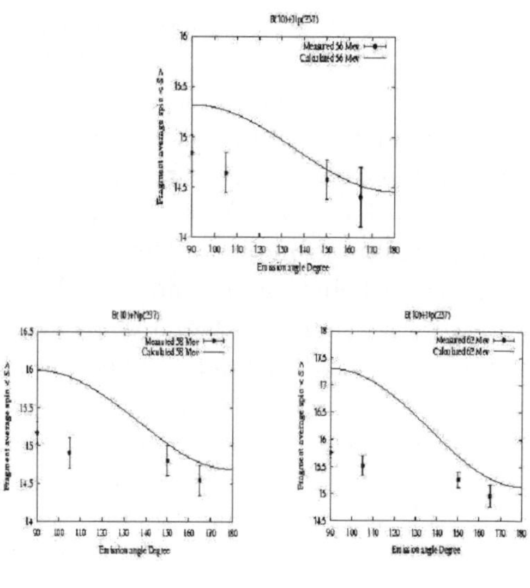

Figure 1. Fission fragments total average spin as a function of emission angle for B^{10} + Np^{237}.

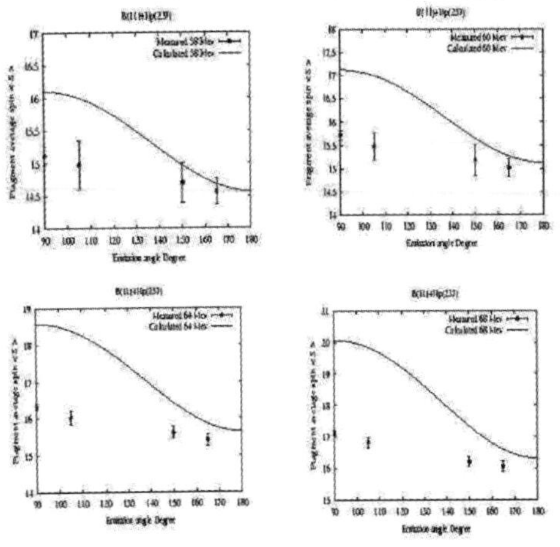

Figure 2. Fission fragments total spin average spin as a function of emission angle for $B^{11} + Np^{237}$.

Measured fission fragments spins in two above mentioned systems at different energy regions as a function of emission angle have shown that the fission fragment spin depends on the emission angle. However, the theoretical calculations based on the TSM show a much stronger emission angle dependence relative to the experiments. It was not possible to consistently explain both the angle dependence of the total fragments spin and fragments angular distribution within the framework of TSM. The K-distribution would have to be considerably distorted from Gaussian to explain the experimentally measured weak emission angle dependence of fission fragment spin. It would be inconsistent to use two different K-distributions to explain the same physical effect. There has been some effort to explain this in terms of the various angular momentum bearing modes (collective modes)[8]. However, a consistent explanation still eludes the nuclear physicists.

Acknowledgments

I would like to thank S. Sinha, R. Varma and V. Kumar from IIT-Bombay and R.K. Choudhoury, B.K. Nayak and A. Saxena from Nuclear Physics Deviation, BARC-Mumbai for their help during the experiments. I also acknowledge the support from the operating staff of the BARC-TIFR Pelletron accelerator facility for providing the required beams during the experiments.

References

1. B.K. Nayak and R.G. Tomas *et al.*, *Phys. Rev.* C **65**, 031601-1 (2000).
2. L. Nowicki and M. Bordene *et al.*, *Phys. Rev.* C **26**, 1114 (1982).
3. D. Volkapic and B. Ivanisevic, *Phys. Rev.* C **52**, 1980 (1995).
4. J.P. Lestone and A.A. Sonozogni *et al.*, *Phys. Rev.* C **52**, 1980 (1995).
5. D.V. Shetty and R.K. Choudhoury *et al.*, *Phys. Rev.* C **56**, 868 (1997).
6. R.K. Schmitt and L. Cook *et al.*, *Nucl. Phys.* A **592**, 130 (1995).
7. C.R. Morton and A.C. Berriman *et al.*, *Phys. Rev.* C **62**, 024607 (2000).
8. D.V. Shetty and R.K. Choudhoury *et al.*, *Phys. Rev.* C **58**, R616 (1998).

SOME REMARKS ON NEUTRINO MASS MATRIX

RIAZUDDIN

National Center for Physics, Quaid-i-Azam University
Islamabad 45320, Pakistan
E-mail: ncp@comsats.net.pk

There is a compelling evidence for neutrino oscillations implying that neutrinos mix and have nonzero mass but without pinning down their absolute mass. The implications of neutrino oscillations and mass squared splitting between neutrinos of different flavor on pattern of neutrino mass matrix is discussed. In particular, a neutrino mass matrix, which shows approximate flavor symmetry where the neutrino mass differences arise from flavor violation in off-diagonal Yukawa couplings is elaborated on. The implications in double β-decay are also discussed.

1 Introduction

[a]Certainly one of the most exciting areas of research at present is neutrino physics. Neutrinos are fantastically numerous in the universe and as such to understand the universe, we must understand neutrinos, particularly their mass. It is fair to say that the results of the last decade on neutrinos from the sun, from the atmospheric interaction of cosmic rays, and from reactors provide a compelling evidence that the neutrinos have nonzero mass and mix.

2 Neutrino Mass

Neutrino occurs in one helicity state (left handed). This together with lepton number L conservation implies $m_\nu = 0$. However there is no deep reason that it should be so. There is no local gauge symmetry and no massless gauge boson coupled to lepton number L, which therefore is expected to be violated. Thus, one may expect a finite mass for neutrino. Moreover, all other known fermions, quarks and charged leptons, are massive. But the intriguing question is: Why $m(\nu_e) \ll m(e)$, which needs to be understood, even though we do not understand why *e.g.* electron mass is what it is and why muons and tauons are heavier than electron. This is the so called flavor problem which has so far eluded us. Neutrino mass has added importance for two other reasons:

- The interesting phenomena of neutrino oscillations are possible if one or more of neutrinos have non-vanishing mass.
- Non-vanishing of neutrino mass has important implications in Astrophysics and Cosmology. It is a candidate for hot dark matter.

2.1 *Astrophysical Constraint on Neutrino Mass*

The total mass-energy of the universe is composed of several constituents, each of which is characterized by its energy density, which is expressed in terms of critical density

$$\rho_{0i} \equiv \Omega_{oi}\rho_{c0}.$$

The critical density is defined as

$$\rho_{c0} = \frac{3H_0^2}{8\pi G_N}, \tag{1}$$

where H_0 is the Hubble constant and G_N is the Newton's gravitational constant. Using the present value of H_0 (Hubble constant), namely $H_0 = 100h_0$ km s^{-1}Mpc^{-1}, $h_0 = 0.72 \pm 0.05$, this gives

$$\begin{aligned}\rho_{c0} &= 1.879h_0^2 \times 10^{-29} gcm^{-3},\\ &= 1.054h_0^2 \times 10^4 eVcm^{-3}.\end{aligned} \tag{2}$$

Neutrinos are fantastically numerous

$$n_\nu = \frac{3}{10} n_\gamma = 112\, cm^{-3},$$

so if they have even a tiny mass, they can outweigh all the stars and galaxies in the universe. The neutrinos contribution to energy density is

$$\rho_{\nu_0} = (112)\left(\sum_i m_{\nu i}\; eV\right) eVcm^{-3},$$

so that the neutrinos contribution to hot dark matter is

$$\Omega_\nu^{HDM} = \frac{\rho_{\nu_0}}{\rho_{c0}}. \tag{3}$$

[a]Invited plenary talk presented at the XIth Regional Conference on Mathematical Physics, May 2004, Tehran, Iran.

Unfortunately, there is no direct particle physics evidence on $\sum_i m_{\nu i}$. We note that $\rho_{\nu_0} \leq \rho_{c0}$, implies that

$$\sum_i m_{\nu i} \leq 93.8 h_0^2 \ (eV) = 49\, eV. \quad (4)$$

This is the astrophysical constraints on light neutrino masses.

2.2 *Double β-Decay*

The double β-decay is another way to look for a finite mass of neutrino. Two kinds of double β-decay can be considered:

$$\begin{aligned} &(2\nu) \ (A,\, Z) \to (A,\, Z+2) + 2e^- + 2\bar{\nu}_e, \\ &(0\nu) \qquad\qquad \to (A,\, Z+2) + 2e^-. \end{aligned}$$

The important physics issues in (0ν) double β-decay are:

(i) Lepton number must not be conserved, which is possible if neutrinos are Majorana particles: $\nu \equiv \bar{\nu}$.

(ii) Helicity of the neutrino cannot be exactly -1, this can be satisfied if $m_\nu \neq 0$.

Thus $(0\nu)\beta\beta$-decay is especially interesting

$$T_{1/2} \propto Q^{-5} \langle m_\nu \rangle^{-2}, \quad (5)$$

where decay Q value $\approx T_{e1} + T_{e2}$. Here

$$\langle m_\nu \rangle = \sum_i \lambda_i \left| U_{ei} \right|^2 m_{\nu_i}, \quad (6)$$

where λ_i is a possible sign since Majorana neutrinos are CP eigenstates; as shown the expectation value is weighted by neutrino's electron couplings. There is a direct evidence of $(2\nu)\,\beta\beta$-decay

$$\begin{aligned} (2\nu)\,\beta\beta \qquad & {}^{82}Se \to {}^{82}Kr, \\ & T_{1/2} = \left(1.1^{+0.8}_{-0.3}\right) \times 10^{20}\ yrs. \end{aligned} \quad (7)$$

For neutrinoless double β-decay

$$\begin{aligned} & {}^{76}Ge \to {}^{76}Se + 2e^-, \\ & T_{1/2} \geq 1.9 \times 10^{25}\ yrs. \end{aligned} \quad (8)$$

One recent result[1] has claimed the evidence for this decay with the best value $T_{1/2} = 1.5 \times 10^{25}$ yrs. This analysis claims $\langle m_\nu \rangle = \left(0.39^{+0.17}_{-0.28}\right)$ eV, which has been commented upon [2]. If the above finding were to be confirmed, it would be the first indication of lepton number violation in nature and that Majorana neutrino can exist in nature. We shall come to other implications of above value of $\langle m_\nu \rangle$ later.

2.3 *Cosmological Constraints*

Recent measurements of the fluctuations by an orbiting observatory called the Wilkinson Microwave Anisotropy Probe (WMAP) and their analysis have settled a number of issues about the universe, its expansion rate and its composition. The results are summarized below:

$$\begin{aligned} \Omega &= \rho/\rho_c = 1.02 \pm 0.02, \\ \rho_{DM} &= (2.25 \pm 0.38) \times 10^{-27} Kg/m^3, \\ \Omega_{DM} &= 0.23 \pm 0.05, \\ \Omega_b &= 0.046 \pm 0.005, \\ \Omega_\nu &< 0.015,\ m_\nu < 0.23\, eV. \end{aligned} \quad (9)$$

Note that visible baryon density is only about 4.6%.

3 Origin of Neutrino Masses

The minimal standard model involves 3 chiral neutrino states, but it does not admit renormalizable interactions that can generate neutrino masses. If there is no $SU_L(2) \times U_Y(1)$-singlet fermion in nature, then neutrino masses are necessarily Majorana

$$\mathcal{L}^{mass} = \frac{1}{2} m \psi^T C^{-1} \psi + h.c. \quad (10)$$

In the SM, Majorana neutrino masses are forbidden by a global Baryon-Lepton $(B-L)$ symmetry, but there is no reason to expect that this symmetry is fundamental. If one allows right-handed neutrinos ν_R which are $SU_L(2) \times U_Y(1)$ singlets, then one can write Yukawa interactions

$$\mathcal{L}_Y = \bar{\ell}_{Li} \phi h_{ij} e_{Rj} + \bar{\ell}_{Li} \tilde{\phi} h_{ij} \nu_{Rj} - \frac{1}{2} \bar{\nu}_R^c M \nu_R + h.c., \quad (11)$$

where the SM places the left-handed components of charged leptons and associated neutrinos into $SU_L(2)$ doublets ℓ_L. ϕ is the usual Higgs doublet under $SU_L(2)$. The lepton number violation is induced by the third term, which is allowed by the gauge symmetry. M is the Majorana mass matrix while h are Yukawa couplings. After spontaneous symmetry breaking the vacuum expectation value of the Higgs field $\langle \phi \rangle \equiv v = 175$ GeV generates the Dirac mass term $(m_D)_{ij} = h_{ij} v$ and 6×6 neutrino mass matrix

$$M_\nu = \begin{pmatrix} 0 & m_D \\ m_D & M \end{pmatrix}. \quad (12)$$

After diagonalization M_ν has 6 mass eigenstates ν_k, that represent Majorana neutrinos ($\nu_k = \bar{\nu}_k$). In the seesaw limit $m_D << M$, there are three light active Majorana neutrinos and the diagonalization of M_ν gives

$$M_\nu = -m_D M^{-1} m_D. \quad (13)$$

This also yields light and heavy neutrino mass eigenstates

$$\begin{aligned} \nu &= V_\nu^T \nu_L + \nu_L^c V_\nu^*, \\ N &\approx \nu_R + \nu_R^c, \\ m_{Ni} &= M_i, \end{aligned} \quad (14)$$

where V_ν is the neutrino mixing matrix. Thus,

$$m_{\nu\ell} \sim \frac{m_D^2}{M} \approx \frac{v^2}{M} \ll m_\ell, \quad (15)$$

by requiring the existence of large scale M, associated with new physics. Indeed, since $v \approx 175$ GeV, $m_\nu \approx 0.03$ eV, for $M \approx 10^{15}$ GeV. Thus, neutrino masses are a probe of physics at grand unification mass scale. We shall see that neutrino oscillations might remarkably provide a mechanism to measure extremely small masses (of order of milli-electron volts and less) and indirectly provide a new scale indicative of new physics.

4 Neutrino Oscillations

4.1 Oscillations in Vacuum

Neutrinos are produced in weak interactions as flavor eigenstates, characterized by e, μ, τ. The flavor eigenstates $|\nu_\alpha\rangle$ need not coincide with mass (energy) eigenstates $|\nu_i\rangle$ and are generally coherent superposition of such states

$$|\nu_\ell\rangle = \sum_i U_{\ell i} |\nu_i\rangle, \quad (16)$$

where the mixing matrix is unitary. This matrix is characterized by 3 angles, $\theta_{12} = \theta_3, \theta_{13} = \theta_2, \theta_{23} = \theta_1$, one CP violating phase δ and two Majorana phases, which we put equal to zero. In vacuum, the mass eigenstates propagate as plane waves

$$|\nu_i(t, \mathbf{x})\rangle = \exp\left(-i\left(E_i t - \mathbf{k}\cdot\mathbf{x}\right)\right)|\nu_i(0)\rangle, \quad (17)$$

where $E_i \approx k + m_i^2/2k^2$, $k \gg m_i$. Thus, flavor eigenstates propagate as

$$|\nu_\ell(t, \mathbf{x})\rangle = U_{\ell i} \exp\left(-i\left(E_i t - \mathbf{k}\cdot\mathbf{x}\right)\right)|\nu_i(0)\rangle. \quad (18)$$

The probability at time t that ν_ℓ is converted into $\nu_{\ell'}$ is

$$P_{\nu_\ell \to \nu_{\ell'}} = |\langle \nu_{\ell'} | \nu_\ell \rangle|^2.$$

For oscillations involving two neutrinos, it takes a simple form

$$P_{\nu_\ell \to \nu_{\ell'}} = \sin^2 2\theta \sin^2\left[\left(\frac{E_1 - E_2}{2}\right) t\right], \quad (19)$$

which can be conveniently written as

$$P_{\nu_\ell \to \nu_{\ell'}} = \sin^2 2\theta \sin^2\left[1.27 \frac{\Delta m^2}{E_\nu} L\right], \quad (20)$$

where L is the distance (measured in meters) traveled after ν_ℓ is converted into $\nu_{\ell'}$. $\Delta m^2 = m_1^2 - m_2^2$ in units of eV^2 while $E_\nu \simeq k$ is measured in MeV. Thus the oscillations in this simple case are characterized by the oscillation length in vacuum

$$L_v = 4\pi \frac{E_\nu}{\Delta m^2}, \quad (21)$$

and by the amplitude $\sin^2 2\theta$.

4.2 Oscillations in Matter

In traversing, matter neutrinos interact with electrons and nucleons of intervening material and their forward coherent scattering induces an effective potential energy $\sqrt{2} G_F N_e$, so that the corresponding matter oscillation length is

$$L_0 = \frac{2\pi}{\sqrt{2} G_F N_e} = 1.7 \times 10^7\,(m)\,/\rho\,(g/cm^3)\,Y_e. \quad (22)$$

Here, N_e denotes the number of electrons per unit volume

$$N_e = \frac{\rho}{m_N} Y_e, \quad (23)$$

where Y_e is the number of electrons per nucleons $\approx 1/2$ in ordinary matter. The effective oscillation length in matter is

$$\begin{aligned} L_m &= L_v \frac{\sin 2\theta_m}{\sin 2\theta}, \\ &= L_v \left[\left(\frac{L_v}{L_0} - \cos 2\theta\right)^2 + \sin^2 2\theta\right]^{-1/2}, \\ \tan 2\theta_m &= \tan 2\theta \left(1 - \frac{L_v}{L_0 \cos 2\theta}\right)^{-1}, \\ P_{(\nu_e \to \nu_e)} &= 1 - \sin^2 2\theta_m \sin^2\left[1.27 \frac{L}{L_m}\right]. \end{aligned} \quad (24)$$

θ_m is new mixing angle in matter. Thus, resonance $[\sin^2 2\theta_m = 1]$ occurs when $\cos 2\theta$ is equal to

$$\begin{aligned}\frac{L_v}{L_0} &= \frac{2\sqrt{2}G_F N_e E_\nu}{\Delta m^2}, \\ &= 0.22\frac{E_\nu}{1\,MeV}\frac{\rho Y}{100g/cm^3}\frac{7\times 10^{-5}eV^2}{\Delta m^2}, \quad (25)\end{aligned}$$

The transition point between the regime of vacuum and matter oscillations is determined by the ratio L_v/L_0. If it is greater than 1, matter oscillations dominate. If it is less than $\cos 2\theta$, vacuum oscillation dominate. Generally, there is a smooth transition between these two regimes. Matter effects become maximum at resonance $L_v/L_0 = \cos 2\theta$. This is the basis of Mikheyev-Smirnov-Wolfenstein (MSW) effect. The survival probability $P_{(\nu_e\to\nu_e)}$ averaged over the detector position L (from the solar surface)

$$P_{(\nu_e\to\nu_e)} = \frac{1}{2}\left[1 + (1 - 2P_x)\cos 2\theta_m(\rho_{\max})\cos 2\theta\right], \quad (26)$$

where $\theta_m(\rho_{\max})$ is the initial mixing angle, usually

$$\cos 2\theta_m(\rho_{\max}) \simeq -1,$$

and P_x is a finite probability for jumping from one eigenstate to the other one and conversion might be incomplete. The survival probability $\langle P_{(\nu_e\to\nu_e)}\rangle$ as a function of E_ν can be plotted for various mixing angles[3].

For the parameters corresponding to preferred solution for neutrino oscillations (see below) $\sin^2 2\theta \approx 0.8$, $\Delta m^2 \approx 7\times 10^{-5}$ eV^2 and $\rho = 100$ g/cm^3 at the center of the sun, $Y_e \approx 1/2$, $L_v/L_0 = 0.11E_\nu/1$ MeV, $\cos 2\theta = 0.45$.

Due to different reaction thresholds, solar neutrinos with energy $E_\nu > 0.814$ MeV can be detected in ^{37}Cl and those with $E_\nu > 0.233$ MeV in ^{71}Ga. Note that for pp neutrinos ($E_\nu < 0.42$ MeV) and ^7Be neutrinos ($E_\nu \approx 0.86$ MeV), $L_v/L_0 < \cos 2\theta$ and they undergo vacuum oscillations, while the neutrinos with $E_\nu > 4.5$ MeV, (^8B neutrinos) undergo MSW matter oscillations.

5 Evidence for Oscillations

One looks for oscillations in two types of experiments.

5.1 *Appearance Experiments*

Here, one searches for a new neutrino flavor, absent in the initial beam, which can arise from oscillations.

Atmospheric Neutrino Anomaly

Atmospheric neutrinos are produced in decays of pions (kaons) that are produced in the interaction of cosmic rays with the atmosphere

$$\begin{aligned}p + A &\to \pi^\pm + A', \\ \pi^\pm &\to \mu^\pm\nu_\mu(\bar\nu_\mu) \\ &\to e^\pm\nu_e(\bar\nu_e)\,\nu_\mu(\bar\nu_\mu).\end{aligned}$$

These neutrinos are detected in and beneath underground detectors through the reactions

$$\begin{aligned}\nu_\mu + n &\to \mu^- + p, \\ \bar\nu_\mu + p &\to \mu^+ + n,\end{aligned}$$

and

$$\begin{aligned}\nu_e + n &\to e^- + p, \\ \bar\nu_e + p &\to e^+ + n.\end{aligned}$$

These are respectively called μ-like and e-like events. The observed ratios of these events was found to be substantially reduced from the expected value ~ 2. There is a compelling evidence that atmospheric neutrinos change flavor as the Super-Kamiokande experiment clearly indicated a deficit of up-ward μ-like events (produced about 10^4 km away at the opposite side of earth) relative to the down-ward going events (produced about 20 km above). The e-like events showed a normal zenith angle dependence. The data is described by $\nu_\mu \to \nu_\tau$ oscillations. The conversion probability $P_{\nu_\mu\to\nu_\tau}$ fits the data quite well for[4]

$$\Delta m^2_{23} = 2.0\times 10^{-3}eV^2,\ \sin^2 2\theta_{23} \approx 1.0. \quad (27)$$

Solar Neutrinos

Particularly compelling evidence that the solar neutrinos change flavor has been reported by the Sudbury Neutrino Observatory (SNO). SNO measures the high energy part of the solar neutrino flux (^8B neutrinos). The reactions

$$\begin{aligned}\nu d &\to \nu n p \\ &\to e p p, \\ \nu e &\to \nu e,\end{aligned}$$

were studied by SNO. SNO measured arriving $\nu_e + \nu_\mu + \nu_\tau$ flux, $\phi_e + \phi_{\mu\tau}$, and the ν_e flux, ϕ_e. From the observed rates for the first two reactions, which involve respectively neutral current and charge current, SNO finds that the ratio of the two fluxes ϕ_e and $\phi_e + \phi_{\mu\tau}$ is 0.306 ± 0.050. This implies that the flux $\phi_{\mu\tau}$ is not zero. Since all the neutrinos are born in nuclear reactions that produce only electron neutrinos, it is clear that neutrinos change flavor. Corroborating information comes from the direct reaction $\nu e \to \nu e$, studied by both SNO and Super-Kamiokande. The strongly favored explanation of solar neutrino flavor change is the Large Mixing Angle version of the MSW effect, with the best fit parameters[5]

$$\Delta m_{12}^2 = 7.1 \times 10^{-5} eV^2, \quad \sin^2 2\theta_{12} = 0.8. \tag{28}$$

5.2 Disappearance Experiments

Reactors are source of $\bar{\nu}_e$'s through the neutron β-decay

$$n \to p + e^- + \bar{\nu}_e,$$

and experiment looks for a possible decrease in the $\bar{\nu}_e$ flux as a function of distance from the reactor, $\bar{\nu}_e \to X$ (if converted to $\bar{\nu}_\mu$, say, one would see nothing, $\bar{\nu}_\mu$ could have produced μ^+, but does not have sufficient energy to do so). Kamland experiment[6] confirms that $\bar{\nu}_e$ do indeed disappear when the reactor $\bar{\nu}_e$ have traveled ≈ 200 km. $\bar{\nu}_e$ flux is only $0.611 \pm 0.085 \pm 0.041$ of what it would be if none of it were disappearing. Interestingly, this reactor $\bar{\nu}_e$ disappearance and the solar neutrino results can be described by the same neutrino mass and mixing parameters indicating that the physics of both phenomena has been correctly identified.

6 Neutrino Mass Matrix

As discussed, the data from solar and atmospheric neutrino and reactor antineutrinos experiments provide evidence for neutrino mass and mixing with two different mass scales and large mixing angles

$$\begin{aligned}
\Delta m_{atm}^2 &\equiv \Delta m_{23}^2 = (2.0 \pm 0.5) \times 10^{-3} eV^2, \\
\sin^2\theta_{23} &\equiv \sin^2\theta_1 = 1.00 \pm 0.4, \\
\Delta m_{solar}^2 &\equiv \Delta m_{12}^2 = (7.1 \pm 0.6) \times 10^{-5} eV^2, \\
\tan^2\theta_{12} &\equiv \tan^2\theta_3 = 0.45 \pm 0.06.
\end{aligned} \tag{29}$$

Further, the CHOOZ experiment[7] gives

$$|U_{e3}|^2 \equiv \sin^2\theta_2 < 4 \times 10^{-2}. \tag{30}$$

We would interpret these results in terms of small off-diagonal perturbations of a degenerate diagonal mass matrix in flavor basis for light Majorana neutrinos[8]. In this approach, there is no fundamental distinction between masses of neutrinos of different flavors; the mass differences arise from small flavor violation of off-diagonal Yukawa coupling constants. Further, the neutrino mass differences do not in anyway constraint the absolute value of neutrino mass. The constraint on it will come from neutrinoless double β-decay experiments, cosmology and direct laboratory experiments, *e.g.* tritium β-decay.

Let us consider a Majorana mass matrix in (e, μ, τ) basis

$$m_\nu = m_0 \begin{pmatrix} a_{ee} & a_{e\mu} & a_{e\tau} \\ a_{e\mu} & a_{\mu\mu} & a_{\mu\tau} \\ a_{e\tau} & a_{\mu\tau} & a_{\tau\tau} \end{pmatrix}. \tag{31}$$

It is convenient to define the neutrino mixing angles as follows

$$\begin{pmatrix} \nu_e \\ \nu_\mu \\ \nu_\tau \end{pmatrix} = U \begin{pmatrix} \nu_1 \\ \nu_2 \\ \nu_3 \end{pmatrix}, \tag{32}$$

where U is the mixing matrix. We shall put CP violating phase δ as well as Majorana phases to be zero. In view of mixing angles given above, we shall take $s_{13} \equiv s_2 = 0$, $c_{13} \equiv c_2 = 1$, and $c_1 = 1/\sqrt{2}$, $s_1 = \mp 1/\sqrt{2}$. The diagonalization gives

$$\begin{aligned}
a_{e\mu} &= \pm a_{e\tau} = \frac{1}{\sqrt{2}} s_3 c_3 \left(-m_1 + m_2\right), \\
a_{\mu\tau} &= \pm\frac{1}{2}\left[\left(m_1 s_3^2 + m_2 c_3^2\right) - m_3\right], \\
a_{\mu\mu} &= a_{\tau\tau} = \frac{1}{2}\left[m_1 s_3^2 + m_2 c_3^2 + m_3\right], \\
a_{ee} &= m_1 s_3^2 + m_2 c_3^2.
\end{aligned} \tag{33}$$

In view of

$$\Delta m_{12}^2 = m_2^2 - m_1^2 \ll \Delta m_{23}^2 = m_3^2 - m_2^2, \tag{34}$$

we can take

$$m_1 \simeq \pm m_2.$$

Thus, we have two possibilities for mass matrix $m_1 = m_2 = m_0$; $m_1 = -m_2 = m_0$:

$$m_\nu = m_0 \begin{pmatrix} 1 & 0 & 0 \\ 0 & \frac{1}{2}(1+a) & \pm\frac{1}{2}(1-a) \\ 0 & \pm\frac{1}{2}(1-a) & \frac{1}{2}(1+a) \end{pmatrix}, \tag{35}$$

$$m_\nu = m_0 \begin{pmatrix} C_3 & -S_3 & \mp S_3 \\ -S_3 & B & C \\ \mp S_3 & C & B \end{pmatrix}, \qquad (36)$$

where $a = m_3/m_0$, $C_3 = \cos 2\theta_3$, $S_3 = \sin 2\theta_3$, $B = \frac{1}{2}(C_3 + a)$, $C = \pm\frac{1}{2}(\cos 2\theta_3 - a)$. If we do not want to commit to any particular value of θ_3, then we have the first case with the following subcases corresponding to $m_0 = 0$, $a = -1, 1, -2, 2, 0$.[9] In order to generate Δm_{12}^2 and Δm_{23}^2, we will now concentrate on the choice $a = 1$, which preserves flavor and add to it a small perturbation which violates flavor in off-diagonal matrix elements

$$m_\nu = m_0 \begin{pmatrix} 1 & \varepsilon_{12} & \varepsilon_{13} \\ \varepsilon_{12} & 1 & \varepsilon_{23} \\ \varepsilon_{13} & \varepsilon_{23} & 1 \end{pmatrix}, \qquad (37)$$

where $\varepsilon_{ij} \ll 1$. The diagonalization gives

$$m_i = m_0 (1 - x_i), \qquad (38)$$

where $x_i\ (i = 1,2,3)$ are roots of cubic equation

$$x^3 - \left(\varepsilon_{12}^2 + \varepsilon_{13}^2 + \varepsilon_{23}^2\right) x + 2\varepsilon_{12}\varepsilon_{13}\varepsilon_{23} = 0. \qquad (39)$$

The choice $\varepsilon_{12} = \varepsilon_{13} = \varepsilon_{23} = \varepsilon$ will give the roots $(\varepsilon, \varepsilon, -2\varepsilon)$ and thus will not lift the degeneracy between m_1 and m_2. To lift this degeneracy we take $\varepsilon_{12} = \varepsilon_{13} = \varepsilon + \delta$, $\varepsilon_{23} = \varepsilon$ with $\delta/\varepsilon \ll 1$. Then, the roots to the first order in δ/ε are $\varepsilon\left(1 + \frac{4}{3}\frac{\delta}{\varepsilon}\right)$, ε, $-2\varepsilon\left(1 + \frac{2}{3}\frac{\delta}{\varepsilon}\right)$, so that

$$\begin{aligned} m_1 &= m_0 \left[1 - \varepsilon - \frac{4}{3}\delta\right], \\ m_2 &= m_0 [1 - \varepsilon], \\ m_3 &= m_0 \left[1 + 2\varepsilon + \frac{4}{3}\delta\right], \\ \Delta m_{12}^2 &\approx \frac{8}{3} m_0^2 \delta (1 - \varepsilon) \simeq \frac{8}{3} m_0^2 \delta, \\ \Delta m_{23}^2 &\approx 6\varepsilon m_0^2. \end{aligned} \qquad (40)$$

This gives

$$\begin{aligned} \frac{\delta}{\varepsilon} &= \frac{9}{4}\frac{\Delta m_{12}^2}{\Delta m_{23}^2} \simeq 5.9 \times 10^{-2} \\ \sqrt{\varepsilon} m_0 &\simeq 2.1 \times 10^{-2}\ eV. \end{aligned} \qquad (41)$$

Thus, m_0 is not constrained. However, m_0 is constrained by WMAP data, $3m_0 < 0.71$ eV. When analyzed in conjunction with neutrino oscillation, it is found that mass eigenvalues are essentially degenerate with $3m_0 > 0.4$ eV. The above limits put limits on ε: $7.9 \times 10^{-3} < \varepsilon < 2.5 \times 10^{-2}$.

For the degenerate neutrino mass pattern $m_1 \sim m_2 \sim m_3 \gg \sqrt{\Delta m_{32}^2} = 0.045$, the effective mass in neutrino-less double β-decay is larger than ~ 0.05 eV, constrained from above by the mass limit from tritium β-decay. If the effective Majorana mass is confirmed to be $\left(0.39^{+0.17}_{-0.28}\right)$ eV,[1] it would strongly indicate that neutrinos follow degenrate mass pattern[3], when

$$\frac{\Delta m^2}{m^2} \ll 1.$$

Finally, for two modest extensions of the standard model in which the neutrino mass matrix advocated in this section can be embedded, see Ref. [8].

7 Conclusion

To conclude various neutrino mass patterns and corresponding neutrino mass matrix types are possible. Further, the absolute value of neutrino mass is not yet determined. However, one thing is certain that neutrinos are providing an evidence for new physics but the scale of new physics is not yet pinned down. The heavy right handed neutrinos at new physics scale may provide an explanation for baryogenesis through leptogenesis. If past is of any guide, neutrinos will enrich physics still further.

Acknowledgements

The author would like to acknowledge the warm hospitality provided by IPM and a travel grant from COMSTECH.

References

1. H.V. Klapdor-Kleingrothaus *et al.*, *Mod. Phys. Lett.* **16**, 2409 (2001).
2. C.E. Aalseth *et al.*, *Mod. Phys. Lett.* **17**, 1475 (2002); F. Feruglio, A. Strumia and F. Vissani, *Nucl. Phys.* B **637**, 345 (2002); H.V. Klapdor-Kleingrothaus, A. Dietz and I. Krivosheina, *Found. Phys.* **32**, 1181 (2002); A.M. Bakalyarov *et al.*, arXiv: hep-ph/0309016.
3. See for example, a recent review: R.D. McKeon and P. Vogel, arXiv: hep-ph/0402025, where detailed references can be found.
4. K. Nishikawa, *Phys. Rev. Lett.* **90**, 041801 (2003).

5. S.N. Ahmed *et al.*, arXiv: nucl-ex/0309004.
6. K. Eguchi *et al.*, *Phys. Rev. Lett.* **90**, 021802 (2003).
7. M. Apollonio *et al.*, *Phys. Lett.* **466**, 415 (1999).
8. Riazuddin, *JHEP* **10**, 009 (2003).
9. See Ref. [8] for original references for various cases.

CHAPTER 4: GENERAL RELATIVITY & QUANTUM GRAVITY

VACUUM ENERGY INSIDE A CAVITY WITH TRIANGULAR CROSS SECTION

H. AHMEDOV and I. H. DURU
Feza Gursey Institute
P.O. Box 6, 81220, Cengelkoy, Istanbul, Turkey
E-mail: hagi@gursey.gov.tr

For certain class of triangles, we construct Green function in a triangle region by making use of the group generated by reflections with respect to the three lines which form the triangle under consideration. Using this result, we calculate Casimir energy for a cavity of infinite height with triangular cross section for scalar massless fields.

1 Introduction

[a]The method of images is a useful tool to obtain the energy-momentum tensor for some systems with plane boundaries. The original parallel plate geometry, and in general rectangular prisms are of that type. For parallel plates the Green function is computed as an infinite sum of the "free" propagators to the image points[1]. The images are due to infinite reflections between the planes, a fact that is best visualized in the path integral formulation of the quantum mechanics[2]. Similarly, for multiply connected geometries (*i.e.* for Aharonov-Born case), one sums over all homotopically non-equivalent paths. As we see in the next sections the groups generated by the reflections, too, provide useful guides in the construction of the Green functions. For parallel plate the group is isomorphic to Z, for the three dimensional rectangular prism, for example, it is Z^3. However, if the rectangularity condition is dropped, the groups generated by reflections becomes non-commutative which is the case for present work.

We first investigate the structure of the group generated by the reflections with respect to the three lines of a triangle under consideration. This group will play central role in the construction of the Green function satisfying the Dirichlet boundary condition. Finally, we give the general expression for the energy-momentum tensor in terms of the sum over elementary power functions.

2 Energy-Momentum Tensor

Reflection Group: For $N = 3,4,5,\cdots$ and $k = 1,2,\cdots,N-2$ consider the triangles $\triangle_k^N$ in x^1x^2-plane formed by the lines

$$L_1 = \{\vec{x} \in R^2 : x^2 = 0\}, \quad (1)$$
$$L_2 = \{\vec{x} \in R^2 : x^2 = x^1 \tan \upsilon\}, \quad (2)$$
$$L_3 = \{\vec{x} \in R^2 : x^2 = (b - x^1)\tan(k\upsilon)\}, \quad (3)$$

where b is the length of the side laying on the line L_1 and $\upsilon = \frac{\pi}{N}$ is the angle between L_1 and L_2. The actions of the reflections Q_j with respect to the lines L_j, $j = 1,2,3$ on the vector

$$\vec{x} = \begin{pmatrix} x^1 \\ x^2 \end{pmatrix}, \quad (4)$$

are given by

$$Q_1\vec{x} = p\vec{x}, \quad Q_2\vec{x} = rp\vec{x}, \quad Q_3\vec{x} = pr^k\vec{x} + \vec{x_0}, \quad (5)$$

where

$$r = \begin{pmatrix} \cos 2\upsilon & -\sin 2\upsilon \\ \sin 2\upsilon & cos 2\upsilon \end{pmatrix}, \quad p = \begin{pmatrix} 1 & 0 \\ 0 & -1 \end{pmatrix},$$
$$\vec{x_0} = (1 - pr^k)\begin{pmatrix} b \\ 0 \end{pmatrix}. \quad (6)$$

The group generated by the these reflections, which we denote by G_N, is a free one with relations. Relations can be obtained from the realization (5) and the properties

$$r^N = 1, \quad p^2 = 1, \quad pr = r^{N-1}p, \quad r^k\vec{x_0} = -p\vec{x_0}. \quad (7)$$

Some of the obvious relations are

$$Q_j^2 = 1, \quad (Q_1Q_2)^N = 1, \quad (8)$$

from which we conclude that the reflections Q_1 and Q_2 generate the finite subgroup

$$D_N = \{r^s, pr^s, s = 0,1,\ldots N-1\}, \quad (9)$$

[a]Talk presented by H.A. at the XIth Regional Conference on Mathematical Physics, May 2004, Tehran, Iran.

which is the dihedral group of dimension $2N$. Consider the linear space V_N which consists of the vectors

$$\vec{\xi} = \sum_{s=0}^{N-1} n_s \vec{x}_s, \tag{10}$$

where n_s are integers and

$$\vec{x}_s = r^s \vec{x}_0. \tag{11}$$

The equalities

$$r\vec{x}_s = \vec{x}_{s+1}, \qquad p\vec{x}_s = \vec{x}_{N-s+k}, \tag{12}$$

imply that D_N is the automorphism group of the linear space V_N. G_N is the semidirect product group of D_N and translation group V_N. In particular,

$$Q_1 = (p, 0), \qquad Q_2 = (rp, 0), \qquad Q_3 = (pr^k, \vec{x}_0). \tag{13}$$

Since V_N is a vector space over the integer numbers the dimension $\mid V_N \mid$ is not necessarily equal to the dimension of the vectors $\vec{x}_s$. It may be larger, that is in our case may be greater than two. For example, the dimensions of V_5 and V_8 are four; while the dimensions of V_3, V_6 and V_4 are two[b]. In general, for $N = M_1^{l_1} M_2^{l_2} \cdots M_f^{l_f}$, where M_j are different prime numbers and l_j are positive integers, we have the following result (for a detailed discussion of this problem see Ref. [3])

$$\mid V_N \mid = N - rank(A_N), \tag{14}$$

where

$$A_N = \begin{pmatrix} I_1 & I_1 & \cdots & I_1 \\ I_2 & I_2 & \cdots & I_2 \\ \cdots & \cdots & \cdots & \cdots \\ I_f & I_f & \cdots & I_f \end{pmatrix}, \tag{15}$$

and I_j is $\frac{N}{M_j} \times \frac{N}{M_j}$ unit matrix.

Casimir Energy: Consider the following function

$$K(x,y) = \sum_{n \in V_N} \sum_{s=0}^{N-1} (G(r^s x + \xi, y) - G(pr^s x + \xi, y)), \tag{16}$$

where

$$G(x,y) = \frac{1}{2\pi^2 \mid x - y \mid^2}, \tag{17}$$

is the Green function in the Minkowski space. Here x is four-vector and action of $g \in G$ is trivial on the components x_0 and x_3. Using the representation (13), one can verify that $K(Q_j x, y) = -K(x,y)$, *i.e.*, $K(x,y)$ vanishes on the boundaries of the triangles. Let us now find out the conditions under which the domain inside a triangle will be the fundamental domain for the reflection group G_N. The orbits of the coset space R^2/D_N are

$$[\vec{x}] = \{r^s \vec{x}, pr^s \vec{x} : \quad s = 0, \ldots, N-1\}. \tag{18}$$

It is clear that we can identify this coset space with region X between two lines L_1 and L_2 including the boundaries. For any orbit in R^2/D_N, there exists a unique representative in X. Since the group G_N is generated by the elements of D_N and Q_3, the problem of constructing the coset space R^2/G_N reduces to finding the subspaces Y of X such that the reflection Q_3 maps Y into X. Consider the area between three lines L_j, which is the triangle under consideration. The previous condition implies that the two angles kv and sv of the triangle between the lines L_1, L_3 and L_2, L_3 must be less than or equal to $\frac{\pi}{2}$. The restrictions

$$kv \le \frac{\pi}{2}, \qquad sv \equiv \pi - (k+1)v \le \frac{\pi}{2}, \tag{19}$$

with solutions

$$k = \begin{cases} \frac{N}{2}, & \text{for even } N, \\ \frac{N-1}{2}, & \text{for odd } N, \end{cases} \tag{20}$$

imply that for triangles without obtuse angle the function $K(x,y)$ is indeed the Green function. Note that the equations (19) have also been solved by $k = \frac{N-2}{2}$ for even N. In this case $s = \frac{N}{2}$. For $k = \frac{N}{2}$ we have $s = \frac{N-2}{2}$. This solution is therefore congruent to the previous one; that is $\Delta^N_{\frac{N}{2}}$ goes to $\Delta^N_{\frac{N-2}{2}}$ when the length b goes to $b\cos v$.

Now we are ready to calculate the energy-momentum tensor density. The energy-momentum tensor for conformally coupled massless scalar field is given by[1]

$$T_{\mu\nu} = \frac{2}{3}\partial_\mu\phi\partial_\nu\phi - \frac{1}{6}\eta_{\mu\nu}\partial_\rho\phi\partial^\rho\phi - \frac{1}{3}\phi\partial_\mu\partial_\nu\phi + \frac{1}{12}\eta_{\mu\nu}\phi\partial_\rho\partial^\rho\phi. \tag{21}$$

Since the vacuum expectation value of the product of two scalar fields is the Green function, we can express the energy-momentum tensor in terms of the Green

[b]The vector spaces $V_3 \equiv V_6$ and V_4 are known as the hexagonal and square lattices[4].

function. The vacuum energy density in particular is given by the following expression

$$T_{00} = \frac{1}{6\pi^2} \sum_{(n,s)} [\frac{|\, (pr^s + 1)\overrightarrow{\xi}\, |^2}{|\, (pr^s - 1)\overrightarrow{x} + \overrightarrow{\xi}\, |^6} - \frac{2 + \cos(2sv)}{|\, (r^s - 1)\overrightarrow{x} + \overrightarrow{\xi}\, |^4}]. \tag{22}$$

References

1. N.D. Birrel and P.C.V. Davies in *Quantum Fields in Curved Spaces*, (Cambridge University Press, 1982).
2. See for example L.S. Schulman in *Techniques and Applications of Path Integration* (John Wiley and Sons, New York, 1981); C. Grosche and F. Steiner in *Handbook on Feynman Path Integrals* (Springer-Verlag, 1998).
3. H. Ahmedov and I.H. Duru, *J. Math. Phys.* **46**, 965 (2004).
4. See for example M.A. Armstrong in *Groups and Symmetry*, (Springer-Verlag, 1997).

CURVATURE COLLINEATIONS OF SOME PLANE SYMMETRIC STATIC SPACETIMES

A. H. BOKHARI*, A. R. KASHIF[b] and A. QADIR*,†

*Department of Mathematical Sciences
King Fahd University of Petroleum and Minerals
Dhahran, 13261 Saudi Arabia

†Centre for Advanced Mathematics and Physics
National University of Sciences and Technology
Peshawar Road, Rawalpindi, Pakistan

[b]National University of Sciences and Technology
E&ME College
Peshawar Road, Rawalpindi, Pakistan
E-mail: kashmology@yahoo.com

A complete classification of plane symmetric static spacetimes by curvature collineations had been obtained earlier[1]. It was found that all cases led to homothetic curvature collineations unless some constraints were met. There were four such cases that were non-trivial. In all these cases their Lie algebra was infinite dimensional. Of these four, three led to specific metrics and hence the constraints had a unique solution. In the fourth case, there was no example provided of a metric satisfying the constraints. In this paper, it is shown, by explicit construction, that the fourth class is non-empty.

[a]In relativity, there exists a large body of literature on the classification of spacetimes according to their isometries or Killing Vectors (KVs) and the group admitted by them (*e.g.* see Ref. [2], and the references therein). Each independent KV gives rise to a conservation law for the spacetime. The geometric symmetries of a spacetime are given by KVs, $\mathbf{k}$, satisfying the Killing equations for the metric coefficients, g_{ab} *viz.*

$$L_{\mathbf{k}} g_{ab} = g_{ab,c}k^c + g_{bc}k^c{}_{,a} + g_{ac}k^c{}_{,b} = 0, \tag{1}$$

where $L_{\mathbf{k}}$ represents the Lie derivative. If the right side of the equation is replaced by βg_{ab}, where β is a real non-zero constant, the particular solution is called a homothetic vector (HV). HVs have been found to be of great physical significance[3]. It is often useful to look for the symmetries of the tensors giving the curvature of the spacetime. The Riemann curvature tensor is an important quantity in general relativity whose trace is the Ricci tensor. The symmetries of the Ricci tensor, R_{ab}, given by

$$L_\xi R_{ab} = R_{ab,c}\xi^c + R_{bc}\xi^c{}_{,a} + R_{ac}\xi^c{}_{,b} = 0, \tag{2}$$

are called Ricci Collineations (RCs) and of the Riemann curvature tensor, R^a_{bcd} given by

$$\begin{aligned} L_\xi R^a_{bcd} = R^a_{bcd,e}\xi^e + R^a_{ecd,b}\xi^e + R^a_{bed,c}\xi^e \\ + R^a_{bce,d}\xi^e - R^e_{bcd}\xi^a{}_{,e} = 0, \end{aligned} \tag{3}$$

are called Curvature Collineations (CCs).

It is clear that every KV is a CC and an RC. It was proved[4] that every CC is an RC but there can be RCs that are not CCs and CCs that are not KVs. Unlike the metric tensor, these tensors can be degenerate, *i.e.* have $\det(R_{ab}) = 0$ or $\det(R_{AB}) = 0$ where A, B are the compound indices [ab] and [cd], respectively, having a range of 1 to 6. Now, any symmetric non-degenerate matrix of rank n can be regarded as the metric of an n-dimensional manifold. The maximum number of isometries[2] of that manifold (for a space of constant curvature) is $n(n+1)/2$. Therefore the number of collineations for a non-degenerate tensor is finite and hence yields a finite dimensional Lie algebra. However, degenerate tensors can have infinite-dimensional Lie algebras. In general, the Lie algebra of KVs is of dimension $n \leq 10$ and of HVs (including KVs) of dimension $(n+1)$ (if there exists a proper homothety). For example, the Schwarzschild metric has a zero Ricci tensor and a non-degenerate curvature tensor. Thus, it has a finite-dimensional Lie algebra of CCs and HVs but an infinite-dimensional Lie algebra of RCs – every vector being an RC. Similarly, for Minkowski space there is a ten-dimensional Lie algebra of KVs, an eleven-dimensional Lie algebra of HVs and an

[a]Talk presented by A.R.K. at the XIth Regional Conference on Mathematical Physics, May 2004, Tehran, Iran.

infinite-dimensional Lie algebra of CCs and of RCs, consisting of all possible vectors. It has also been demonstrated that $\{HVs\} \sqsubseteq \{RCs\}$.

The procedure adopted earlier[5] involved solving the 10 Killing equations as a system of quasi-linear first order partial differential equations for the 10 metric coefficients and the 4 components of the KV as functions of the 4-spacetime variables, subject to the minimal symmetry constraints on the metric coefficients. Thus, for plane symmetry, we can write all metrics or classes of metrics associated with each algebra $A \sqsupseteq so(3)$. A complete classification of plane symmetric static spacetimes

$$ds^2 = e^{\nu(x)}dt^2 - e^{\lambda(x)}dx^2 - e^{\mu(x)}(dy^2 + dz^2), \quad (4)$$

by curvature collineations was obtained[1]. It was found that all cases led to homothetic curvature collineations unless constraints were met. There were four such cases that were non-trivial. In the fourth case, there was no example provided of a metric satisfying the constraints. A solution compatible with the fourth case is presented here. The constraint requirements came form the conditions that

$$2\nu' + \nu''^2 - \nu'\mu' \neq 0,$$

unless $\mu' = 0$. Thus the metric can be written as

$$ds^2 = e^{\nu(x)}dt^2 - dx^2 - (dy^2 + dz^2), \quad (5)$$

subject to the constraints

$$\left.\begin{aligned} &\left(e^{\nu/2}\right)' \neq 0, \\ &a^2\left[\frac{\nu' + 2\nu'\upsilon' + \frac{1}{2}\nu^8}{2\nu' + \nu^2}\right]' = \frac{e^{-\nu}}{(2\nu' + \nu^2)^{1/2}}. \end{aligned}\right\} \quad (6)$$

These conditions are met by the choice

$$e^{\nu/2} = \cosh(x/a). \quad (7)$$

In this case the homotheties, which are the same as the isometries

$$\left.\begin{aligned} \xi^0 &= -\tfrac{1}{2}a\tanh(x/a)[c_4\cosh(t/a) + c_5\sinh(t/a) + c_6, \\ \xi^1 &= \tfrac{1}{2}[c_4\cosh(t/a) + c_5\sinh(t/a), \\ \xi^2 &= c_3 z + c_1, \\ \xi^3 &= -c_3 z + c_2, \end{aligned}\right\} \quad (8)$$

satisfy the Lie algebra $so(1,2) \otimes_s \mathbf{R}^3$. Here $\otimes_s$ stands for the semi-direct product. The CCs have the same values as the isometries for the first two components (ξ^0 and ξ^1) but the second two components (ξ^2 and ξ^3) are arbitrary functions of y and z. Hence, this class of spacetimes is non-empty as it has at least one element (given by Eq. (8)). The components of the energy-momentum tensor for above spacetime are $T_0^0 = 0 = T_1^1$, $T_2^2 = -\frac{2\nu' + \nu^2}{4} = T_3^3$. Thus it represents an electromagnetic non-null field.

Acknowledgments

The presenting author (A.R.K.) thanks Pakistan Science Foundation for providing travel support to participate in the conference. The authors are also thankful to K. Saifullah and A. A. Siddiqui for their help in preparing the manuscript.

References

1. A.H. Bokhari, A.R. Kashif and A. Qadir, *J. Math. Phys.* **44**, 2167 (2000).
2. H. Stephani, D. Kramer, M.A.H. MacCallum, C. Hoenselaers and E. Herlt in *Exact Solutions of Einstein's Field Equations* (Cambridge University Press, 2003); G.S. Hall in *Symmetries and Curvature Structure in General Relativity* (World Scientific, Singapore, 2004).
3. C.B.G. McIntosh, *Phys. Lett.* 50 **A**, 429 (1975); A.H. Taub in *General Relativity: Papers in honor of J.L. Synge, Chap. VIII*, ed. L. O'Raifeartaigh (Oxford University Press, London, 1972); M.E. Cahill and A.H. Tuab, *Commun. Math. Phys.* **21**, 1 (1971); J. Carot and A.M. Santos, *Class. Quantum Grav.* **14**, 1183 (1997).
4. G.H. Katzin, J. Levine and W.R. Davis, *J. Math. Phys.* **10**, 617 (1969).
5. A.H. Bokhari and A. Qadir, *J. Math. Phys.* **28**, 1019 (1998); *J. Math. Phys.* **31**, 1463 (1990); *J. Math. Phys.* **34**, 3543 (1993); R. Bertolotti, G. Contoreros, L.A. Nunez, U. Percoco and J. Carot, *J. Math. Phys.* **37**, 1086 (1996); A. Qadir and M. Ziad, *J. Math. Phys.* **29**, 2473 (1988); *J. Math. Phys.* **31**, 254 (1990); *Nuovo Cimento* B **110**, 317 (1995); *Nuovo Cimento* B **110**, 277 (1995).

PROBING UNIVERSALITY OF GRAVITY

N. DADHICH
Inter-University Centre for Astronomy & Astrophysics (IUCAA)
Post Bag 4, Ganeshkhind, Pune 411 007, India
E-mail: nkd@iucaa.ernet.in

I wish to expound a novel perspective of probing universal character of gravity. To begin with, inclusion of zero mass particle in mechanics leads to special relativity, while its interaction with a universal force shared by all particles leads to general relativity. The universal nature of force further suggests that it is intrinsically attractive, self-interactive and higher-dimensional. I argue that the principle of universality could serve as a good guide for future directions.

[a]The moment we admit the existence of zero mass particle, we need a new mechanics for it moves with the same universal speed relative to all observers. Newtonian mechanics cannot accommodate a universal speed. The new mechanics that emerges is the special relativity which synthesizes space and time into one whole – spacetime. Alongside come synthesis of energy, momentum and mass, and the light-cone structure of spacetime events. It is the principle of universality that has led us to SR.

A universal entity is the one which is freely accessible and shared by all particles. Space and time are without question the most primary universal entities we know of. There cannot in principle exist two unconnected universal things. How is one universal entity is distinguished from the other universal entity? This distinguishing feature will contradict the universal nature of them. Thus all universal things must be related and this relation would have also to be universal. It is hence pertinent to ask for a relation between the two universal entities; space and time. That will require a universal speed which is what is precisely provided by the zero mass particle – photon, a quantum of light.

It is geometry that makes a universal statement. For instance, motion of free particles on a straight line is a universal property which is expressed as a geometric statement. A universal relation or property must therefore be expressible as a geometrical statement. The universality of speed of light must also be a geometric statement and the Minkowski geometry of spacetime is that statement. From this we wish to propound a general guiding principle, *any universal physical property/force must be expressible as a property of universal entities, space and time, through a geometric statement.*

What we have done is that, in conformity with universality, we have asked for incorporation of massless particles in mechanics, which is not admissible in the Newtonian mechanics. Taking into account the characteristic property of motion of massless particles, we are forced to enlarge the Newtonian framework to Minkowski framework which admits both massive and massless particles. In our further exploration, this would be the guiding finger: *Ask a question which is not admitted in the existing framework, then enlarge the framework as indicated/suggested by the question itself such that the question is answered.*

Of the four basic forces in Nature, gravitation distinguishes itself from the rest by the property of universality. As argued above, universal force must be unique. Let us hence begin by considering a universal force, and then show that it can be nothing other than gravity. Since the force is universal, it must be long range and act on all particles including massless ones as well. But, massless particle propagates with universal constant speed which cannot be changed. Then, how do we make it feel the universal force, and feel it must? We thus face the contradiction that massless particle must feel the force but its speed must not change. This is certainly not possible in the existing framework. How do we enlarge the framework? Two suggestions come forward from our guiding principle, one since the force is universal, it must be expressible as a geometric property of spacetime and two, what do we really want massless particle to do is that when it is skirting the source of the universal force it should acknowledge its presence

[a]Invited plenary talk presented at the XIth Regional Conference on Mathematical Physics, May 2004, Tehran, Iran.

by bending rather than going straight. To illustrate this point, let us consider a piece of wood floating in a river. It floats freely and bends as the river bends. No force really acts on it to bend its course, it simply follows the flow of the river. This suggests that bend the *river* in which massless particle propagates freely. Thus, we arrive at the profound insight into the nature of the universal force, that it must bend/curve space – rather spacetime in which massive as well as massless particles propagate freely. That is the envisaged universal force can only be described by curvature of spacetime and in no other way. Isn't it a remarkable and insightful conclusion that follows from the very simple common sense arguments?

Thus, it is the universality of force which demands spacetime to respond to it by being its *potential*. It can no longer remain as inert background, but, instead it now has all the dynamics of the universal force. This is the unique and distinguishing property of universality. It then ceases to be an external force and its dynamics has to be fully governed by the spacetime curvature. This automatically incorporates the first suggestion mentioned above that universal force should become a property of spacetime. The equation of motion for the force does not have to be prescribed, but must rather follow from the spacetime curvature all by itself. The curvature of spacetime is given by the Riemann curvature tensor for the metric g_{ab} and it satisfies the Bianchi differential identity. The contraction of which yields the second rank symmetric tensor, constructed from the Ricci tensor, and it is divergence free. That is,

$$\nabla_b G^{ab} = 0, \tag{1}$$

where

$$G_{ab} = R_{ab} - \frac{1}{2} R g_{ab}. \tag{2}$$

The above equation implies

$$G_{ab} = -\kappa T_{ab} - \Lambda g_{ab}, \tag{3}$$

with

$$\nabla_b T^{ab} = 0, \tag{4}$$

where T_{ab} is a symmetric tensor, and κ and Λ are constants. On the left, we have a differential expression involving the second derivative and square of the first derivative of the metric which now acts as a potential for the universal force. Thus, on the right should be the source of the force. What should be the source for a universal force? Something which is shared by all particles – matter energy. That identifies T_{ab} with the energy momentum tensor of matter distribution and vanishing of its divergence ensures conservation of energy and momentum. In the weak field and slow motion limit, this equation for small Λ takes the form $\nabla^2\phi = \frac{1}{2}\kappa\rho$ where ρ is energy density of matter distribution. Hence, the force in this limit is inverse square law. Further, since the force is self-interacting, which is indicated by the presence of square of first derivative of the metric in the Riemann curvature, it must have charge. What should be its polarity? We all know that for stability, it is essential that total charge is zero. The matter energy is positive and hence charge neutrality will require field energy to have opposite charge; *i.e.* it should be negative. Since the potential is negative for a long range force, it must hence be attractive – yet another remarkable result from simple common sense arguments. We thus have a inverse square law attractive universal force which must agree with the Newtonian gravitational law in the first approximation. That determines $\kappa = 8\pi G/c^2$. The above equation is nothing but the Einstein equation for gravitation with a new constant Λ. The universal force we began with could be nothing other than gravitation and thus gravity is the unique universal interaction[1]. The universality of force also tells us that all its properties are self-determined and we have no freedom to put any constraints on it. For instance, the Einstein equation is valid in all dimensions where Riemann curvature can be defined ($D \geq 2$). *The dimensionality of space would have to be determined by the dynamics of gravitational field itself.*

Note that the constant Λ enters into the equation naturally and is in fact at the same footing as the energy-momentum tensor. It all follows from the Riemann curvature which contains the whole dynamics. It appears as a constant of integration in the field equation, and its Newtonian analogue would be $\nabla^2\phi = k = const.$ This gives rise to radially symmetric harmonic oscillator potential. It is addition of a constant to the field equation which has come about because now dynamics of the field cannot be prescribed but it follows from the spacetime geometry. It is the geometry which brings in this new constant which is a true new constant of the Einstein gravity. What value should it have to be determined by observation/experiment? It makes a very important statement that even when space is free of all non-gravitational matter energy distribution, empty space has non-trivial dynamics. The above equation refers to spacetime in entirety, which means whatever can be done in it has to be included in it. It is well known that vacuum can suffer quantum fluctuations which produce stress-energy tensor precisely of the form Λg_{ab} and it must be included in the equation. Had Einstein followed this chain of arguments, he would have anticipated gravitational effect of quantum fluctuations of vacuum and would have made a profound prediction rather than a profound blunder.

Classically a constant scalar field with a constant potential has no dynamics while in GR, it again generates precisely the stress tensor of the type Λg_{ab} due to constant potential. That of course has non-trivial dynamics. In other perspective, we can thus say that Λ is the measure of the constant potential to which vacuum has been raised[b].

That means setting $\Lambda = 0$ is to put vacuum to absolute zero potential. Note that spacetime responds to gravity by curving, and it also mediates motion of massless particles. It should therefore possess micro-structure which can facilitate bending (curvature) due to gravity and propagation of massless particles as a property of spacetime. It could then naturally suffer quantum fluctuations. That means, it would thus be natural for vacuum to have the inherent dynamics of Λg_{ab}. The big question however is what value should Λ have? The answer to this question is that its value has to be determined empirically by experiment that probes the micro-structure of spacetime.

Let us now turn to dimensionality of the gravitational interaction. Since the equation is valid for any D, it is therefore inherently a higher-dimensional interaction. However, let us go by the principle of minimum requirement. For the full realization of dynamics of gravity, the minimum dimension required is $4 = 3 + 1$. Let us keep time aside, so we require 3-space which could be inhibited by its source, matter energy distribution. Would gravity be confined to 3-space/brane or could it propagate off the brane? To establish the attractive character of the universal force, we had earlier argued that the field energy should have charge and its polarity be negative (opposite of positive polarity of matter energy). This was invoked to have overall charge neutrality. Here negative charge is spread all over the space and hence complete charge neutrality could only be achieved when integration is taken all over the space. So long as the total charge is non-zero on any finite surface, then the field must propagate off the surface. For instance, if net electric charge on a sphere is non-zero, the electric field does propagate off the sphere into 3-space. Similarly, since total gravitational charge in a finite region of space cannot be fully neutralized, the field will have to propagate off the 3-brane into $(5 = 4 + 1)$-D bulk spacetime. However, as we include larger and larger region on the brane, charge strength diminishes to zero asymptotically. This is perhaps suggestive of the fact that gravity cannot go deep enough into the bulk but rather remains confined near to the brane. If the matter fields are confined to n-space, the field will leak into the $(n + 1)$th dimension but will not propagate deep enough. That is, massless graviton will have ground state on the brane and hence will remain confined to the brane. This is exactly what is required to happen for the braneworld gravity[3,4]. It is remarkable that we have here motivated the higher-dimensional and the braneworld nature of the gravitational field purely from classical standpoint. In particular we can have the Randall-Sundrum braneworld model[4]. In that case, the confinement of gravity on the brane requires that bulk spacetime must be Anti-deSitter (AdS) with negative Λ in the bulk. It has been shown for the AdS bulk and flat brane system, that the Newtonian gravity can be recovered on the brane with high energy $1/r^3$ correction to the potential[4,5]. The most interesting case is of Schwarzschild – AdS bulk harboring FRW brane. In that case localization of gravity on the FRW brane requires non-negative effective Λ on the brane[6,7].

The universality of gravitation demands that field is self-interactive and spin 2 non-Abelian field. Self-interaction is however an iterative process and the Riemann curvature includes the first iteration through square of first derivative of the metric. The Riemann curvature led to the Einstein equation. How about the next order of iteration involving higher powers of the derivative of the metric? That is

[b] In the Schwarzschild field, we have the Newtonian potential, $\phi = -M/r$, note that a non-zero constant, which is classically inert, cannot be added to it. The Einstein equation determines the potential of an isolated body absolutely with its zero fixed at infinity and nowhere else[2]. That is for the local situation and here we have constant potential in a global setting.

inclusion of higher powers of the Riemann curvature. At any rate, the resulting equation should be second order quasi-linear (*i.e.* second derivative must occur linearly). Then the question is, what is the most general second order quasi-linear equation which can be derived from polynomial in Riemann curvature. It turns out that there exists the specific Gauss-Bonnet (GB) combination involving the square of Riemann, Ricci and scalar curvature (and there is the Lovelock generalization for higher order polynomial, see for instance in Ref. [8,9]), which still yields the second order quasi-linear equation for the field. However, the GB term is topological for $D < 5$; *i.e.* it makes non-trivial contribution only for $D > 4$. That is higher order iteration effects of self-interaction will therefore be non-trivial only in the higher dimensions $D > 4$. The high energy gravity would naturally involve higher order iterative GB term, which can have non-trivial meaning only in higher dimensions. This is therefore an independent new argument for higher dimensionality of gravity.

Now, if we adhere to our minimum requirement principle, the minimum number of spatial dimension required for full realization of gravitational dynamics is 3 and hence there is no compelling reason for non-gravitational matter energy to exist in dimension greater than 3. Gravity could however leak into the fourth (bulk) spatial dimension. In the 5-D bulk, the GB term attains dynamics which accounts for higher order self-interaction effects and hence must be included. It turns out that dS or AdS automatically solves the GB term in the equation and it simply redefines the bulk Λ. It suggests that the dynamics of the bulk spacetime may be governed by the GB term; *i.e.* the source for the AdS bulk in the Randall-Sundrum braneworld model may fully or partially be provided by the GB term[10]! As a matter of fact the GB term will generate effective negative Λ and thereby AdS in the bulk naturally, because it is the measure of gravitational field energy in higher order iteration. As argued elsewhere[11], the positive energy condition for the field energy is that it be negative.

It is however well-known that charge neutrality is a necessary condition for stable equilibrium but not sufficient. The Earnshaw's theorem states that it is impossible to attain stable equilibrium purely under electromagnetic force. This is because the field has two kinds of charges which are isolated and localized. On the other hand, in the case of gravity the other (negative) charge is distributed all over the space and hence is not isolated and localizable. A gravitational situation could be envisioned as follows: A positive charge (body) sitting in its own field which has negative charge spread around it in space like a net. This system is obviously stable. It is the distributed nature of the other charge that provides the stability. It is thus no surprise that systems bound by gravity are always stable.

Let us for a moment digress to quantum theory. What question should we ask which can lead us from classical to quantum mechanics? We have two kinds of motion, particle and wave. Like particle, wave must also carry energy and momentum with it. So it must like particle have a 4-momentum vector while on the other hand its motion is completely determined by the 4-wave vector. Since both these vectors refer to the same wave, they must be proportional. This gives the basic quantum mechanical relations between energy and frequency, and momentum and 3-(wave)vector. From this, it is easy to get to the uncertainty and commutation relations which form the basic quantum principle.

First of all, we have not yet succeeded in writing the uncertainty principle in a spacetime covariant form. Further, the quantum principle is universal. Going by the guiding principle of universality, it must, like the speed of light, be expressible as a property of spacetime. This has unfortunately not happened. What is required is exactly what Minkowski did to SR by synthesizing the speed of light into the spacetime structure. This is, however, very difficult because synthesis of quantum principle with the spacetime would ask for discrete structure, which is in contradiction with the inherent continuum of spacetime. However, so long as this does not happen, quantum theory will remain incomplete. Thus, for completion of quantum theory, it would be required that spacetime must have micro-structure which could accord to quantum principle. It is the geometry of that which would synthesize quantum principle with the spacetime. This is an open question of over 100 years standing.

The same question is also coming up from the gravity side as well. Unlike quantum theory, GR is complete like classical electrodynamics. However, at high energy we have quantum electrodynamics. We know that at high energy matter attains quantum character, *i.e.* T_{ab} on the right becomes quantum. On the left, is the spacetime which would now have to become quantum – discrete. This is what being pursued in the canonical approach of loop quantum gravity[12]. It is a program of quantizing gravitational field; *i.e.* quantizing spacetime in 4-dimensions. On the other hand, string theory begins with the quantum principle and SR to construct a consistent theory of matter. That naturally leads first to 26- and then to 10- or 11-dimensions. Gravity is considered as a spin 2 gauge field in higher-dimensional flat spacetime. There is, however, no unique way of getting down to the usual 4-dimensions. It makes connection with GR as a low energy effective theory[13]. The two approaches appear to be complementary, either catching some aspect of the problem. The strong point of the former is spacetime background independent formulation, while that of the latter is the gauge field framework which is shared by other fields and thereby its strong orientation to unification of the fields. However, asymptotically the two would have to converge when the complete theory comes about.

Both gravity at high energy, as well as completion of quantum theory require discrete microstructure for spacetime. This suggests that there may perhaps be one and the same answer to both the questions. In a sense the two approaches, string theory and loop quantum gravity anchor respectively to quantum field theoretic and geometric gravitational aspects.

Adhering to our guiding principle of universality, what question should we ask and how should we enlarge the existing framework to answer the question? Since spacetime is the fundamental universal entity, let us ask, do there still remain some properties of it which have yet remained untapped? One is dimensionality of space and the other is its (non)-commutativity. The former is, however, essential for the string theory and is also quite in vogue in the braneworld models. We have, however, attempted to articulate a new and simple minded motivation purely based on classical consideration for higher dimensions. On the other hand, noncommutativity of space has also attracted some attention recently and it is hoped that it might facilitate in building up discrete structure for space.

Returning to the enigmatic Λ, we would like to argue that it is really anchored on the microstructure of spacetime and hence might hold the ultimate key to the problem. It may, in a deep sense, connect through some duality relation micro- with macro-structure of the Universe. It defines a new scale given by the Einstein gravity. On the other hand, we already have the Planck length which is not given by any theory but constructed by using three universal constants. I believe that it is not a wise thing to take the Planck length a priori fundamental but instead we should attempt to deduce it in a fundamental manner. Λ, on the other hand, is a scale provided by a fundamental theory and hence should be respected. We thus have two length scales, where we require only one and hence it is natural to expect that there should exist a relation between them, which will perhaps encode a profound physical truth.

Let me come back to the guiding question one should ask? In the gravitational field equation, we have the curvature of spacetime on the left, and matter stress tensor and the vacuum energy Λ on the right. If T_{ab} lives only in 3-brane which becomes quantum at high energy, while vacuum energy can still support a continuum bulk spacetime. Taking the cue from what we have discussed so far what question should we ask and how should we enlarge the framework to admit and answer the question? The universality demands that the Einstein equation should remain valid whether the matter field source is classical or quantum. For the quantum case, the spacetime curvature will be required to have discrete quantum character. Should that mean that spacetime itself becomes quantum? This is what is being explored by the canonical loop quantum gravity approach[12]. In this approach, we remain bound to the 4-dimensional spacetime and there is no background spacetime relative to which spacetime is being quantized. In the string theory approach, we are already in higher dimensions and gravity is being considered as massless spin 2 field and GR appears as the low energy effective theory in 4-dimensions along with plethora of other fields[13]. In either case, it is adaption of radically new framework which cannot be seen as enlargement of the existing framework.

One possible enlargement of the framework could be, for matter fields confined to 3-brane, the 5-dimensional bulk spacetime with Λ and GB support could provide the continuum background for gravity (spacetime curvature) to be quantized on the 3-

brane. This is the suggestion that crops up in the spirit of what has been done in going from Newton to Einstein. However, the pertinent question is whether it is technically and conceptually workable? That has to be investigated. This suggestion is in the spirit of the guiding principle we have propounded. Perhaps it may lead to some insight.

Let me reiterate that we have here enunciated a method of asking question which is motivated by the principle of universality, and the question also suggests enlargement of the framework such that it gets answered. And, we arrive at a new framework. One of the remarkably interesting applications of this method is to establish why the universal force has to be attractive? This is perhaps the simplest and most direct demonstration of why gravity is attractive. Equally illuminating is the physical realization of the higher order iteration of self-interaction through the GB contribution, and thereby the higher-dimensional nature of gravity. Following this train of thought what we need to do is to ask the right kind of question which will show us the way beyond GR or quantum theory. The main problem is to identify the right question.

We have argued that one of the most challenging problems is to understand Λ in terms of the basic building blocks of spacetime. Any quantization scheme should have to address to it[14]. We believe that there must exist some fundamental relation between Λ and the Planck length, which needs to be discovered (recently there have come up a couple of such proposals[15]). That may hold the key to the problem.

These are some of the rumbling thoughts which I wanted to share[16].

Acknowledgements

I wish to thank warmly the organizers of the Regional conference for their kind invitation and wonderful hospitality which gave me the opportunity to share and propound a new perspective. I also thank Sudhendu Rai Choudhury for helpful discussions.

References

1. N. Dadhich, arXiv: gr-qc/0102009; arXiv: physics/0203004.
2. N. Dadhich, arXiv: gr-qc/9704068.
3. N. Arkani-Hamed, S. Dimopolous and G. Dvali, *Phys. Lett. B* **429**, 263 (1998).
4. L. Randall and R. Sundrum, *Phys. Rev. Lett.* **83**, 4690 (1999).
5. J. Garriga and T. Tanaka, *Phys. Rev. Lett.* **84**, 2778 (2000).
6. A. Karch and L. Randall, *JHEP* **0105**, 008 (2001).
7. P. Singh and N. Dadhich, to appear in *Mod. Phys. Lett. A*, arXiv: hep-th/0204190; arXiv: hep-th/0208080.
8. N. Deruelle and J. Madore, arXiv: gr-qc/0305004.
9. D. Lovelock, *J. Math. Phys.* **12**, 498 (1971).
10. M. Sami and N. Dadhich, arXiv: hep-th/0405016.
11. N. Dadhich, *Phys. Lett.* B **492**, 357 (2000); arXiv: hep-th/0009178.
12. C. Rovelli in *Living Review* **1**, http://www.livingreviews.org/articles; T. Thiemann, arXiv: gr-qc/0110034.
13. M. B. Green, J. H. Schwarz and E. Witten in *Superstring Theory*, (Cambridge University Press 1988); J. Polchinski in *String Theory*, (Cambridge University Press 1998).
14. R. Gambini and J. Pullin, *Class. Quant. Grav.* **17**, 4515 (2000).
15. T. Padmanabhan, arXiv: hep-th/0406060; S. Hsu and A. Zee, arXiv: hep-th/0406142.
16. N. Dadhich, arXiv: gr-qc/0311028; arXiv: gr-qc/0405115.

MAGNETIC ROTATING SOLUTIONS IN GAUSS-BONNET GRAVITY AND THE COUNTERTERM METHOD

M. H. DEHGHANI

Physics Department and Biruni Observatory
Shiraz University, Shiraz 71454, Iran
and
Institute for Studies in Theoretical Physics and Mathematics (IPM), School of Physics
P.O. Box 19395-5531, Tehran, Iran
Email: mhd@shirazu.ac.ir

We present a static magnetic solution in Einstein-Maxwell-Gauss-Bonnet gravity without a cosmological constant. This solution yields an asymptotically AdS-5-dimensional spacetime with a longitudinal magnetic field generated by a static magnetic brane. We also generalize this solution to the case of a spinning magnetic brane with two rotation parameters and find that the spinning magnetic brane has a net electric charge proportional to the magnitude of rotation parameters. We also find that these solutions present naked singularities. Finally, we use the counterterm method in the Gauss-Bonnet gravity and compute the conserved quantities of these spacetimes.

1 Introduction

[a]Recent astrophysical data have created a great deal of attention to the asymptotically de Sitter (dS) spacetimes. On the other hand asymptotically anti-de Sitter (AdS) spacetimes continue to attract more attention due to the fact that there is a correspondence between supergravity (the low-energy limit of string theory) in $(n+1)$-dimensional asymptotically AdS spacetimes and conformal field theory living on an n-dimensional boundary, known as the AdS/CFT correspondence.

The simplest way of having an asymptotically (A)dS spacetime is to add a cosmological constant term to the right hand side of Einstein equation. However, the cosmological constant meets its well-known cosmological, fine-tuning and coincidence problems. Thus, it seems natural to get rid of the cosmological constant and look for an alternative candidate for it. In the context of classical theory of gravity, the second way of having an asymptotically (A)dS spacetime is to add higher curvature terms to the left hand side of Einstein equation. Also, from the point of view of brane world cosmology, one should investigate classical gravity in higher dimensions, and again it seems natural to reconsider the left hand side of Einstein equation.

In this context, one may use another consistent theory of gravity in any dimension with a more general action. This action may be written, *e.g.*, through the use of string theory. The effect of string theory on classical gravitational physics is usually investigated by means of a low energy effective action which describes gravity at the classical level. This effective action consists of the Einstein-Hilbert action plus curvature-squared terms and higher powers as well, and in general, gives rise to fourth order field equations and bring in ghosts. However, if the effective action contains the higher powers of curvature in particular combinations, then only second order field equations are produced and consequently no ghosts arise. The effective action obtained by this argument is precisely of the form proposed by Lovelock[1].

In this paper, we want to restrict ourself to the first two terms of Lovelock gravity, which are the Einstein-Hilbert and the Gauss-Bonnet terms. From a geometric point of view, the combination of the Einstein-Gauss-Bonnet terms constitutes, for five-dimensional spacetimes, the most general Lagrangian producing second order field equations, as in the four-dimensional gravity, where the Einstein-Hilbert action is the most general Lagrangian producing second order field equations. Of course, one may note that the Gauss-Bonnet term is topological in 4-dimensions, and hence has no dynamics.

The outline of our paper is as follows: In Sec. 2, we give a brief review of the field equations. In Sec. 3, we introduce the static and rotating solutions of Gauss-Bonnet gravity without a cosmological constant term in the presence of electromagnetic field.

[a]Talk presented at the XIth Regional Conference on Mathematical Physics, May 2004, Tehran, Iran.

Sec. 4 will be devoted to the use of the counterterm method to compute the conserved quantities of these spacetimes. We finish our paper with some concluding remarks.

2 Field Equations

The most fundamental assumption in the standard general relativity is the requirement that the field equations be generally covariant and contain at most second order derivative of the metric. Based on this principle, the most general classical theory of gravitation in five-dimensions is the Einstein-Gauss-Bonnet gravity. The gravitational action for the spacetime $(\mathcal{M},\, g_{\mu\nu})$ with a flat boundary $(\Sigma,\, \gamma_{ab})$ can be written as

$$I_G = \frac{1}{2}\int_{\mathcal{M}} dx^{n+1}\sqrt{-g}(R+\alpha\mathcal{L}_{GB}+\mathcal{L}_m) - \int_{\Sigma} dx^{n}\sqrt{-\gamma}(K+2\alpha J), \tag{1}$$

where α is the Gauss-Bonnet coefficient with dimension (length)^2, K is the trace of $K_{\mu\nu}$ which is the extrinsic curvature of the hypersurface Σ with induced metric γ_{ab} and J is the trace of

$$J_{ab} = \frac{1}{3}\{\ 2KK_{ac}K^c_b + K_{cd}K^{cd}K_{ab} -2K_{ac}K^{cd}K_{db} - K^2K_{ab}\}. \tag{2}$$

In (1), $\mathcal{L}_{GB}$ is the Gauss-Bonnet Lagrangian given by

$$\mathcal{L}_{GB} = R_{\mu\nu\gamma\delta}R^{\mu\nu\gamma\delta} - 4R_{\mu\nu}R^{\mu\nu} + R^2, \tag{3}$$

where R, $R_{\mu\nu\rho\sigma}$ and $R_{\mu\nu}$ are the Ricci scalar, Riemann and Ricci tensors of the spacetime, $\mathcal{L}_m$ is the Lagrangian of the matter field and the last term is the boundary term, which is chosen such that the variational principle is well-defined[2]. In this paper, we want to consider the magnetic rotating solutions of Gauss-Bonnet gravity. Thus, the Lagrangian of the matter action is $F_{\mu\nu}F^{\mu\nu}$, where $F_{\mu\nu} = \partial_\mu A_\nu - \partial_\nu A_\mu$ is the electromagnetic tensor field, and A_μ is the vector-potential. Varying the total action over the metric tensor $g_{\mu\nu}$ and electromagnetic field $F_{\mu\nu}$, the equations of gravitational and electromagnetic fields are obtained as

$$\begin{aligned} &R_{\mu\nu} - \frac{1}{2}g_{\mu\nu}R - \frac{1}{2}\alpha\{\mathcal{L}_{GB}g_{\mu\nu} + 4R^{\rho\sigma}R_{\mu\rho\nu\sigma} \\ &\quad +4R_{\mu\lambda}R^\lambda_\nu - 2RR_{\mu\nu} - 2R_\mu{}^{\rho\sigma\lambda}R_{\nu\rho\sigma\lambda}\} = T_{\mu\nu}, \\ &\nabla_\mu F_{\mu\nu} = 0, \end{aligned} \tag{4}$$

where $T_{\mu\nu}$ is the electromagnetic stress tensor

$$T_{\mu\nu} = 2F^\lambda{}_\mu F_{\lambda\nu} - \frac{1}{2}F_{\lambda\sigma}F^{\lambda\sigma}g_{\mu\nu}. \tag{5}$$

3 Magnetic Solutions

First, we obtain the 5-dimensional static solution of Eq. (4), which produce a longitudinal magnetic field normal to the $(r-\phi_1)$-plane. These types of solutions in Einstein gravity have been considered in Ref. [3] by Dehghani. We assume that the metric has the following form

$$ds^2 = -\frac{r^2}{\alpha}dt^2 + \frac{dr^2}{f(r)} + \alpha f(r)d\phi_1^2 + r^2 d\phi_2^2 + \frac{r^2}{\alpha}dx^2.$$

The coordinate x has the dimension of length, while the angular coordinates ϕ_i are dimensionless as usual and range in $0 \le \phi_i < 2\pi$. The motivation for this metric gauge $[g_{tt} \propto -r^2$ and $(g_{rr})^{-1} \propto g_{\phi_1\phi_1}]$ instead of the usual Schwarzschild gauge $[(g_{rr})^{-1} \propto g_{tt}$ and $g_{\phi\phi} \propto r^2]$ comes from the fact that we are looking for a magnetic solution instead of an electric one[4,5]. Since, we want to have a magnetic field, one may assume that $A_\mu = h(r)\delta^{\phi_1}_\mu$. Using the field equation (4), one obtains

$$f(r) = \frac{r^2}{4\alpha} \pm \sqrt{\frac{r^4}{16\alpha^2} + \frac{m}{6\alpha} + \frac{q^2}{6\alpha r^2}}, \tag{6}$$

$$A_\mu = \frac{q}{2\sqrt{\alpha}r^2}\delta^{\phi_1}_\mu. \tag{7}$$

The only non-vanishing component of electromagnetic field is $F_{r\phi_1} = \sqrt{\alpha}q/r^3$, which is a longitudinal magnetic field normal to the $(r-\phi_1)$-plane. In order to study the general structure of these solutions, we first look for curvature singularities. It is easy to show that the Kretschmann scalar $R_{\mu\nu\lambda\kappa}R^{\mu\nu\lambda\kappa}$ diverges at $r=0$ and therefore, there is an essential singularity located at $r=0$. As one can see from (6), the solution has two branches with "−" and "+" signs. Since the "−" signs branch goes to zero as r goes to infinity, therefore it cannot be accepted. The "+" signs branch is always positive, and therefore this spacetime has no horizon. Thus, it presents a naked singularity.

Now we consider the most general magnetic rotating solution in five-dimensions. The rotation group in $n+1$-dimensions is $SO(n)$ and therefore the number of independent rotation parameters is $[(n+1)/2]$, where $[x]$ is the integer part of x. Thus, the most general rotating solution in five-dimensions has two rotation parameters. It is easy to show that the following metric with the two rotation parameters a_1 and a_2 satisfies the field equations (4)

$$ds^2 = -\frac{r^2}{\alpha}\left(\Xi dt - \sum_{i=1}^{2} a_i d\phi_i\right)^2 + \frac{dr^2}{f(r)}$$
$$+\frac{f(r)}{\Xi^2-1}\sum_{i=1}^{2}\left((\Xi^2-1)dt - \Xi\sum_{i=1}^{2} a_i d\phi_i\right)^2$$
$$-\frac{r^2}{\alpha(\Xi^2-1)}(a_1 d\phi_2 - a_2 d\phi_1)^2 + \frac{r^2}{\alpha}dx^2, \quad (8)$$

$$\Xi^2 = 1 + \alpha^{-1}\sum_{i=1}^{2} a_i^2, \quad (9)$$

where $f(r)$ is the same as $f(r)$ given in (6). The gauge potential is

$$A_\mu = \frac{q}{2r^2}((\Xi\delta_\mu^0 - a_i\delta_\mu^i), \quad \text{(no sum over } i). \quad (10)$$

Again this spacetime has an essential naked singularity at $r = 0$. It is worthwhile to mention that in this spacetime, in addition to the longitudinal magnetic field, one encounters with an electric field too. Using the Gauss law, one can show that the magnetic spinning brane has a total electric charge which is proportional to the magnitude of the rotation parameter given by $\sqrt{\alpha(\Xi^2-1)}$.

4 Conserved Quantities

Having the action, one can use the Brown-York[6] definition of stress-energy tensor to construct a stress-energy tensor as

$$T^{ab} = \sqrt{-\gamma}\{(K\gamma^{ab} - K^{ab}) + 2\alpha(J\gamma^{ab} - 3J^{ab})\}. \quad (11)$$

One may note that when α goes to zero, the stress-energy tensor (11) reduces to that of Einstein gravity. As in the case of Einstein gravity, the conserved quantities computed through the use of stress-energy tensor of (11) are divergent. For 5-dimensional magnetic rotating solution (9) with flat boundary, the Riemann tensor is zero and therefore the divergence can be removed by adding the counterterm

$$T^{ab}_{\text{coun}} = \frac{5}{\sqrt{2\alpha}}\gamma^{ab}. \quad (12)$$

The conserved quantity associated with a Killing vector ξ^a is

$$\mathcal{Q}(\xi) = \int_{\mathcal{B}} d^n x\sqrt{\sigma}u^a(T_{ab} + T_{ab}^{\text{coun}})\xi^b, \quad (13)$$

where $\mathcal{B}$ is the hypersurface of fixed r and t, u^a is the unit normal vector on $\mathcal{B}$, and σ is the determinant of the metric σ_{ij}, appearing in the ADM-like decomposition of the boundary metric,

$$ds^2 = -N^2dt^2 + \sigma_{ij}(dx^i + N^i dt)(dx^j + N^j dt). \quad (14)$$

In (14), N and N^i are the lapse and shift functions, respectively. For the spacetimes introduced in this paper, the 5-dimensional boundaries $\mathcal{B}$ have timelike Killing vector ($\xi = \partial/\partial t$) and rotational Killing vectors ($\zeta_i = \partial/\partial\phi^i$). Thus, one obtains the conserved mass and angular momentum densities as

$$M = \frac{11m}{6\alpha}\left[4(\Xi^2-1)+1\right], \quad (15)$$
$$J_i = -\frac{22}{3\alpha}\Xi m a_i. \quad (16)$$

5 Closing Remarks

In this paper, we investigated the classical theory of gravity without cosmological constant. Indeed, we added the Gauss-Bonnet term to the Einstein action and introduced a static asymptotically AdS solution of the field equations in the presence of an electromagnetic field. This solution yields a 5-dimensional spacetime with a longitudinal magnetic field [the only nonzero component of the vector potential is $A_\phi(r)$] generated by a static magnetic brane. We found that this solution presents a naked singularity at $r = 0$. We also generalized this solution to the case of rotating magnetic brane and found that when all the rotation parameters are zero (static case), the electric field vanishes, and therefore the brane has no net electric charge. For the spinning brane, when one or more rotation parameters are nonzero, the brane has a net electric charge density which is proportional to the magnitude of the rotation parameter.

The counterterm method inspired by the AdS/CFT correspondence conjecture has been widely applied to the case of Einstein gravity. Here, we applied this method to the case of Gauss-Bonnet gravity and calculated the conserved quantities of the two classes of solutions.

References

1. D. Lovelock, *J. Math. Phys.* **12**, 498 (1971).
2. S.C. Davis, *Phys. Rev.* D **67**, 024030 (2003).
3. M.H. Dehghani, *Phys. Rev.* D **69**, 044024 (2004).
4. M.H. Dehghani, *Phys. Rev.* D **67**, 064017 (2003).
5. M.H. Dehghani and A. Kodam-Mohammadi, *Phys. Rev.* D **67**, 084006 (2003).
6. J.D. Brown and J.W. York, *Phys. Rev.* D **47**, 1407 (1993).

OBSERVING BLACK HOLES

F. DE PAOLIS, G. INGROSSO and A. A. NUCITA
Dipartimento di Fisica, Universita degli Studi di Lecce, and INFN,
Sezione di Lecce, Via Arnesano, CP 193, 73100 Lecce, Italy
E-mail: depaolis@le.infn.it

A. QADIR
Centre for Advanced Mathematics and Physics,
National University of Science and Technology, Rawalpindi, Pakistan
E-mail: aqadirs@comsats.net.pk

Black holes are predicted by General Relativity and there has been a long search for them. Now they have been observed. General Relativity also predicts effects of high spin of the black hole. There is good reason to believe that the best known black hole is spinning more or less as rapidly as it is allowed to. Now, the search is on for the spin of the black hole. In this paper, we briefly review the theoretical reasons for expecting black holes to exist and the earlier searches for them. We then explain how the supermassive black holes at galactic centers have been observed. We conclude with some work suggesting how the spin of the black holes can be observed by the next generation of space borne telescopes.

1 Introduction

[a]The first suggestion for the existence of what we now call a black hole came independently from Mitchell and Laplace. Using Newton's laws of motion and gravitation, they noted that if a star became sufficiently dense that its radius is less than $2GM/c^2$, where M is the mass of the star, G Newton's gravitational constant and c the speed of light, the escape velocity from its surface would exceed the speed of light. This distance was later called the *Schwarzschild radius.* As such it would be a "dark star", observable by its gravitational effects but not *visible.* The full significance of this observation could not be appreciated till after Einstein stated his restricted theory of relativity in 1905. Since no signal can travel faster than light, according to his theory, not only would the "dark star" not be visible, *nothing could emerge from it.* However, its static gravitational field would still cause the orbits of bodies in its vicinity to be modified.

Of course, the analysis could not be undertaken meaningfully in the context of the restricted theory, as the restriction imposed is of unaccelerated motion, and the effect of gravity is to accelerate. This analysis had to await the advent of the unrestricted, or general, theory of relativity (GR) in 1916. Soon thereafter, the rigorous analysis of orbits about a point source of gravitation (given by the Schwarzschild metric) proved that the escape velocity in GR, for a point gravitational source, is the classical one.

Chandrasekhar[5] had obtained a mass limit of 1.4 $M_\odot$ for a normal star to collapse to a *white dwarf* star (having a density of a ton/cc). This limit was based on the assumption that the Fermi degeneracy pressure of the electron gas in the star stops further gravitational collapse. Later, Landau[20] and Chandrasekhar[6] independently found the limit for the star to collapse to a *neutron star* (having a density of a billion tons/cc) and a mass limit of $0.7M_\odot$, if it is cold and 1.4 $M_\odot$ if it is hot. They replaced the electron degeneracy by the neutron degeneracy and then refined the calculation by taking the star to consist of neutrons, protons and electrons.

In 1935 Oppenheimer and Volkov[26] extended the analysis to show that, in principle, there could be a mass for a star such that it would exceed the Fermi degeneracy of nucleons as well, so that no known force could stop endless gravitational collapse of the star. Wheeler dubbed such stars *black holes*[24], as they would pull matter in but be totally black. For a black hole of the mass of the Sun, the Schwarzschild radius is about 3 km and the density about 100 times that of a corresponding neutron star! As the mass increases, the density *decreases* quadratically (since the

---

[a]Invited plenary talk presented by A.Q. at the XIth Regional Conference on Mathematical Physics, May 2004, Tehran, Iran.

volume *increases* cubically). In 1971 Christodoulou and Ruffini[8] found that there is a certain portion of the mass of any black hole that cannot be reduced. If this is enough, the star would have to collapse. Later, Rhoades and Ruffini[28] demonstrated that the maximum limit is 3.2 $M_\odot$, beyond which there can be no way to stop endless gravitational collapse. This is because the mass equivalent of any energy put in to stop the collapse, due to the non-linearity of GR, would actually act to *enhance* the collapse. Independently, during the "cultural revolution" in China, at about the same time, Fang Li Zhi found the same limit but could not publish it in a reasonable journal because of restrictions in China[13]. (It is interesting that in the Conference, when Fang was asked what he was doing at the time, he said that he was "working in the fields". It was assumed that his English was weak and someone corrected him to say "You mean working in field *theory*". "No", he replied, "I mean I was working in the fields" and he made motions of digging and hoeing).

Seen from the outside of the black hole, it appears that the matter of the star has collapsed to a point. However, analyzing what would appear from the inside, using the relevant Carter-Penrose coordinates and diagram[17], it can be seen that the collapse is on to an infinite line of infinite energy density. The new dimension, in which the line lies, looks like the time dimension as seen from outside. This is one of the weird aspects of the (enforced) change of coordinates in GR. It has been shown that in terms of the relevant geometrically defined time parameter K, the length of the line increases with the time as[27] $K^{1/3}\ln K$.

In 1963 Kerr[19] derived the metric for a spinning gravitational source of mass M and angular momentum per unit mass $a = J/Mc$. It was later shown that this is the most general end result of collapse of neutral matter. Since most stars spin, conservation of angular momentum would lead one to expect that most actual black holes would be Kerr, and not Schwarzschild. Analysis of the metric[24] showed that it would not be physical for $a > GM/c^2$. The rotating black hole has a radius of $GM/c^2 + \sqrt{G^2M^2/c^4 - a^2}$. It is used to believe that the spin of such holes would be small compared with the mass, but there is no reason why this should be so.

2 Early Observational Attempts

At first sight it may seem to be a contradiction in terms to talk of observing what should be unobservable. However, it is not only the change in the orbits of nearby bodies but the strong non-linear gravitational effects of GR on photon trajectories that could be looked for. They would be specially strong near the surface of the black hole. If matter is falling into the black hole, it would be highly accelerated and would give a characteristic radiation at some given frequency. This was the basis of Ruffini's identification of Cygnus X-1 as a black hole[21]. As he was the first to identify one, he deserves credit as the discoverer of black holes. This X-ray source is associated with a normal star. Matter from an accretion disc about the black hole was being pulled into it. Working out the Kepler problem for the binary system, one could obtain the orbits of both stars about their common center of mass. The X-ray source switched off as the black hole was occluded by the normal star, confirming the modelling[29]. However, there were caveats about the identification and astronomers wanted some confirmation of the existence of such bizarre objects as black holes. The resistance to the claimed observation of black holes may have had more to do with a rejection of the *singularity*, a region of infinite energy density, at the core implicit in their existence than with actual observational problems.

One of the most significant reservations about the observation had to do with the supposed dynamical association of the visual star with the X-ray source, on account of the poor accuracy of identifying the location of X-ray sources at the time. As such, there was a search for "visual black holes" that would radiate in the visible band instead of the X-ray band. This severely constrained the mass of the black hole and the associated visual star. Further, one needed to catch them at just the right time and see them at just the right angle. It is not surprising that this confirmation was not so rapidly forthcoming.

There had been speculations that black holes may form most readily at galactic cores on account of the high density of stars there. Further, it had been speculated that many black holes of a few $M_\odot$ may coalesce to form larger black holes of $\sim 10^3 M_\odot$. Liang[22] pointed out that there could be a significant electron-positron pair production in accretion disks around black holes. Stern[32] suggested that these could explain the observed radiation peak at 1 MeV around $\sim 10^3\ M_\odot$. Later Chakrabarti[7] worked out a more precise tail of the spectrum of pairs produced, which fitted observations very well. One would have thought that the sceptics may have been silenced by now. However, it was pointed out that one could conceive of other processes providing those signals.

3 Supermassive Black Holes

There had long been the proposal that *QUASARs*, optical sources with a pair of radio sources on either side of them, were black holes eating up matter in the early stages of galaxy formation. This view was supported by the discovery of *active galactic nuclei* (AGNs), which seemed to be "half-way to quasars", in that they were closer and could be regarded as later stages of quasars evolving into usual galaxies – remember that the further away we look the further back in time do we see the object. All such suggestions relied on *supermassive* black holes with mass $\sim 10^6 M_\odot$ – $10^9 M_\odot$. The $\sim 10^3 M_\odot$ black holes are now referred to as "intermediate mass" black holes. There is now extremely strong evidence for there being supermassive black holes at the centers of all galaxies. In fact, some galaxies have more than one. These arise because of collision of galaxies, so that the resulting larger galaxy has the central black holes of each of the galaxies. It would be possible that, given time, the central black holes would merge and we would get one larger black hole.

Even though the existence of supermassive black holes appears very likely, at present we have no *direct* test of the presence of such compact massive objects in AGNs, QUASARs and, in general, in galactic cores. In fact, the size of a black hole with mass $\simeq 10^8$ $M_\odot$ has a diameter $\simeq 3 \times 10^{13}$ cm which, for a distance of about 10 Mpc, corresponds to a subtended angle of only $\simeq 2 \times 10^{-7}$ arc-seconds ruling out the possibility, with the present instruments, to directly form an image of a hole that may possibly be hosted in the galactic cores.

Nevertheless, there exists a lot of compelling evidence for central supermassive black holes. In fact, it is expected that the strong gravitational potential affects the density distribution of the stellar cluster itself. In this case, dynamical considerations about black hole adiabatic growth show that a spike of matter – following a density profile $\rho \propto r^{-3/2}$ – forms[23]. The observation of such a cusp will give the proof of the existence of a supermassive black hole. Also the stellar velocity distribution around the hole is expected to change. From the virial theorem, one gets the black hole mass estimate[2] $M_{vir} = 16 < v^2 r > /\pi G$. Further, all available observations show the existence of a strong correlation between the black hole mass and the velocity dispersion of the stars, *i.e.* $M \propto \sigma^4$ which seems to hold for all the known galactic centres[14]. Also the reverberation mapping technique, consisting of monitoring the variability and spectrum of AGNs, allows us to get information about the mass of the putative supermassive black holes hosted in distant QUASARs[18].

Final evidence came from infrared and microwave observations of the center of our own galaxy, which seems to host the closer supermassive black hole. These had not been possible by Earth-bound observations as most of the radiation gets absorbed by the atmosphere. Further, there is a dust cloud between us and the galactic core, which occludes all observation in the visible and higher frequency ranges. However, once space-borne telescopes became operational the entire electromagnetic spectrum opened up for observation. About 14 years ago observations of the galactic center became possible[15]. Eight stars were found in orbit about the putative central black hole. The one that approaches closest, called S1, has a highly eccentric orbit with closest approach distance of about 1.5×10^{10} km and a mass close to 15 times that of the Sun. The next closest, S2, also has a highly eccentric orbit, but spends more time closer to the central source than S1. Its closest approach is 3×10^{10} km and its mass about 15 $M_\odot$. Its orbital period is about 15 years and the first completed orbit will be seen by 2005! It will provide the possibility to measure the perihelion shift of S2 and obtain an enormous amount of information about the distribution of matter about the hole. Solving the Kepler problem gives the mass of the central source as $4.07 \times 10^6\ M_\odot$. It is totally dark, so it can not be a normal star. Further, if it extended out to $\sim 10^8$ km tidal forces would rip out the material from it into the S2. This means that it has to be a collapsed object. Since no collapsed object of that mass can be stable under gravitational collapse, it has to be a black hole.

4 Retro-Lensing by Supermassive Black Holes

It is well known that a dark object moving close to the source-observer line of sight acts as a *gravitational lens*, causing the formation of two or more images of the source[30]. If the images can not be resolved, the lensing phenomenon results in an overall source image amplification, to which one refers as a *gravitational microlensing event.* The usual lensing is taken to occur with the lens between the source and the observer. However, we can also have the observer between the source and the lens, or the source between the observer and the lens. In this case, there have to be large angle deflections[5]. Of course, the luminosity of such images will be sharply reduced, but with a sufficiently powerful lens (sufficiently large black hole) it may be detectable. This is known as *strong lensing.*

Holz and Wheeler[17] considered the case that a Schwarzschild black hole bends the light *all the way back*, so that it is in effect *reflected.* In other words, the ray is deflected through an angle π. They further point out that there can be repeated images formed by deflection through $3\pi, 5\pi, \cdots, (2n-1)\pi$. The geometry required, is that the light source, observer and lens are collinear. They called this a *retro-lens.* This forms when the incoming photons have impact parameters comparable to the photosphere of the black hole, which is $3GM/2c^2$ for a Schwarzschild black hole. The shape and the magnitude of the images formed, arcs becoming rings in the perfect alignment case, depends strongly on the mass of the black hole and on the relative distances between the observer, the lens and the source. In the perfect alignment case, after π rotations around the black hole, photons if collected form a sequence of circular rings.

Holz and Wheeler considered the Sun as the light source which emits photons towards a stellar black hole at distance D_L and then evaluated the magnitude of the retro-lensed images. For perfect alignment between the source, the observer and the lens, the maximum distance at which the retro-images can be seen by an instrument with limiting magnitude $\bar{m}$ is $D_L = 0.02\text{pc} \times \left[10^{(\bar{m}-30)/2.5}(M/10\ M_\odot)^2\right]^{1/3}$, or about 6×10^{11} km for an assumed image baseline of magnitude $m = 30$. For comparison, the distance from the Earth to the Sun is 1.5×10^8 km and the Oort cloud of comets is only a quarter the proposed distance. However, since we do not expect massive black holes lying so close to the Earth (a significant black hole there should seriously perturb the Oort cloud and send in many more comets), it is not worth while to search for retro-lenses illuminated by the Sun.

It was proposed[9] that the black hole at the galactic center, Sgr A* with mass[15] $M \simeq 4.07 \times 10^6\ M_\odot$ can be considered as the lens and a really bright star close to Sgr A* as the light-source, namely, the above mentioned S2 star. We can now have direct lensing (or lensing with large angle deflections of $2n\pi$) or Holz-Wheeler retro-lensing. Obviously, the true shape and magnitude of the retro-lensing rings due to the black hole at the galactic center can only be obtained by carefully studying the S2 orbit. It was found that the expected infrared magnitude of the retro-lensing images are in the range $m_{\text{K min}} \simeq 33.3$ and $m_{\text{K max}} \simeq 36.8$, depending on the S2 periastron ($D_{LS} \simeq 2 \times 10^8$ km) and apastron ($D_{LS} \simeq 3 \times 10^9$ km) positions, respectively. Of course, present instruments can not detect such faint images. However, the attainable limiting magnitude will increase with the Next Generation Space Based Telescope (NGST)[25], planned to work in the wavelength range $0.6 - 27\ \mu$ m, is expected to detect (within an integration time of 3 hours) an energy flux (in the K band) of $\Phi \simeq 2 \times 10^{-19}$ erg cm^2 s^{-1} corresponding to a limiting magnitude $\bar{m} \sim 32$. This limiting magnitude is very close to that necessary to observe S2 retro-lensing images that could be effectively observed by NGST increasing the integration time to about 27 hours.

5 Observing the Spin of Black Holes

The central black hole may well have a significant spin. In fact, recent estimates suggest that its spin is 0.996 of the maximum value allowed for a physical black hole[1]. *It would be fascinating to not only observe the black hole but measure its spin as well!* The non-zero spin breaks the spherical symmetry of the Schwarzschild geometry and affects the gravitational field around the black hole. This leads to a modification of the retro-images. For such large spins as are expected, can we actually detect the change due to the spin?

Due to the loss of symmetry of the geometry,

it becomes difficult to trace out the paths of light rays in general. The reason for this is that in the Kerr geometry angular momentum is not conserved. Only the azimuthal component and the total angular momentum squared are conserved[16]. Thus particle orbits or light ray paths are not generally coplanar. However, since the angular momentum for a path in the equatorial plane is simply the azimuthal part, such paths *will* be coplanar. One effect is a distortion of the ring due to the change in the geometry[5]. The reason is that "co-rotating photons" – in the sense that their angular momentum relative to the black hole is in the direction of its spin – can come much closer, up to twice as close, to the spinning black hole than to the Schwarzschild black hole, while "counter-rotating photons" are kept much further away, up to twice as far. This can be calculated by computing the "impact parameter" b of the photons[3]. Using the same formalism another effect of the modified geometry, to slightly enhance the energy of co-rotating photons and decrease the energy of the counter-rotating photons, was also worked out[10]. Unfortunately, it does not seem so likely to actually be able to see these effects with the NGST, even for the central black hole illuminated by the S2 star.

The most exciting possibility is provided by adverting to the implications of the energy enhancement (loss) for co-rotating (counter-rotating) photons. There must be a corresponding frequency shift for the photons. Since measuring frequency shifts is much easier than measuring the shape of (extremely small) rings or their (extremely small) intensity changes, one could hope to actually see the effect if it is real. It turns out[11] that the effect is genuine. The spinning black hole acts like an enormous *photon synchrotron*. For the nth-ring we have[11] $(\delta\nu/\nu)_\nu = 4GMa(2n-1)/b^2c^2$. For co-rotating and counter-rotating photons, in the cases $a = 0.5$ and $n = 1$, the frequency shifts turn out to be 0.026 and -0.013, respectively. While, for a maximally rotating black hole ($a = 1$) and $n = 1$, the frequency shifts are 0.13 and -0.02, respectively. These should be detectable by the NGST!

A word of caution is in place here. The above calculations apply only in the equatorial plane. The S2 star does not have just the right geometry for our observation to give retro-lensing[9]. It is simply a strong lensing image that would be seen. It is unlikely that at the time of observation the star and the observer (us) lie in the equatorial plane. We need to calculate off this plane as well. The general calculation can be done using the procedure of Bozza and Mancini[4]. This is being done[12] using the additional observation that the generalization of coplanar motion in the spherically symmetric case is motion on the surface of a *cone* whose axis lies along the spin axis of the black hole. The crucial point to note is that the frequency shift is damped by a factor of $sin^2\theta$, where θ is the half angle of the cone. The reason for this factor is that the Kerr metric, giving the relativistic proper time or length, has a term that mixes the time and the azimuthal directions due to the dragging of inertial frames by the rotating gravitational source. It is this rotation that brings the spin into the observable effects. This "cross term" includes the mass, the spin parameter, a, an inverse square of the effective distance from the source $(r^2 + a^2 sin^2\theta)$, and a $sin^2\theta$. In the equatorial plane the value of the last term is unity. However, off the equatorial plane this term will also enter. Of course, there will be a further modification by the dependence of the denominator on the polar angle, but that will be a higher order effect. There will be a corresponding diminution of the intensity by the $sin^2\theta$ factor for those rays that come on the cone. Naturally, there will continue to be other rays that are not so diminished. The frequency shifts, coupled with the intensity profiles of the rings at different observation times could even provide the spin direction.

6 Conclusion

This is an exciting time. There was time when little of astrophysics or cosmology could be taken too seriously, because of the tremendous paucity of data and the lack of control on the objects to be observed. The laboratory method of *experimentation* cannot be used. Now some of the most precise measurements come from astrophysics and cosmology. The current record for accuracy is held by the indirect observation of gravitational waves[33] in the Taylor-Hulse binary pulsar system[31]. We have also come a long way from the time when it was taken to be tautological that black holes could not be observed. No longer do we only have to rely on "black hole *candidates*". Black holes are a reality. We now have to get down to the serious task of actually *measuring* the parameters of the black hole. It is hoped that within a few years we will *know* what the spin and the axis of our central black hole is. *Science fiction cannot begin to match the excitement of the next decade of Astrophysics.*

Acknowledgments

One of us (A.Q.) is most grateful to COMSTECH for providing travel support, and to the organizers for providing local hospitality, to attend the Conference. He is also indebted to INFN and the Department of Physics of the University of Lecce, Italy, where much of the work was done, for support for his visit.

References

1. B. Aschenbach, N. Grosso, D. Porquet and P. Predhel, *A&A* **417**, 71 (2004).
2. D.C. Bachall and S. Tremaine, *ApJ* **244**, 805 (1981).
3. V. Bozza, *Phys. Rev.* D **67**, 103006 (2002).
4. V. Bozza and L. Mancini, arXiv: astro-ph/0404526.
5. S. Chandrasekhar, *ApJ* **74**, 81 (1931).
6. S. Chandrasekhar, *Observatory* **57**, 373 (1934).
7. S.K. Chakrabarti, *ApJ* **464**, 623 (1996).
8. D. Christodoulou and R. Ruffini, *Phys. Rev.* D **4**, 3552 (1971).
9. F. De Paolis, G. Ingrosso, A. Geralico and A.A. Nucita, *A&A* **409**, 809 (2003).
10. F. De Paolis, A. Geralico, G. Ingrosso, A.A. Nucita and A. Qadir, *A&A* **415**, 1 (2004).
11. F. De Paolis, A. Geralico, G. Ingrosso, A.A. Nucita and A. Qadir, submitted to *A&A* (2004).
12. F. De Paolis, G. Ingrosso, A.A. Nucita and A. Qadir, in preparation (2004).
13. L.Z. Fang, A. Qadir and R. Ruffini in *Physics and Contemporary Needs*, Eds. Riazuddin and A. Qadir (Plenum Publishers, Vol. 5, 1983).
14. L. Ferrarese and D. Merritt, *ApJ* **539**, L29 (2000).
15. A.M. Ghez *et al.*, *ApJ* **586**, L127 (2003).
16. S.W. Hawking and G.F.R. Ellis, *ApJ* **152**, 25 (1968).
17. D.E. Holz and J.A. Wheeler, *ApJ* **57**, 330 (2002).
18. S. Kaspi *et al.*, *ApJ* **533**, 631 (2000).
19. R.P. Kerr, *Phys. Rev. Lett.* **11**, 237 (1963).
20. L.D. Landau, *Phys. Z. Sowjetunion* **1**, 285 (1932).
21. R.W. Leach and R. Ruffini, *ApJ* **180**, L15 (1973).
22. E.P.T. Liang, *ApJ* **234**, 1105 (1979).
23. D. Merritt and G.D. Quinlan, *ApJ* **498**, 625 (1998).
24. C.S. Misner, K.S. Thorne and J.A. Wheeler in *Gravitation*, (Freeman Press, San Francisco, 1973).
25. National Research Council, *Astronomy and Astrophysics in the New Millennium*, (Nat. Ac. Press, Washington D.C., 2001).
26. J.R. Oppenheimer and G.M. Volkoff, *Phys. Rev.* **55**, 374 (1939).
27. A. Qadir and A.A. Siddiqui, *preprint form QAU* (2003).
28. C.E. Rhoades Jr. and R. Ruffini, *Phys. Rev. Lett.* **32**, 324 (1974).
29. R. Ruffini in *Physics and Contemporary Needs Vol. 1*, Eds. Riazuddin (Plenum Publishers, 1977).
30. P. Schneider, J. Ehlers and E.E. Falco in *Gravitational Lenses*, (Springer-Verlag, Cambridge, 1992).
31. S.L. Shapiro and S.A. Teukolsky in *Black Holes, White Dwarfs and Neutron Stars*, (John Wiley and Sons, 1983).
32. B.E. Stern, *Astronomicheskii Zhurnal* **62**, 529 (1985).
33. J.M. Weisberg and J.H. Taylor in *Binary Radio Pulsars, Proc. Aspen Conference, ASP Conf. Series*, eds. F.A. Rasio and I.H. Stairs (2004).

RICCI CONFORMAL COLLINEATIONS FOR STATIC SPACETIMES

K. SAIFULLAH
Department of Mathematics, Quaid-i-Azam University, Islamabad, Pakistan
E-mail: saifullah@qau.edu.pk

In general relativity the Lie derivatives of different geometric quantities have been studied extensively. Invariance under the Lie transport of the metric, Ricci and Riemann tensors, respectively defines the Killing vectors (KVs), Ricci collineations (RCs) and curvature collineations (CCs). These symmetries have important applications not only for classifying spacetimes but also from the physical point of view. However, if the Lie derivative of the Ricci tensor is proportional to the tensor itself (by some arbitrary function of the coordinates), it gives rise to Ricci conformal collineations (RCCs). In this paper, the static cylindrically symmetric spacetimes have been studied for their RCCs when the Ricci tensor is degenerate.

[a]The theory of Lie derivatives has important applications in general relativity as well as other branches of mathematics. For each point p in a manifold M, a vector field $\mathbf{V}$ on M determines a unique curve $\alpha_p(t)$ such that $\alpha_p(o) = p$ and $\mathbf{V}$ is the tangent vector to the curve. Consider a mapping h_t dragging each point p, with coordinates x^i, along the curve $\alpha_p(t)$ through p into the image point $q = h_t(p)$, with coordinates $y^i(t)$. If t is very small, the map h_t is a one-one map and induces a map $h_t^*\mathbf{T}$ of any tensor $\mathbf{T}$. The Lie derivative of $\mathbf{T}$ with respect to $\mathbf{V}$ is defined by[1,2]

$$\pounds_{\mathbf{V}}\mathbf{T} = \lim(t \to 0)\frac{1}{t}(h_t^*\mathbf{T} - \mathbf{T})\,. \qquad (1)$$

If $\mathbf{T}$ is of type (r,s), its Lie derivative is also a tensor of type (r,s). Using the coordinate bases $\{\partial_{x^i}\}$ and $\{\partial_{y^i}\}$, the Lie derivative of a vector $\mathbf{U}$ with respect to $\mathbf{V}$ in component form can be written as

$$\begin{aligned}\pounds_{\mathbf{V}}\mathbf{U} &= v^m\frac{\partial}{\partial x^m}\left(u^i\frac{\partial}{\partial x^i}\right) - u^m\frac{\partial}{\partial x^m}\left(v^i\frac{\partial}{\partial x^i}\right)\\ &= [\mathbf{V}\,,\mathbf{U}]\,, \end{aligned}\qquad (2)$$

where the commutator $[\mathbf{V}\,,\mathbf{U}]$ is the Lie bracket. It has the following properties

(a) it is bilinear;

$$[a\mathbf{U}+b\mathbf{V},\mathbf{W}] = a\,[\mathbf{U},\mathbf{W}] + b\,[\mathbf{V},\mathbf{W}]\,,$$

a and b are scalars;

(b) it is antisymmetric;

$$[\mathbf{V}\,,\mathbf{U}] = -\,[\mathbf{U},\mathbf{V}]\,;$$

(c) it satisfies the Jacobi identity;

$$[\mathbf{U},[\mathbf{V},\mathbf{W}]] + [\mathbf{V}\,,[\mathbf{W}\,,\mathbf{U}]] + [\mathbf{W}\,,[\mathbf{U},\mathbf{V}]] = 0\,.$$

A vector space together with a product [,] satisfying (a), (b) and (c) is a Lie algebra. If $\{\mathbf{X}_i, i = 1,\cdots,n\}$ is a basis for the Lie algebra, then we can always write

$$[\mathbf{X}_k,\mathbf{X}_l] = C^j_{kl}\mathbf{X}_j, \qquad C^j_{kl} = -C^j_{lk}, \qquad (3)$$

where we have assumed the summation convention. The Lie algebra is completely characterized by C^j_{kl}, the structure constants. If all the structure constants vanish, we get the Abelian Lie algebra. Every Lie algebra defines a unique simply connected Lie group and *vice versa*. Naturally the basis $\{\mathbf{X}_i\}$ is not unique, and under a change of basis the numbers C^j_{kl} transforms as components of a tensor. Every Lie group and algebra has a unique "structure tensor" $\mathbf{C}$.

Lie derivatives are used to express the invariance of a tensor field under some transformation. We say that a tensor field $\mathbf{T}$ is invariant under a vector field $\mathbf{V}$ if the tensors $h_t^*\mathbf{T}$ and $\mathbf{T}$ coincide for t in some interval around 0, *i.e.*, the Lie derivative vanishes

$$\pounds_{\mathbf{V}}\mathbf{T} = 0. \qquad (4)$$

The manifolds of interest in theoretical physics have metric tensors. It is of interest to know when the metric is invariant with respect to some vector field. The vector fields along which the metric remains invariant are called Killing vector (KV) fields or isometries[1,3]. After the spacetime metric, the Ricci tensor, $\mathbf{R}$, and the stress-energy tensor, $\mathbf{T}$, are other important candidates which play a significant role in

[a]Talk presented at the XIth Regional Conference on Mathematical Physics, May 2004, Tehran, Iran.

understanding the geometric structure of spacetimes by virtue of the Einstein field equations (EFEs)

$$R_{ab} - \frac{1}{2} R g_{ab} = \kappa T_{ab}, \qquad (a, b = 0, 1, 2, 3). \qquad (5)$$

For a tensor **A** of rank two, the vanishing of the Lie derivative $\pounds_{\mathbf{B}} \mathbf{A} = 0$, can be written in component form as

$$B^c A_{ab,c} + A_{ac} B^c_{,b} + A_{bc} B^c_{,a} = 0 \,. \qquad (6)$$

If the tensor is the metric tensor, **g**, then the last equation is called the Killing equation and the vector $\mathbf{B} = (B^0, B^1, B^2, B^3)$ a KV or an isometry. If the tensor in Eq. (6) is the Ricci tensor or the stress-energy tensor the vector **B** is called the Ricci or the matter collineation (RC or MC), respectively. Similarly, the vanishing of the Riemann curvature tensor defines curvature collineations (CCs)[3,4,5]. While the isometries provide information of the symmetries inherent in the spacetime, RCs, vector fields along which the Ricci tensor is invariant under Lie transport, are important from the physical point of view as well. These symmetry properties are described by continuous groups of motions or collineations which lead to conservation laws and these vectors generate a Lie algebra. As the metric tensor cannot be degenerate, *i.e.*, cannot have zero determinant, the Lie algebra for the KVs can only be finite dimensional. However, for RCs and MCs, it can be finite- as well as infinite-dimensional as the Ricci and the stress-energy tensors can be degenerate as well as non-degenerate. As the Ricci tensor can be written in terms of the metric tensor, we note that every KV is an RC but the converse is not true, in general. The physical significance of these symmetries has been discussed in the literature[6].

Now, we consider the case when the Lie derivative of a tensor is not equal to zero but is proportional to the tensor itself so that Eq. (6) becomes non-homogeneous. For example, for the metric tensor we have

$$\pounds_{\mathbf{B}} \mathbf{g} = \alpha \mathbf{g}.$$

If α is a constant, then **B** gives the homothetic vector and if it is not a constant but a function of the coordinates, then we get the conformal vector[1,3]. These vectors play an important role in the study of self-similar spacetimes. Analogously, this definition can be extended to the Ricci tensor *i.e.*,

$$\pounds_{\mathbf{B}} \mathbf{R} = \psi(x^c) \mathbf{R},$$

in which case **B** is called the Ricci conformal collineation (RCC)[5]. Similarly, we can define matter conformal collineation (MCC) for the stress-energy tensor as

$$\pounds_{\mathbf{B}} \mathbf{T} = \psi(x^c) \mathbf{T}.$$

Here, we study the RCC equations for the general cylindrically symmetric static spacetime written in (t, ρ, θ, z) coordinates as

$$ds^2 = e^{\nu(\rho)} dt^2 - d\rho^2 - a^2 e^{\lambda(\rho)} d\theta^2 - e^{\mu(\rho)} dz^2. \quad (7)$$

For this metric, the non-vanishing components of the Ricci tensor are given as

$$\begin{aligned}
R_{00} &= \tfrac{e^{\nu}}{4}\left(2\nu'' + \nu'^2 + \nu'\lambda' + \nu'\mu'\right), \\
R_{11} &= -\left(\tfrac{\nu''}{2} + \tfrac{\lambda''}{2} + \tfrac{\mu''}{2} + \tfrac{\nu'^2}{4} + \tfrac{\lambda'^2}{4} + \tfrac{\mu'^2}{4}\right), \\
R_{22} &= -\tfrac{a^2 e^{\lambda}}{4}\left(2\lambda'' + \nu'\lambda' + \lambda'^2 + \lambda'\mu'\right), \\
R_{33} &= -\tfrac{e^{\mu}}{4}\left(2\mu'' + \nu'\mu' + \lambda'\mu' + \mu'^2\right).
\end{aligned} \qquad (8)$$

Here ′ denotes differentiation with respect to ρ. In component form for the range of four on the indices we get the following RCC equations

$$\begin{aligned}
&R'_{00} B^1 + 2R_{00} B^0_{,0} = 2\psi(x^c) R_{00}, \\
&R'_{11} B^1 + 2R_{11} B^1_{,1} = 2\psi(x^c) R_{11}, \\
&R'_{22} B^1 + 2R_{22} B^2_{,2} = 2\psi(x^c) R_{22}, \\
&R'_{33} B^1 + 2R_{33} B^3_{,3} = 2\psi(x^c) R_{33}, \\
&R_{00} B^0_{,1} + R_{11} B^1_{,0} = 0, \\
&R_{00} B^0_{,2} + R_{22} B^2_{,0} = 0, \\
&R_{00} B^0_{,3} + R_{33} B^3_{,0} = 0, \\
&R_{11} B^1_{,2} + R_{22} B^2_{,1} = 0, \\
&R_{11} B^1_{,3} + R_{33} B^3_{,1} = 0, \\
&R_{22} B^2_{,3} + R_{33} B^3_{,2} = 0.
\end{aligned} \qquad (9)$$

The solution of these equations can broadly be divided into two cases depending upon whether the Ricci tensor is degenerate or non-degenerate. The case for the degenerate Ricci tensor can further be divided into the following fifteen cases depending upon which one or more of the four components of the Ricci tensor is/are zero

Case	R_{00}	R_{11}	R_{22}	R_{33}
I(a)	$=0$	$\neq 0$	$\neq 0$	$\neq 0$
I(b)	$\neq 0$	$=0$	$\neq 0$	$\neq 0$
I(c)	$\neq 0$	$\neq 0$	$=0$	$\neq 0$
I(d)	$\neq 0$	$\neq 0$	$\neq 0$	$=0$
II(a)	$=0$	$=0$	$\neq 0$	$\neq 0$
II(b)	$=0$	$\neq 0$	$=0$	$\neq 0$
II(c)	$=0$	$\neq 0$	$\neq 0$	$=0$
II(d)	$\neq 0$	$=0$	$=0$	$\neq 0$
II(e)	$\neq 0$	$=0$	$\neq 0$	$=0$
II(f)	$\neq 0$	$\neq 0$	$=0$	$=0$
III(a)	$=0$	$=0$	$=0$	$\neq 0$
III(b)	$=0$	$=0$	$\neq 0$	$=0$
III(c)	$=0$	$\neq 0$	$=0$	$=0$
III(d)	$\neq 0$	$=0$	$=0$	$=0$
IV	$=0$	$=0$	$=0$	$=0$

We see that if R_{00} is zero then B^0 gets eliminated from the RCC equations, resulting in the arbitrariness in the value of the RCC vector, thus, giving rise to infinite dimensional Lie algebra. Similarly, when R_{22} or R_{33} is zero B^2 or B^3, respectively, are eliminated from these equations and we get Lie algebra of infinite dimensions. However, when R_{11} is zero no component of the vector **B** is eliminated, thus giving rise to the possibility of obtaining finite dimensional Lie algebras.

To conclude we observe that Case I(b) from the above table is the only case which can admit finite dimensional Lie algebras. In all other cases the Lie algebra has infinite dimensions.

Acknowledgments

Useful discussions with Professor Asghar Qadir and Ugur Camci are gratefully acknowledged. The author is also thankful to the organizers of the Regional Conference for travel support and hospitality.

References

1. H. Stephani, D. Kramer, M.A.H. MacCallum, C. Hoenselaers and E. Herlt in *Exact Solutions of Einstein's Field Equations*, (Cambridge University Press, 2003).
2. K. Yano in *The Theory of Lie Derivatives and its Applications*, (North-Holland Publishing Co., 1955).
3. G.H. Katzin, J. Levine and W.R. Davis, *J. Math. Phys.* **10**, 617 (1969); *J. Math. Phys.* **11**, 1518 (1970).
4. A.H. Bokhari and A. Qadir, *J. Math. Phys.* **34**, 3543 (1993); M. Ziad, *Gen. Rel. Grav.* **35**, 915 (2003); A. Qadir, K. Saifullah and M. Ziad, *Gen. Rel. Grav.* **35**, 1927 (2003).
5. M. Tsamparlis and P.S. Apostolopoulos, *Gen. Rel. Grav.* **36**, 47 (2004).
6. W.R. Davis and G.H. Katzin, *American J. Phys.* **30**, 750 (1962); W.R. Davis, L.H. Green and L.K. Norris, *Nuovo Cimento* B **24**, 256 (1976).

KINEMATIC SELF-SIMILAR SOLUTIONS

M. SHARIF and S. AZIZ
Department of Mathematics, University of the Punjab
Quaid-i-Azam Campus, Lahore-54590, Pakistan
E-mail: hasharif@yahoo.com

We review some kinematic self-similar solutions of spherically symmetric spacetime. This analysis has been used to investigate the kinematic self-similar solutions of cylindrically symmetric spacetime. These have been discussed for the tilted case only. We have attempted solutions for the first, second, zeroth and infinite kind. The governing equations for perfect fluid cosmological models are introduced and a set of integrability conditions for the existence of a kinematic self-similar solutions are derived.

1 Introduction

[a]There is no characteristic scale in Newtonian gravity or General Relativity (GR). A set of field equations is invariant under a scale transformation if we assume appropriate matter field. This implies the existence of scale invariant solutions to the field equations. Such solutions are called self-similar solutions. A characteristic of self-similar solutions is that, by a suitable coordinate transformations, the number of independent variables can be reduced by one, thus allowing a reduction of field equations. In the broadest sense, self-similarity refers to an invariance which simply allows the reduction of a system of partial differential equations to ordinary differential equations.

Similarity solutions were first studied in GR by Cahill and Taub[1], who did so in the cosmological context and under the assumption of spherically symmetric distribution of a self-gravitating perfect fluid. They assumed that the solution was such that the dependent variables are essentially functions of a single independent variable constructed as a dimensionless combination of the independent variables and that the model contains no other dimensionful constants. They showed that the existence of a similarity of the first kind in this situation could be invariantly formulated in terms of the existence of a homothetic vector.

In GR, self-similarity is defined by the existence of a homothetic Killing vector field. Such similarity is called the first kind (or homothety). There exists a natural generalization of homothety called kinematic self-similarity, which is defined by the existence of a kinematic self-similar vector field. The basic condition characterizing a manifold vector field ξ as a self-similar generator[2] is given by

$$\pounds_\xi A = \lambda A, \tag{1}$$

where λ is constant and A is independent physical field. This field can be scalar (*e.g.* μ), vector (*e.g.* u_a), or tensor (*e.g.* g_{ab}). In GR, the gravitational field is represented by the metric tensor g_{ab}, and an appropriate definition of geometrical self-similarity is necessary. The work by Cahill and Taub[3], the simplest generalization was effected, whereby the metric itself satisfies an equation of the form (1).

A kinematic self-similarity satisfies the condition

$$\pounds_\xi u_a = \alpha u_a, \tag{2}$$

with

$$\pounds_\xi h_{ab} = 2\delta h_{ab}, \tag{3}$$

where α and δ are constants and h_{ab} is the projection tensor. Carter and Henriksen[3] defined the other kinds of self-similarity, namely second, zeroth and infinite kind. In the context of kinematic self-similarity, homothety is considered as the first kind. Kinematic self-similar perfect fluid solutions have been explored by several authors[4,5,6,7,8,9,10,11].

In this paper, we are attempting to find self-similar solution for cylindrical symmetric spacetime for the tilted case only. The paper has been organized as follows: In the next section, we shall review briefly some results about kinematic self-similar solutions of spherically symmetric spacetime. In section 3, we shall formulate the governing equations for perfect fluid cosmological models and possible kinematic self-similar solutions are derived for cylindrical symmetric spacetime. We shall conclude the results in the last section.

---

[a]Talk presented by M.S. at the XIth Regional Conference on Mathematical Physics, May 2004, Tehran, Iran.

2 Spherically Symmetric Kinematic Self-Similar Solutions

The line element in a spherically symmetric spacetime is given by

$$ds^2 = -e^{2\phi(t,r)}dt^2 + e^{2\psi(t,r)}dr^2 + R^2(t,r)d\Omega^2, \quad (4)$$

where $d\Omega^2 = d\theta^2 + \sin^2\theta d\varphi^2$.

We consider a perfect fluid as a matter field

$$T_{ab} = p(t,r)g_{ab} + [\mu(t,r) + p(t,r)]u_a u_b, \quad (a,b = 0,1,2,3), \quad (5)$$

where $u_a = (-e^{\phi}, 0, 0, 0)$ is the four-velocity of the fluid element in the co-moving coordinates.

The Einstein field equations and the equations of motion for the perfect fluid are reduced to the following simple form

$$-8\pi G\mu e^{2\phi} = e^{2\phi-2\psi}\left(\frac{R_r^2}{R^2} + \frac{2R_{rr}}{R} - \frac{2R_r\psi_r}{R}\right) - \frac{2R_t\psi_t}{R} - \frac{R_t^2}{R^2} - \frac{e^{2\phi}}{R^2}, \quad (6)$$

$$-8\pi Gpe^{2\psi} = e^{2\psi-2\phi}\left(\frac{R_t^2}{R^2} + \frac{2R_{tt}}{R} - \frac{2R_t\phi_t}{R}\right) - \frac{2R_r\phi_r}{R} - \frac{R_r^2}{R^2} + \frac{e^{2\psi}}{R^2}, \quad (7)$$

$$-8\pi Gp = e^{-2\phi}\left(\psi_{tt} - \psi_t\phi_t + \psi_t^2 - \frac{R_t\phi_t}{R} + \frac{R_t\psi_t}{R} + \frac{R_{tt}}{R}\right) - e^{-2\psi}\left(\phi_{rr} - \psi_r\phi_r + \phi_r^2 + \frac{R_r\phi_r}{R} - \frac{R_r\psi_r}{R} + \frac{R_{rr}}{R}\right), \quad (8)$$

$$0 = R_t\phi_r + R_r\psi_t - R_{tr}. \quad (9)$$

The conservation of energy-momentum, ${T^b}_{a;b} = 0$, implies that

$$\phi_r = -\frac{p_r}{\mu + p}, \quad (10)$$

$$\psi_t = -\frac{\mu_t}{\mu + p} - \frac{2R_t}{R}, \quad (11)$$

$$p = p(r). \quad (12)$$

Substituting Eqs. (14) and (15) in (13), we obtain

$$(pR^2R_t)_r + (\mu R^2 R_r)_t = 0. \quad (13)$$

This implies that there exists a function $m(t,r)$ such that

$$m_t = -4\pi pR^2R_t, \quad m_r = 4\pi\mu R^2 R_r. \quad (14)$$

If we make use of Eqs. (10), (11), (13) and (17), we finally arrive at

$$m = \frac{1}{2G}R[1 - e^{-2\psi}(R_r)^2 + e^{-2\phi}(R_t)^2]. \quad (15)$$

For different spherically symmetric self-similar solutions one can see the literature given in the references.

3 Cylindrically Symmetric Kinematic Self-Similar Solutions

In this section we first discuss the self-similar variables for the cylindrical case. We shall restrict ourselves only for the tilted case. The line element in a cylindrically symmetric spacetime is given by

$$ds^2 = -e^{2\phi(t,r)}dt^2 + dr^2 + e^{2\mu(t,r)}d\theta^2 + e^{2\nu(t,r)}dz^2. \quad (16)$$

The kinematic self-similar vector field in the case of tilted fluid flow, *i.e.*, neither parallel nor orthogonal to the fluid flow, is given by (only t and r are independent variables)

$$\xi^a\frac{\partial}{\partial x^a} = (\alpha t + \beta)\frac{\partial}{\partial t} + r\frac{\partial}{\partial r}. \quad (17)$$

For the tilted case, the similarity index $\frac{\alpha}{\delta}$ yields the following two possibilities according to $\delta = 0$ or $\delta \neq 0$. We find that self-similar variable for cylindrically symmetric spacetime remains the same in the case of first kind, zeroth kind and second kind as for the spherically symmetric spacetime. However, for the infinite kind, the self-similar variable is different. In this case, self-similar variable is given by $\xi = \frac{e^{r/c}}{t}$.

When $\delta \neq 0$ and self-similar variable is ξ, the metric functions for different kinds can be written as

$$\phi = \bar{\phi}(\xi), \quad e^{\mu} = re^{\bar{\mu}(\xi)}, \quad e^{\nu} = re^{\bar{\nu}(\xi)}. \quad (18)$$

When $\delta = 0$, *i.e.*, for self-similarity of infinite kind, the kinematic self-similar vector ξ is given by

$$\xi^\mu\frac{\partial}{\partial x^\mu} = t\frac{\partial}{\partial t} + r\frac{\partial}{\partial r}. \quad (19)$$

The metric functions can be written as

$$\phi = \bar{\phi}(\xi), \quad \mu = \bar{\mu}(\xi), \nu = \bar{\nu}(\xi). \quad (20)$$

Now we discuss the cylindrically symmetric self-similar solutions. The Einstein field equations for cylindrically symmetric spacetime will become

$$k\mu e^{2\phi} = e^{2\phi}(-e^{2\phi}\nu_t\mu_t - \nu_r\mu_r - \mu_{rr} - \mu_r^2 - \nu_{rr} - \nu_r^2), \quad (21)$$

$$kp = e^{-2\phi}(-\nu_{tt} + \phi_t\mu_t + \phi_t\nu_t - \nu_t\mu_t - \mu_t^2 - \mu_{tt} - \nu_t^2) + \phi_r\nu_r + \phi_r\mu_r + \nu_r\mu_r, \quad (22)$$

$$kpe^{2\mu} = e^{2\mu-2\phi}(-\nu_{tt} + \phi_t\nu_t - \nu_t^2)$$

$$+ e^{2\mu}(\phi_{rr} + \phi_r^2 + \nu_{rr} + \nu_r^2 + \phi_r\nu_r), \quad (23)$$

$$kpe^{2\nu} = e^{2\nu-2\phi}(-\mu_{tt} + \phi_t\mu_t - \mu_t^2)$$

$$+ e^{2\nu}(\phi_{rr} + \phi_r^2 + \mu_{rr} + \mu_r^2 + \phi_r\mu_r), \quad (24)$$

$$0 = \mu_{tr} + \mu_t\mu_r - \mu_t\phi_r + \nu_{tr} + \nu_t\nu_r - \nu_t\phi_r. \quad (25)$$

The conservation of energy momentum tensor yields the following results

$$\phi_r = -\frac{p_r}{\mu + p}, \quad (26)$$

$$\mu_t = -\frac{\mu_t}{\mu + p} - \nu_t, \quad (27)$$

$$p = p(r). \quad (28)$$

Since there is no direct way available to form the exact differential any of Eqs. (21)-(25), we could not follow the way of spherically symmetric case here to find self-similar solutions. Recently, there are other attempts[12,13] which provide solution with some other assumptions. Here, we are taking vacuum situation together with the special cylindrical symmetry when $\phi = \mu = \nu$. We consider infinite kind for which the self-similar variable is given by $\xi = \frac{e^{r/c}}{t}$. When we convert the non-zero components of the Ricci tensor in the form of kinematic self-similar variable solve simultaneously, we obtain only one solution which satisfies all the equations. This is $\mu = constant$ and consequently we obtain Minkowski spacetime.

4 Summary

The aim of this paper is to find out some cylindrical symmetric self-similar solutions. We have attempted for this purpose. For the cylindrically symmetric spacetime, the field equations have been written in full detail. The way of spherically symmetric could not provide fruitful results due to the lack of exact differential. We take the assumption of vacuum case and further it is assumed that all metric coefficients are the same. In doing so, we finally have a solution in the infinite kind which turns out to be the Minkowski metric. For the special case $\phi = \mu = \nu$, first, zeroth and second kinds of cylindrically symmetric spacetime, we have problems with kinematic condition. However, self-similar solution is possible. We are working on other aspects to find kinematic self-similar solutions, which is in progress and will appear somewhere else[14].

Acknowledgments

One of the authors (S.A.) acknowledges the enabling role of the Higher Education Commission Islamabad, Pakistan and appreciate its financial support through *Merit Scholarship Scheme for Ph.D. Studies in Science and Technology (200 Scholarships)*. M.S. would like to thank Pakistan Science Foundation for providing traveling grant and IPM for providing local hospitality.

References

1. M.E. Cahill and A.H. Taub, *Commun. Math. Phys.* **21**, 1 (1971).
2. B. Carter and R.N. Henriksen, *Ann. de Phys.* **14**, 47 (1989).
3. B. Carter and R.N. Henriksen, *J. Math. Phys.* **32**, 2580 (1991).
4. B.J. Carr, *Phys. Rev.* D **62**, 044022 (2000).
5. B.J. Carr and A.A. Coley, *Phys. Rev.* D **62**, 044023 (2000).
6. A.A. Coley, *Class. Quantum. Grav.* **14**, 87 (1997).
7. C.B.G. McIntosh, *Gen. Rel. Grav.* **7**, 199 (1975).
8. P.M. Benoit and A.A. Coley, *Class. Quant. Grav.* **15**, 2397 (1998).
9. A.M. Sintes, P.M. Benoit and A.A. Coley, *Gen. Rel. Grav.* **33**, 1863 (2001).
10. H. Maeda, T. Harada, H. Iguchi and N. Okuyama, *Phys. Rev.* D **66**, 027501 (2002).
11. H. Maeda, T. Harada, H. Iguchi and N. Okuyama, *Prog. Theor. Phys.* **108**, 819 (2002); *ibid.* **110**, 25 (2003).
12. J. Bick, T. Ledvinka, B.G. Schmidt and M. Zofka, *Class. Quant. Grav.* **21**, 1583 (2004).
13. A.Z. Wang, Y.M. Wu and Z.C. Wu, *Gen. Rel. Grav.* **36**, 1225 (2004).
14. M. Sharif and S. Aziz, work in progress.

CONSTRAINT ALGEBRA IN CAUSAL LOOP QUANTUM GRAVITY

F. SHOJAI and A. SHOJAI

Physics Department, University of Tehran
North Karegar Ave., 14395 Tehran, Iran
and
Institute for Theoretical Physics and Mathematics (IPM), School of Physics
P.O. Box 19395-5531, Tehran, Iran
Emails: fatimah, shojai@theory.ipm.ac.ir

de Broglie-Bohm causal interpretation of canonical quantum gravity in terms of Ashtekar new variables is built. The Poisson brackets of (de Broglie-Bohm) constraints are derived and it is shown that the Poisson bracket of Hamiltonian with itself would change with respect to its classical counterpart.

1 Introduction

[a]Recently[1], it is shown that using de Broglie-Bohm causal interpretation of quantum mechanics[2], one can derive meaningful relations for constraint algebra and the equations of motion. This is done using the *old* variables, *i.e.* the dynamical variable is chosen to be the metric on spatial slices in an ADM 3+1 decomposition. The new algebra is a clear projection of general coordinate transformation into the spatial and temporal diffeomorphisms.

In Ref. [1] it is shown that the diffeomorphism subalgebra does not change with respect to the classical one. The Poisson bracket of the quantum Hamiltonian and the diffeomorphism constraints, which represents the fact that the quantum Hamiltonian is a pseudo-scalar under diffeomorphisms, is also the same as in the classical case. Finally, the Poisson bracket of the quantum Hamiltonian constraint with itself differs dramatically with its classical counterpart. In fact, this Poisson bracket would be zero weakly, *i.e.* using the equations of motion. This result is just what one expects for the Hamiltonian.

The quantum Hamiltonian is the classical one added with the quantum potential and gives the system quantum trajectories. In the de Broglie-Bohm interpretation of quantum mechanics, one deals with trajectories and thus with Poisson brackets (not commutators). The system has a well-defined trajectory in the phase space obtained from quantum Hamilton-Jacobi equation. Thus the de Broglie-Bohm quantum mechanics represents a deterministic picture of particle trajectory consistent of the statistical predictions of the standard quantum mechanics. In the gravitational case, the dynamics of the metric is determined as a modification of classical Einstein's equation by the quantum potential and quantum force. These are covariant under spatial and temporal diffeomorphisms[1].

It is proved that the *new* variables of gravity[3] are more useful in making quantum gravity. So a natural question is how the causal interpretation of canonical quantum gravity in terms of new variables looks like? One can also ask about the constraints algebra and equations of motion. In this paper we shall answer to these questions.

2 Causal interpretation in terms of new variables

In terms of new variables, gravity consists of three constraints, gauge, diffeomorphism and Hamiltonian constraints. The dynamical variables are the self-dual connection A_a^i and the canonical momenta are $\widetilde{E}_i^a$. The constraints are given by

$$G_i = \mathcal{D}_a \widetilde{E}_i^a, \tag{1}$$
$$C_b = \widetilde{E}_i^a F_{ab}^i, \tag{2}$$
$$H = \epsilon_k^{ij} \widetilde{E}_i^a \widetilde{E}_j^b \mathcal{F}_{ab}^k, \tag{3}$$

in which $\mathcal{D}_a$ represents self-dual covariant derivative and F_{ab}^i is self-dual curvature.

Canonical quantization of these constraints can be achieved in the connection representation via changing $\widetilde{E}_i^a$ into $-\hbar\delta/\delta A_a^i$ and acting them on the

[a]Talk presented by F.S. at the XIth Regional Conference on Mathematical Physics, May 2004, Tehran, Iran.

wavefunctional $\Psi(A)$. We have

$$\hbar \mathcal{D}_a \frac{\delta \Psi(A)}{\delta A_a^i} = 0, \quad (4)$$

$$\hbar F_{ab}^i \frac{\delta \Psi(A)}{\delta A_a^i} = 0, \quad (5)$$

$$\hbar^2 \epsilon_k^{ij} F_{ab}^k \frac{\delta^2 \Psi(A)}{\delta A_a^i \delta A_b^j} = 0. \quad (6)$$

Note that we have chosen the specific ordering that the triads act at left. In order to get the causal interpretation, one should put the definition $\Psi = R \exp(iS/\hbar)$ into these relations. The result is

$$\mathcal{D}_a \frac{\delta R(A)}{\delta A_a^i} = 0, \quad (7)$$

$$\mathcal{D}_a \frac{\delta S(A)}{\delta A_a^i} = 0, \quad (8)$$

$$F_{ab}^i \frac{\delta R(A)}{\delta A_a^i} = 0, \quad (9)$$

$$F_{ab}^i \frac{\delta S(A)}{\delta A_a^i} = 0, \quad (10)$$

$$\epsilon_k^{ij} F_{ab}^k \frac{\delta}{\delta A_a^i} \left(R^2 \frac{\delta S(A)}{\delta A_b^j} \right) = 0, \quad (11)$$

$$-\epsilon_k^{ij} F_{ab}^k \frac{\delta S(A)}{\delta A_a^i} \frac{\delta S(A)}{\delta A_b^j} + Q = 0, \quad (12)$$

where the *quantum potential* is defined as

$$Q = -\hbar^2 \epsilon_k^{ij} F_{ab}^k \frac{1}{R} \frac{\delta^2 R(A)}{\delta A_a^i \delta A_b^j}. \quad (13)$$

Equations (7)-(10) show the gauge and diffeomorphism invariance of the norm and the phase of the wavefunctional. Equation (11) is the continuity equation, while (12) is the quantum Einstein-Hamilton-Jacobi equation. The quantum effects as it is always the case in causal interpretation, are introduced via the quantum potential. These are the classical equations[4] corrected by the quantum potential.

The quantum trajectories would be achieved via the guidance relations

$$\widetilde{E}_i^a = i \frac{\delta S(A)}{\delta A_a^i}. \quad (14)$$

3 Constraints algebra

In this section, we shall study the quantum version of the constraints algebra. In terms of the smeared out Gauss, vector, and scalar constraints

$$\mathcal{G}(\Lambda_i) = -i \int d^3x \Lambda_i \mathcal{D}_a \widetilde{E}_i^a, \quad (15)$$

$$\mathcal{C}(\vec{N}) = i \int d^3x N^b \widetilde{E}_i^a F_{ab}^i - \mathcal{G}(N^a A_a^i), \quad (16)$$

$$\mathcal{H}(\underset{\sim}{N}) = \frac{1}{2} \int d^3x \, \underset{\sim}{N} \epsilon_k^{ij} \widetilde{E}_i^a \widetilde{E}_j^b \mathcal{F}_{ab}^k, \quad (17)$$

the classical algebra is given by

$$\{\mathcal{G}(\Lambda_i), \mathcal{G}(\Theta_j)\} = \mathcal{G}(\epsilon_{jk}^i \Lambda^j \Theta^k), \quad (18)$$

$$\{\mathcal{C}(\vec{N}), \mathcal{C}(\vec{M})\} = \mathcal{C}(\mathcal{L}_{\vec{M}} \vec{N}), \quad (19)$$

$$\{\mathcal{C}(\vec{N}), \mathcal{G}(\Lambda_i)\} = \mathcal{G}(\mathcal{L}_{\vec{N}} \Lambda_i), \quad (20)$$

$$\{\mathcal{C}(\vec{N}), \mathcal{H}(\underset{\sim}{M})\} = \mathcal{H}(\mathcal{L}_{\vec{N}} \underset{\sim}{M}), \quad (21)$$

$$\{\mathcal{G}(\Lambda_i), \mathcal{H}(\underset{\sim}{N})\} = 0, \quad (22)$$

$$\{\mathcal{H}(\underset{\sim}{N}), \mathcal{H}(\underset{\sim}{M})\} = \mathcal{C}(\vec{K}) + \mathcal{G}(K^a A_a^i), \quad (23)$$

where $K^a = \widetilde{E}_i^a \widetilde{E}^{bi} (\underset{\sim}{N} \partial_b \underset{\sim}{M} - \underset{\sim}{M} \partial_b \underset{\sim}{N})$.

In the previous section, we saw that the quantum Einstein-Hamilton-Jacobi equation is just the classical one added with the quantum potential. So the quantum trajectories can be obtained from the quantum Hamiltonian given by

$$H_Q = H + Q. \quad (24)$$

The smeared out gauge and diffeomorphism constraints would not change but the Hamiltonian constraint is now given by

$$\mathcal{H}_Q(\underset{\sim}{N}) = \frac{1}{2} \int d^3x \underset{\sim}{N} \epsilon_k^{ij} \widetilde{E}_i^a \widetilde{E}_j^b \mathcal{F}_{ab}^k + \mathcal{Q}(\underset{\sim}{N}), \quad (25)$$

where $\mathcal{Q}(\underset{\sim}{N}) = \int d^3x \underset{\sim}{N} Q$. The constraint Poisson bracket (18), (19) and (20) would not change. The Poisson bracket (21) is still valid for the quantum Hamiltonian, because the quantum potential is a scalar density. So we have

$$\{\mathcal{C}(\vec{N}), \mathcal{H}_Q(\underset{\sim}{M})\} = \mathcal{H}_Q(\mathcal{L}_{\vec{N}} \underset{\sim}{M}). \quad (26)$$

The same is true for the Poisson bracket (22) as the quantum potential is gauge invariant

$$\{\mathcal{G}(\Lambda_i), \mathcal{H}_Q(\underset{\sim}{N})\} = 0. \quad (27)$$

The difference comes in evaluation of the quantum version of Poisson bracket (23). We have:

$$\begin{aligned} &\{\mathcal{H}_Q(\underset{\sim}{N}), \mathcal{H}_Q(\underset{\sim}{M})\} = \\ &\{\mathcal{H}(\underset{\sim}{N}), \mathcal{H}(\underset{\sim}{M})\} + \{\mathcal{Q}(\underset{\sim}{N}), \mathcal{H}(\underset{\sim}{M})\} + \\ &\{\mathcal{H}(\underset{\sim}{N}), \mathcal{Q}(\underset{\sim}{M})\} + \{\mathcal{Q}(\underset{\sim}{N}), \mathcal{Q}(\underset{\sim}{M})\}, \end{aligned} \quad (28)$$

the fourth term is zero identically, since the quantum potential is a functional of the connection only. The sum of the second and the third terms is

$$\{\mathcal{Q}(\underset{\sim}{N}),\mathcal{H}(\underset{\sim}{M})\}+\{\mathcal{H}(\underset{\sim}{N}),\mathcal{Q}(\underset{\sim}{M})\}\sim$$
$$i\int d^3z\left(\underset{\sim}{N}\epsilon^{ij}{}_kF^k_{ab}\widetilde{E}^b_j\mathcal{D}_c(\underset{\sim}{M}\epsilon_i^{lm}\widetilde{E}^a_l\widetilde{E}^c_m)-\right.$$
$$\left.\underset{\sim}{M}\epsilon^{ij}{}_kF^k_{ab}\widetilde{E}^b_j\mathcal{D}_c(\underset{\sim}{N}\epsilon^{lm}{}_i\widetilde{E}^a_l\widetilde{E}^c_m)\right), \tag{29}$$

where we have used the symbol $\sim$ in order to show that this equality is valid weakly. That is, the equation of motion - functional derivative of the Hamiltonian constraint with respect to the connection - is used in its evaluation. A simple calculation then shows that the Poisson bracket of the quantum Hamiltonian with itself is given by

$$\{\mathcal{H}_Q(\underset{\sim}{N}),\mathcal{H}_Q(\underset{\sim}{M})\}\sim 0, \tag{30}$$

which is a result very similar to the one in terms of the old variables[1].

At this end it may be useful to obtain the quantum equations of motion by making use of the Hamilton equations. We have

$$\dot{A}^i_a=-i\epsilon^{ijk}\underset{\sim}{N}\widetilde{E}^b_jF_{abk}-N^bF^i_{ab}, \tag{31}$$

$$\dot{\widetilde{E}}{}^a_i=i\epsilon_i^{jk}\mathcal{D}_b(\underset{\sim}{N}\widetilde{E}^a_j\widetilde{E}^b_k)$$
$$-2\mathcal{D}_b(N^{[a}\widetilde{E}^{b]}_i)+\frac{i}{2}\int d^3z\frac{\delta\mathcal{Q}(\underset{\sim}{N})}{\delta A^i_a(z)}. \tag{32}$$

Also to recover the real quantum general relativity one must set the reality conditions. These are

$$\widetilde{E}^a_i\widetilde{E}^{bi},\qquad \textit{must be real,} \tag{33}$$

and

$$i\epsilon^{ijk}\widetilde{E}^{(a}_i\mathcal{D}_c\left(\widetilde{E}^{b)}_k\widetilde{E}^c_j\right)+\frac{i}{2}\int d^3z\frac{\delta\mathcal{Q}}{\delta A^i_{(a}(z)}\widetilde{E}^{b)i}(x), \tag{34}$$

must be real.

4 Conclusion

We saw that one can successfully construct a causal version of canonical quantum gravity in terms of new variables using the de Broglie-Bohm interpretation of quantum mechanics. As it is usual in this theory, all the quantum behavior is coded in the *quantum potential.* Since the theory is a constrained one, one should calculate the constraints algebra and check it for consistency. As in the de Broglie-Bohm theory any quantum system has a well-defined trajectory in the configuration space and one has no operator, the algebra action is in fact the Poisson bracket. We have shown that only the Poisson bracket of Hamiltonian with itself would change with respect to the classical algebra. This Poisson bracket is *weakly,* that is on the equations of motion, equal to zero, different from the classical case, where it is *strongly* equal to sum of a gauge transformation and a 3-diffeomorphism. The result is just like to that of the theory in terms of old variables[1]. This enables one to give the meaning of time generator to the Hamiltonian constraint. At the end the equations of motion are written out and as it is expected the quantum force appears in them.

It must be noted here that all the above results are formal, that is to say, in evaluation of the Poisson brackets and other things we have not regularized the ill-defined terms. For having a rigor result, one should evaluate them using a regulator. Introduction of a regulator in general needs to use a background metric and one must show at the end that the result is independent of that background metric. We shall do this in a forthcoming paper.

References

1. A. Shojai and F. Shojai, *Class. Quant. Grav.* **21**, 1 (2004); F. Shojai and A. Shojai, *Pramana J. Phys.* 58, **1**, 13 (2002).
2. D. Bohm, *Phys. Rev.* **85**, 166 (1952); *ibid.* **85**, 180 (1952); T. Horiguchi, *Mod. Phys. Lett.* A **9**, 1429 (1994).
3. A. Ashtekar in *Lectures on Non-perturbative Canonical Gravity,* (World Scientific Publishing co. Pte. Ltd., Singapore, 1991).
4. C. Rovelli in *Quantum Gravity,* the draft version may be found at: http://www.cpt.univ-mrs.fr/~rovelli/rovelli.html, 2003.

THE COSMIC CENSORSHIP HYPOTHESIS AND THE NAKED REISSNER-NORDSTROM SINGULARITY

A. QADIR
Centre for Advanced Mathematics and Physics
National University of Sciences and Technology
Peshawar Road, Rawalpindi, Pakistan
and
Department of Mathematical Sciences
King Fahd University of Petroleum and Minerals
Dhahran, 13261, Saudi Arabia

A. A. SIDDIQUI
National University of Sciences and Technology
E&ME College
Peshawar Road, Rawalpindi, Pakistan
E-mail: azadasj@yahoo.com

Penrose's cosmic censorship hypothesis excludes the physical existence of naked singularities, as they could otherwise introduce unpredictable influences in their future null cones. In this paper, we have analyzed timelike geodesics for a naked Reissner-Nordstrom singularity. It is found that the singularity is effectively clothed by its own repulsive nature, making the hypothesis redundant in this case as it provides "censorship without censorship".

1 Introduction

[a]It has been argued[1] that special significance attaches to the frame of observers falling freely from rest at infinity. In this frame the gravitational force deduced for a Schwarzschild source would simply be the Newtonian force. As such, for a more general source the relativistic *correction* to the Newtonian gravitational force could be computed in this frame. For the Reissner-Nordstrom (RN) geometry[2], one finds a repulsion due to the charge and for the Kerr-Newmann spacetime[3] a rich structure of forces. Such frames have been called pseudo-Newtonian frames and have been identified[4] as a special class of Fermi-Walker frames. In this paper, free fall and other timelike geodesics in a naked RN singularity background are analyzed in the context of the cosmic censorship hypotheses[5].

Penrose proposed the cosmic censorship hypothesis so as to avoid the possibility of unpredictable influences emerging from the singularity, where physical laws break down. As he put it[6], "it is as if there is a cosmic censor board that objects to *naked* singularities being seen and ensures that they only appear suitably clothed by an event horizon". Due to this conjecture, naked singularities are seldom studied seriously in themselves, though various discussions focus on the possibility of finding counter-examples to it even for singularities that arise from realistic gravitational collapse processes. There *is* a paper[7] that investigates geodesics of arbitrarily charged particles in a naked RN singularity background, but it concentrates on calculating the geodesics only and not on deducing any consequences from them. In the following sections, after presenting the RN spacetime and the geodesic equation, timelike geodesics for the RN spacetime are given. The significance of these geodesics is then discussed.

2 The RN Spacetime

Reissner[8] and Nordstrom[9] obtained the solution of Einstein field equations, with a non-vanishing energy-momentum tensor arising form the sourceless electromagnetic field ($j^{\mu} = 0$), which describes the field outside a spherically symmetric massive charged point, called the RN black hole, is given (in gravita-

[a]Talk presented by A.A.S. at the XIth Regional Conference on Mathematical Physics, May 2004, Tehran, Iran.

tional units $G = c = 1$) by the metric,

$$ds^2 = e^{\upsilon(r)}dt^2 - e^{-\upsilon(r)}dr^2 - r^2 d\theta^2 - r^2\sin^2\theta d\phi^2 \,, \quad (1)$$

where

$$e^{\upsilon(r)} = \frac{r^2 - 2mr + Q^2}{r^2}, \quad (2)$$

Q being the charge (in gravitational units).

For $m > Q$ the RN metric, (1) becomes singular at $r = 0$ and

$$r_\pm = m \pm \sqrt{m^2 - Q^2}. \quad (3)$$

The curvature invariants,

$$\left.\begin{array}{l} R_1 = R^{ab}{}_{ab} \\ R_2 = R^{ab}{}_{cd}R^{cd}{}_{ab} \end{array}\right\}, \quad (4)$$

etc., associated with the curvature tensor, $R^a{}_{bcd}$, are regular on the singular surfaces $r_\pm$. Thus there are no physical singularities there, but the coordinate system is singular. The latter singularities are called the *outer* and *inner horizons*, respectively. There are appropriate coordinates available which remove these singularities[10]. At $r = 0$, the curvature invariants are not regular. The singularity at this point is a physical singularity, called a *curvature* or *essential singularity*.

For $r = m$, (3) gives $r_- = r_+ = m$, so there is only one horizon in this case. This singularity can be removed by using the Carter coordinates[10]. In the case when $r < m$ there are no coordinate singularities and hence no horizon to shield the essential singularity. This is called the naked RN singularity.

3 Timelike Geodesics (Paths) in RN Spacetime

The geodesic equation, for the extremal path between two points, is

$$\ddot{x}^a + \Gamma^a_{bc}\dot{x}^b\dot{x}^c = 0, \quad (5)$$

where the dot represents the derivative with respect to the arc length parameter, s. The timelike geodesics represent the paths of uncharged test particles. In the RN background, the geodesic equation for t gives

$$\frac{dt}{ds} = \dot{t} = ke^{-\upsilon(r)}. \quad (6)$$

From (1), for $\theta = \phi =$ constant, we have

$$e^{\upsilon(r)}\dot{t}^2 - e^{-\upsilon(r)}\dot{r}^2 = 1. \quad (7)$$

We can re-write the two requirements, (6 and 7) in a single equation for the change of r with t, as

$$\frac{dr}{dt} = \frac{\dot{r}}{\dot{t}} = \pm\frac{\sqrt{k^2 - e^{\upsilon(r)}}}{ke^{-\upsilon(r)}}. \quad (8)$$

On account of the initial attraction of the gravitational source, we must take the negative root. The geodesics can be better represented in terms of a re-scaled radial parameter

$$r^* = \int e^{-\upsilon(r)}dr = \int \left(1 - 2m/r + Q^2/r^2\right)^{-1} dr. \quad (9)$$

The constant of integration must be chosen so that $r^* = 0$ at $r = 0$, to get

$$r^* = r + m\ln\left(\frac{r^2 - 2mr + Q^2}{Q^2}\right) + \frac{2m^2 - Q^2}{\sqrt{Q^2 - m^2}} \times \left[\tan^{-1}\left(\frac{r - m}{\sqrt{Q^2 - m^2}}\right) - \tan^{-1}\left(\frac{m}{\sqrt{Q^2 - m^2}}\right)\right]. \quad (10)$$

In (t, r^*) coordinates, (8) gives

$$\frac{dr^*}{dt} = \pm\frac{\sqrt{k^2 - e^{\upsilon(r)}}}{k}. \quad (11)$$

3.1 Case (i): $k = 1$

This corresponds to the path of a test particle falling freely from rest at infinity, and we have $dr^*/dt = \pm\sqrt{2m/r - Q^2/r^2}$. The term inside the square root becomes negative for $r < r_b$, where

$$r_b = \frac{Q^2}{2m}. \quad (12)$$

Therefore, no timelike geodesics from infinity can enter the region $r < r_b$. That region is protected from view by its repulsive nature!

3.2 Case (ii): $k > 1$

This corresponds to a test particle falling from infinity with a positive velocity. Putting $k^2 = 1 + \varepsilon$, we find that the geodesics will have a barrier not at $r = r_b$, but rather at $r_c = \left[-m \pm \sqrt{m^2 + \varepsilon Q^2}\right]/\varepsilon$. It is easily verified that only the +ve root is valid. Now the boundary moves *back* from r_b to $r_c \approx r_b - \varepsilon Q^4/8m^3$, and the "clothed" singularity appears *smaller* to a faster moving observer. Note that here k is bounded from below by 1, but is not bounded from above. In the limit as k goes to infinity, r_c goes to zero.

3.3 *Case (iii): $k < 1$*

For $k < 1$ we put $k^2 = 1 - \varepsilon$. Now the barriers are at $r_\pm = \left[m \pm \sqrt{m^2 - \varepsilon Q^2}\right] / \varepsilon$. In this case the boundary will move forward from r_b to $r_- \approx r_b + \varepsilon Q^4/8m^3$ while the limit at infinity moves back to $r_+ \approx 2m/\varepsilon$. The "clothed" singularity appears *larger*. Clearly, k is bounded from below by the requirement that $m^2 > \varepsilon Q^2$. Thus, for a given m and Q, $\varepsilon \leq m^2/Q^2$. At $\varepsilon = m^2/Q^2$ we get $r_+ = r_- = 2r_b$. This gives the geodesic as $r = 2r_b$. In the other limit, as k tends to 1, we see that the outer limit, r_+, tends to infinity.

4 Conclusion

Penrose's cosmic censorship hypothesis: "*no naked singularities can form by a physical collapse process but must always be clothed by an event horizon*" has to be assumed separately as a physical requirement. It would be nicer if the theory took care of the problem for us and no additional assumptions were needed. For the naked RN singularity the theory appears to run into problems as there is no horizon. However, our analysis shows that though there is no horizon, timelike geodesics can not reach the essential singularity, nor can any emerge from it. However, null geodesics *can* go up to the singularity and emerge from it. Due to the back-reaction of the test-particle on the background spacetime, null particles would acquire an *effective* mass. Hence the singularity at $r = 0$ *would* be protected from the outside world, making it unnecessary to add on the conjecture. It must be admitted that we have only demonstrated the redundancy of the hypothesis in this one case and a more general analysis is required.

Acknowledgments

The presenting author (A.A.S.) is grateful to the Higher Education Commission of Pakistan for providing travel support to participate in the Regional Conference. The authors also thank the organizers for providing local hospitality. They express their gratitude to K. Saifullah for help in preparing the manuscript.

References

1. S.M. Mahajan, A. Qadir and P.M. Valanju, *Nuovo Cimento* B **65**, 404 (1981); A. Qadir and J. Quamar in *Proceedings of the Thitd Marcel Grossmann Meeting*, ed. Hu Ning (North Holland Publishing Co., 1983).
2. A. Qadir, *Phys. Lett.* A **99**, 419 (1983).
3. A. Qadir, *Europhys. Lett.* **2**, 426 (1986).
4. A. Qadir and I. Zafarullah, *Nuovo Cimento* B **111**, 79 (1996).
5. See, for example, R. Penrose, in *Physics and Contemporary Needs, Vol. 1*, ed. Riazuddin (Plenum Press, 1977).
6. R. Penrose, *private communication.*
7. J.M. Cohen and R. Gautreau, *Phys. Rev.* D **19**, 2273 (1990).
8. H. Reissner, *Ann. Phys.* **50**, 106 (1916).
9. G. Nordstrom, *Proc. Koninkl. Ned. Akad. Wetenschap.* **20**, 1238 (1918).
10. See, for example, A. Qadir in *Einstein's General Theory of Relativity*, in preparation; A.A. Siddiqui in *Ph.D. Thesis* (Quaid-i-Azam University, Islamabad, 2000).

CHAPTER 5: MATHEMATICAL PHYSICS

ON THE ROLE OF NON-NOETHER SYMMETRY IN INTEGRABILITY OF DISPERSIVELESS LONG WAVE SYSTEM

G. CHAVCHANIDZE

Department of Theoretical Physics, A. Razmadze Institute of Mathematics
1 Aleksidze Street, Tbilisi 0193, Georgia
E-mail: gch@rmi.acnet.ge

We show that infinite sequence of conserved quantities and bi-Hamiltonian structure of DLW hierarchy of integrable models are related to the non-Noether symmetry of dispersiveless water wave system.

[a]Symmetries play an essential role in dynamical systems, because they usually simplify analysis of evolution equations and often provide quite elegant solution of problems that otherwise would be difficult to handle. In the present paper, we show how knowing just single generator of non-Noether symmetry one can construct infinite involutive sequence of conserved quantities and bi-Hamiltonian structure of one of the remarkable integrable models – dispersiveless long wave system. In fact, among nonlinear partial differential equations that describe propagation of waves in shallow water, there are many interesting integrable models. And most of them seem to have non-Noether symmetries leading to the infinite sequence of conservation laws and bi-Hamiltonian realization of these equations. In dispersiveless long wave system, such a symmetry appears to be local, that in some sense simplifies investigation of its properties and calculation of conserved quantities.

The evolution of dispersiveless long wave system is governed by the following set of nonlinear partial differential equations

$$\begin{aligned} v_t &= v_x w + v w_x, \\ w_t &= v_x + w w_x. \end{aligned} \tag{1}$$

Each symmetry of this system must satisfy linear equation

$$\begin{aligned} E(v)_t &= (wE(v))_x + (vE(w))_x, \\ E(w)_t &= E(v)_x + (wE(w))_x, \end{aligned} \tag{2}$$

obtained by substituting infinitesimal transformations

$$\begin{aligned} v &\to v + aE(v) + O(a^2), \\ w &\to w + aE(w) + O(a^2), \end{aligned} \tag{3}$$

into equations of motion (1) and grouping first order (in a) terms. One of the solutions of this equation yields the following symmetry of dispersiveless water wave system

$$\begin{aligned} E(v) &= 4vw + 2x(vw)_x + 3t(v^2 + vw^2)_x, \\ E(w) &= w^2 + 4v + 2x(ww_x + v_x), \\ &\quad + t(6vw + w^3)_x. \end{aligned} \tag{4}$$

Tt is remarkable that this symmetry is local in the sense that $E(v), E(w)$ in point x depend only on v and w and their derivatives are evaluated in the same point (this is not the case in Korteweg-de Vriez, modified Korteweg-de Vriez and non-linear Schrödinger equations where similar symmetries appear to be non-local[3]).

Before we proceed, let us note that dispersive water wave system is actually an infinite dimensional Hamiltonian dynamical system. Assuming that v and w fields are subjected to zero boundary conditions

$$v(\pm\infty) = w(\pm\infty) = 0, \tag{5}$$

it is easy to verify that equations (1) can be represented in Hamiltonian form

$$\begin{aligned} v_t &= \{h, v\}, \\ w_t &= \{h, w\}, \end{aligned} \tag{6}$$

with Hamiltonian equal to

$$h = \frac{1}{2}\int_{-\infty}^{+\infty} (vw^2 + v^2)dx, \tag{7}$$

and Poisson bracket defined by the following Poisson bivector field

$$W = \int_{-\infty}^{+\infty} \left[\frac{\delta}{\delta v} \wedge \left[\frac{\delta}{\delta w}\right]_x\right] dx. \tag{8}$$

[a]Talk presented at the XIth Regional Conference on Mathematical Physics, May 2004, Tehran, Iran.

Now using our symmetry that appears to be non-Noether, one can calculate the second Poisson bivector field involved in the bi-Hamiltonian realization of dispersiveless long wave system

$$\hat{W} = [E, W] = -2\int_{-\infty}^{+\infty}\left[v\frac{\delta}{\delta v}\wedge\left[\frac{\delta}{\delta v}\right]_x\right.$$
$$\left.+w\frac{\delta}{\delta v}\wedge\left[\frac{\delta}{\delta w}\right]_x+\frac{\delta}{\delta w}\wedge\left[\frac{\delta}{\delta w}\right]_x\right]dx. \quad (9)$$

Note that $\hat{W}$ gives rise to the second Hamiltonian realization of the model

$$v_t = \{h^\bullet, v\}_\bullet,$$
$$w_t = \{h^\bullet, w\}_\bullet, \quad (10)$$

where

$$h^\bullet = -\frac{1}{4}\int_{-\infty}^{+\infty} vw dx, \quad (11)$$

and $\{,\}_\bullet$ is the Poisson bracket defined by bivector field $\hat{W}$.

Now let us pay attention to conservation laws. By integrating the third equation of dispersive water wave system (1), it is easy to show that

$$J^{(0)} = \int_{-\infty}^{+\infty} w dx, \quad (12)$$

is a conservation law. Using non-Noether symmetry one can construct other conservation laws by taking Lie derivative of $J^{(0)}$ along the generator of symmetry and in this way entire infinite sequence of conservation laws of dispersive water wave system can be reproduced

$$J^{(0)} = \int_{-\infty}^{+\infty} w dx,$$

$$J^{(1)} = L_E J^{(0)} = 2\int_{-\infty}^{+\infty} v dx,$$

$$J^{(2)} = L_E J^{(1)} = (L_E)^2 J^{(0)} = 4\int_{-\infty}^{+\infty} vw dx,$$

$$J^{(3)} = L_E J^{(2)} = (L_E)^3 J^{(0)}$$
$$= 12\int_{-\infty}^{+\infty} (vw^2 + v^2) dx,$$

$$J^{(4)} = L_E J^{(3)} = (L_E)^4 J^{(0)}$$
$$= 48\int_{-\infty}^{+\infty} (3v^2 w + vw^3) dx,$$

$$J^{(n)} = L_E J^{(n-1)} = (L_E)^n J^{(0)}. \quad (13)$$

As we see, non-Noether symmetry (4) naturally leads to an infinite sequence of conserved quantities (13) and second Hamiltonian realization (9) of dispersiveless water wave system.

Acknowledgements

This work was supported by INTAS (00-00561).

References

1. G. Bluman and S. Kumei in *Symmetries and Differential Equations*, (Springer-Verlag, New York, 1989).
2. G. Chavchanidze, *J. Geom. Phys.* **48**, 190 (2003), arXiv: math-ph/0211014.
3. G. Chavchanidze, arXiv: math-ph/0405003.
4. P. Olver in *Applications of Lie groups to Differential Equations, GTM 107*, (Springer Verlag, New York, 1986).

CONNECTION BETWEEN GROUP BASED QUANTUM TOMOGRAPHY AND WAVELET TRANSFORM IN BANACH SPACES

M. A. JAFARIZADEH, M. MIRZAEE and M. REZAEE
Department of Theoretical Physics and Astrophysics
Tabriz University, Tabriz 51664, Iran
E-mails: jafarizadeh, mirzaee, karamaty@tabrizu.ac.ir

The intimate connection between the Banach space wavelet reconstruction method for each unitary representation of a given group and some of well-known quantum tomographies, such as, tomography of rotation group, spinor tomography and tomography of unitary group, is established. Also both the atomic decomposition and Banach frame nature of these quantum tomographic examples is revealed in details.

1 Introduction

[a]The mathematical theory of wavelet transform finds nowadays an enormous success in various fields of science and technology, including treatment of large databases, data and image compression, signal processing, telecommunication and many other applications. Recently, another concept, called atomic decomposition, has played a key role in further mathematical development of wavelet theory. As far as the Banach space is concerned, Feichtinger-Grochenig[1] provided a general and very flexible way to construct coherent atomic decompositions and Banach frames for Banach spaces.

The quantum states can be determined completely from the appropriated experimental data by using the well-known technique of quantum tomography or better to say tomographic transformation.

Here, in this manuscript, we are trying to establish the intimate connection between the Banach space wavelet reconstruction method developed by Feichtinger-Grochenig and some of well-known quantum tomographies associated with mixed states.

2 Wavelet Transform, Frame, and Atomic Decomposition on Banach Spaces

The following is a brief recapitulation of some aspects of the theory of wavelets, atomic decomposition and Banach frame. We only mention those concepts that will be needed in the sequel, a more detailed treatment may be found, for example, in Refs. [1-3].

Let G be locally compact group with left Haar measure $d\mu$ and let π be a continuous representation of a group G in a (complex) Banach space $\mathcal{B}$.

A representation for group $G \times G$ in the space $\mathcal{L}(\mathcal{B})$ of bounded linear operators acting on Banach space $\mathcal{B}$ is defined as

$$T : G \times G \to \mathcal{L}(\mathcal{L}(\mathcal{B})) : \hat{O} \to U(g_1^{-1})\hat{O}U(g_2), \quad (1)$$

where if $g_1 = g_2$, the representation is called adjoint representation, and if g_2 is equal to identity operator, the representation is called left representation of group. We will say that the set of vectors $b_g = T(g)b_0$ form a family of coherent states, if there exists a continuous nonzero linear functional $l_0 \in \mathcal{B}^*$, called test functional, and a vector $b_0 \in \mathcal{B}$, called vacuum vector, such that

$$C(b_0, b_0^{'}) = \int_G < T(g^{-1})b_0, l_0 >< T(g)b_0^{'}, l_0^{'} > d\mu(g), \quad (2)$$

is non-zero and finite, which is known as the admissibility relation. For unitary representation in Hilbert spaces, the condition (2) is known as square integrability. Thus, our definition describes an analog of square integrable representation for Banach space.

The wavelet transform $\mathcal{W}$ from Banach space $\mathcal{B}$ to a space of function $F(G)$, that is defined by a representation π of G on $\mathcal{B}$, a vacuum vector b_0 and a test functional l_0, is given by

$$\begin{aligned} \mathcal{W} : \mathcal{B} \to F(G) : O \to \hat{O}(g) = [\mathcal{W}\hat{O}](g) \\ =< T(g^{-1})\hat{O}, l_0 >=< \hat{O}, \pi^*(g) l_0 >, \quad (3) \end{aligned}$$

and, the inverse wavelet transform $\mathcal{M}$ from $F(G)$ to $\mathcal{B}$ is given by

$$\begin{aligned} \mathcal{M} : F(G) \to \mathcal{B} : \hat{O}(g) \to \mathcal{M}[\hat{O}] \\ = \int_G \hat{O}(g)\; b_x \; d\mu(g) = \int_G \hat{O}(g)\; T(g)\; b_0 \; d\mu(g). (4) \end{aligned}$$

[a]Talk presented by M.A.J. at the XIth Regional Conference on Mathematical Physics, May 2004, Tehran, Iran.

The operator $\mathcal{P} = \mathcal{MW} : \mathcal{B} \longmapsto \mathcal{B}$ is a projection of $\mathcal{B}$ into its linear subspace, in which b_0 is cyclic, *i.e.* the set $\{T(g)b_0|g \in G\}$ spans a Banach space $\mathcal{B}$ and $\mathcal{MW}(\hat{O}) = P(\hat{O})$ where the constant P is equal to $\frac{c(b_0,b_0')}{<b_0,l_0'>}$. Especially, if in cases, the left representation U is an irreducible representation, then, the inverse wavelet transform $\mathcal{M}$ is a left inverse operator of wavelet transform $\mathcal{W}$ on $\mathcal{B}$, *i.e*, $\mathcal{MW} = I$.

Frames can be seen as a generalization of basis in Hilbert or Banach space. Banach frames and atomic decomposition are sequences that have basis-like properties but which need not to be bases. Atomic decomposition has played a key role in the recent development of wavelet theory.

Now we define a decomposition of a Banach space as follows:

Definition of atomic decomposition: Let $\mathcal{B}$ be a Banach space, and $\mathcal{B}_d$ be an associated Banach space of scalar-valued sequences indexed by $N = \{1, 2, 3, \cdots\}$, and let $\{y_i\}_{i\in N} \subset \mathcal{B}^*$ and $\{x_i\}_{i\in N} \subset \mathcal{B}$ be given. If

a) $\{< \hat{O}, y_i >\} \in \mathcal{B}_d$ for each $\hat{O} \in \mathcal{B}$,

b) the norms $\|\hat{O}\|_{\mathcal{B}}$ and $\|\{< \hat{O}, y_i >\}\|_{\mathcal{B}_d}$ are equivalent,

c) $\hat{O} = \sum_{i=1}^{\infty} < \hat{O}, y_i > x_i$ for each $\hat{O} \in \mathcal{B}$,

then $(\{y_i\}, \{x_i\})$ is an atomic decomposition of $\mathcal{B}$ with respect to $\mathcal{B}_d$. In cases, the norm equivalence is given by

$$A\|\hat{O}\|_{\mathcal{B}} \leq \|\{< \hat{O}, y_i >\}\|_{\mathcal{B}_d} \leq B\|\hat{O}\|_{\mathcal{B}}, \tag{5}$$

then A, B are a choice of atomic bounds for ($\{y_i\}$ and $\{x_i\}$).

Definition of Banach frame: Let X be a Banach space, and $\mathcal{B}_d$ be an associated Banach space of scalar-valued sequences indexed by $\mathcal{N}$ and let $\{y_i\}_{i\in N} \subset \mathcal{B}^*$ and $S : \mathcal{B}_d \longrightarrow \mathcal{B}$ be given. If

a) $\{< \hat{O}, y_i >\} \in \mathcal{B}_d$ for each $x \in \mathcal{B}$,

b) the norms $\|x\|_{\mathcal{B}}$ and $\|\{< \hat{O}, y_i >\}\|_{\mathcal{B}_d}$ are equivalent,

c) S is bounded and linear, and $S\{< \hat{O}, y_i >\} = \hat{O}$ for each $O \in \mathcal{B}$,

then $(\{y_i\}, S)$ is a Banach frame for $\mathcal{B}$ with respect to $\mathcal{B}_d$. The mapping S is called the reconstruction operator. If the norm equivalence is given by

$$A\|\hat{O}\|_{\mathcal{B}} \leq \|\{< \hat{O}, y_i >\}\|_{\mathcal{B}_d} \leq B\|\hat{O}\|_{\mathcal{B}},$$

then A, B will be a choice of frame bounds for $(\{y_i\}, S)$.

Obviously, one can show that admissibility condition is the same as frame condition.

3 Quantum Tomography via Group Theory with Wavelet Transform on Banach Space

The Group tomography of a compact group G, with an irreducible unitary representation U acting on separable Hilbert space $\mathcal{H}$, means that every element of $\mathcal{B}(\mathcal{H})$, the Banach algebra of bounded linear operators acting on $\mathcal{H}$, can be constructed by the set $\{U(g), g \in G\}$ according to formula (6), where the set $\{U(g), g \in G\}$ is known as tomographic set and Tr $[U^\dagger(g)\hat{O}]$ is sampling set or tomogram set of a given operator[4] $\hat{O}$. When $\mathcal{H}$ is finite-dimensional, the hypothesis that $\{U(g)\}$ is a tomographic set is sufficient to reconstruct any given operator from the tomographic set by using (6). But the case of $\dim(\mathcal{H}) = \infty$ needs a further condition to make sure that every expression converges and that it can be attributed to a precise mathematical meaning. If O is a trace-class operator on $\mathcal{H}$ and $\{U(g)\}$ is a tomographic set, then we have

$$\hat{O} = \int d\mu(g) \text{Tr } [U^\dagger(g)\hat{O}]U(g). \tag{6}$$

Now, we try to obtain the above explained tomography via wavelet transform in Banach space. In order to do so, we need to choose the tomographic set $U(g)$ as a continuous representation of the wavelet transformation and the identity operator as a vacuum vector. Therefore, the corresponding wavelet transformation takes the following form:

$$\mathcal{W} : \mathcal{B} \mapsto F(g) : \hat{O} \mapsto \hat{O}(g) =< \hat{O}, l_g >=$$
$$=< \hat{O}, U(g)l_0 >=< \hat{O}U(g)^\dagger, l_0 >= \text{Tr}(\hat{O}U(g)^\dagger). \tag{7}$$

With this condition, the inverse wavelet transform $\mathcal{M}$ becomes a left inverse operator of the wavelet transform $\mathcal{W}$

$$\mathcal{MW} = I \Rightarrow \mathcal{M} : F(g) \mapsto \mathcal{B} : \hat{O}(g) \mapsto \mathcal{M}[\hat{O}]$$
$$= \mathcal{MW}(\hat{O}) = \hat{O} = \int d\mu(g) < \hat{O}, l_g > b_g. \tag{8}$$

Therefore, with the choice of $b_0 = I$ (identity operator), we obtain the tomography relation (6). At the end, we obtain atomic decomposition and Banach frame with atomic bounds $A = B = 1$.

3.1 *Tomography for Rotation Group*

We try to obtain the rotation group tomography via wavelet transform in Banach space. In order to do so, we need to choose the tomographic set $D(\alpha,\beta,\gamma)$ as a continuous representation of the wavelet transformation and the identity operator as a vacuum vector. Therefore, the corresponding wavelet transformation is given by

$$\mathcal{W}:\mathcal{B}\mapsto F(G):\hat{\rho}(\alpha,\beta,\gamma)=\mathrm{Tr}\ (\hat{\rho}D^{\dagger}(\alpha,\beta,\gamma)).$$

With this condition, inverse wavelet transform is obtained by

$$\mathcal{M}:F(G)\mapsto\mathcal{B}:\mathcal{MW}(\hat{\rho})=$$
$$=\int d\mu(\alpha,\beta,\gamma)<\hat{\rho},l_{(\alpha,\beta,\gamma)}>b_{(\alpha,\beta,\gamma)}.$$

Therefore, the rotation group tomography is given by[5]

$$\hat{\rho}=\int \mathrm{Tr}(\hat{\rho}D^{\dagger}(\alpha,\beta,\gamma))D(\alpha,\beta,\gamma)d\mu(\alpha,\beta,\gamma).$$

At the end, we obtain atomic decomposition and Banach frame with atomic bounds $A=B=1$.

3.2 *Tomography of Quantum Spinor States*

Now we extract quantum spinor states tomography via wavelet transform in Banach space. Then, the the wavelet transform for adjoint representation is defined by $\hat{\rho}(\Omega)=\langle U(\Omega)\hat{\rho}(\Omega)^{\dagger},l_0\rangle$, where U is irreducible representation of $SU(2)$ group for spin J. In this case, by choosing the test functional as $l_0(\hat{\rho})=\mathrm{Tr}\ [\hat{\rho}\mid j,m_1><j,m_2\mid]$, the corresponding wavelet transform becomes

$$\hat{\rho}(\Omega)=<j,m_1\mid U(\Omega)\hat{\rho}U(\Omega)^{\dagger}\mid j,m_2>=\omega(\Omega,m_1,m_2).$$

The inverse wavelet transform is

$$\mathcal{M}(\hat{\rho})=\int d\Omega\ \omega(\Omega,m_1,m_2)U^{\dagger}(\Omega)b_0U(\Omega).$$

Therefore, the quantum spinor states tomography is given by[6]

$$(-1)^k\hat{\rho}_{ik}^{(j)}=\sum_{J'=0}^{2j}\sum_{M=-J'}^{J'}(2J'+1)^2\sum_{m=-j}^{j}(-1)^m\times$$
$$\int w(m,\Omega)D_{0M}^{J'}(\Omega)\frac{d\Omega}{8\pi^2}\begin{pmatrix} j & j & J' \\ m & -m & 0\end{pmatrix}\begin{pmatrix} j & j & J' \\ i & -k & M\end{pmatrix},$$

where $i,k=-j,-j+1,\ldots,j$. At the end, we obtain atomic decomposition and Banach frame with atomic bounds $A=B=\frac{1}{2j+1}$.

3.3 *Discrete Spin Tomography*

For spin $s=1$, it is possible to find a finite group instead of $SU(2)$. The corresponding wavelet transform becomes $\mathcal{W}:\hat{\rho}(n,\Psi)=\mathrm{Tr}\ (R_n^{\dagger}(\Psi)\hat{\rho})$. For the choice of the identity operator as a vacuum vector and the test functional $l_0(\hat{\rho})=\mathrm{Tr}\ [\hat{\rho}]$, the inverse wavelet transform becomes

$$\mathcal{MW}=PI\Rightarrow\mathcal{MW}(\hat{\rho})=\sum_{\Psi,n}\frac{1}{P}<\hat{\rho},l_{(n,\Psi)}>b_{(n,\Psi)}.$$

Hence, the tomography for $S=1$ can be written as[7]

$$\hat{\rho}=\frac{1}{4}\sum_{m=-1}^{1}\sum_{j=1}^{7}P(\vec{n}_j,m)K_j(m-\vec{s}.\vec{n}_j)+\frac{1}{4}I,$$

where $P(\vec{n}_j,m)$ is the probability of having outcome **m** which is the result of measuring the operator $\vec{s}\cdot\vec{n}$ and $K_j(m-\vec{s}\cdot\vec{n}_j)$ is a kernel function representation $e^{i\Psi(\vec{s}\cdot\vec{n})}$. Also, we obtain the atomic decomposition and Banach frame with atomic bounds $A=B=4$.

3.4 *Unitary Group Tomography*

By choosing an irreducible square integrable representation of $SU(d)$ group as $U(\Omega)=e^{i\hat{J}\cdot\hat{n}\psi}$, we can obtain the unitary group tomography via wavelet transform in Banach space. The wavelet transform from Banach space with the selection of a vacuum vector b_0 which is equal to identity is given by the expression[8] $\widehat{\rho}(\Omega)=Tr[U^{\dagger}(\Omega)\rho]$. Therefore, tomography relation for unitary group is given by $\rho=\int d\mu(\Omega)Tr[U^{\dagger}(\Omega)\rho]U(\Omega)$. Finally, we obtain atomic decomposition and Banach frame with atomic bounds $A=B=1$.

References

1. H.G. Feichtinger and K.H. Grochenig, *J. Functional Anal.* **86**, 308 (1989).
2. O. Christensen and C. Heil, *Math. Nachr.* **185**, 33 (1997).
3. W. Miller in *Topics in Hormonic Analysis with Applications to Radar and Sonar*, (Lecture note 23 October 2002).
4. M. Paini, arXiv: quant-ph/0002078.
5. G.M. D'Ariano, *Phys. Lett A.* **268**, 151 (2000).
6. V.V. Dodonov and V.I. Man'ko, *Phys. Lett.* A **229**, 335 (1997).
7. G.M. D'Ariano, L. Maccone and M. Paini, *J. Opt. B: Quantum Semicalss. Opt.* **5**, 77 (2003).
8. G. M. D'Ariano, L. Maccone and M. G. A. Paris, *Phys. Lett.* A **276**, 25 (2000).

DIFFERENTIAL GORMS AND WORMS

D. KOCHAN

Department of Theoretical Physics, Faculty of Mathematics, Physics and Informatics
Comenius University, Mlynská dolina F2, 842 48 Bratislava, Slovakia
E-mail: kochan@sophia.dtp.fmph.uniba.sk

We study "higher-dimensional" generalization of differential forms in the framework of supergeometry. From a more conceptual point of view, forms over manifold M are functions on the superspace of maps $\mathbb{R}^{0|1} \to M$ and the action of $Diff(\mathbb{R}^{0|1})$ on forms is equivalent to deRham differential and to degree of forms. Gorms on M are functions on the superspace of maps $\mathbb{R}^{0|2} \to M$ and we study the action of $Diff(\mathbb{R}^{0|2})$ on gorms, it contains more than just degrees and differentials. By replacing 2 with an arbitrary n, we get differential worms.
This stub paper concerns a very slender part of our work with Pavol Ševera, who mainly originated the concept and properties of differential gorms.

1 Gorms as Functions over Iterated Odd Tangent Bundle

[a] *n-times iterated odd tangent bundle* $(\Pi T)^n M$ is defined as the supermanifold of all maps from $\mathbb{R}^{0|n}$ (Cartesian space with n global Grassmann coordinates θ^α) to smooth manifold M. It is well-known (see Refs. [1-3]) that the algebra of functions on $(\Pi T)M$ corresponds to exterior algebra of differential forms on M and related calculus on the forms is equivalent to special supergroups actions on $(\Pi T)M$. Therefore, for the functions over $(\Pi T)^n M$ we adopted wittily name *differential worms* on M, which in the special cases $n = 1, 2, \cdots$ reduce to the names like *differential forms, gorms*[b], $\cdots$. In this paper we will provide a short description of the supermanifold $(\Pi T)^2 M$ and basic properties of the *gorms*, since all higher iterations ($n > 2$) are only a technically more complicated, mainly, due to the presence of larger number of incoming coordinates.

Using any local coordinates x^i on M, we can express an arbitrary map $F \in (\Pi T)^2 M$ (expanding in θ's) in the form

$$x^i[F(\theta^1, \theta^2)] = x^i + \theta^1 \xi_1^i + \theta^2 \xi_2^i + \theta^2 \theta^1 y^i,$$

where $i = 1, \cdots, m = \dim M$, with immediate consequence that $(\Pi T)^2 M$ is $2m|2m$-dimensional supermanifold with even coordinates x^i, y^i and odd coordinates ξ_1^i, ξ_2^i. The global structure of $(\Pi T)^2 M$ is inherited from the atlas of the manifold M, *i.e.* when $x^i \mapsto \tilde{x}^i(x)$, then

$$\xi_\alpha^i \mapsto \tilde{\xi}_\alpha^i = \frac{\partial \tilde{x}^i}{\partial x^j} \xi_\alpha^j, \tag{1}$$

$$y^i \mapsto \tilde{y}^i = \frac{\partial \tilde{x}^i}{\partial x^j} y^j + \frac{\partial^2 \tilde{x}^i}{\partial x^j \partial x^k} \xi_1^j \xi_2^k. \tag{2}$$

Let us stress that odd coordinates ξ_α^i ($\alpha = 1, 2$) transform exactly like two independent sets of co-vectors dx^i and the body of the supermanifold $(\Pi T)^2 M$, defined by $\xi_\alpha^i = 0$ ($\alpha = 1, 2$), is diffeomorphic to TM.

Since $(\Pi T)^2 M$ is a supermanifold of all maps "running" from $\mathbb{R}^{0|2}$ to M, we have a corresponding action of the (super)groups $Diff(\mathbb{R}^{0|2})$ and $Diff(M)$ on it. An arbitrary element $\Phi \in Diff(\mathbb{R}^{0|2})$ is by definition an invertible map from $\mathbb{R}^{0|2}$ to itself and can be expressed as

$$\theta^\alpha \mapsto \tilde{\theta}^\alpha[\Phi(\theta^1, \theta^2)] = \beta^\alpha + A^\alpha_\omega \theta^\omega + \gamma^\alpha \theta^1 \theta^2,$$

where matrix $A \in \mathrm{Gl}_2(\mathbb{R}) = \mathrm{Sl}_2(\mathbb{R}) \times \mathbb{R}^*$. Translation parameters β^α and non-linear terms γ^α are formal Grassmann numbers. The $Diff(\mathbb{R}^{0|2})$ action is simply a composition of the maps, namely, when $\Phi : \mathbb{R}^{0|2} \to \mathbb{R}^{0|2}$ and $F : \mathbb{R}^{0|2} \to M$, then $F \circ \Phi$ is the "new" element from $(\Pi T)^2 M$. Let us analyze the infinitesimal action, *i.e.* the situation when $A = \mathbb{I} + \delta A$ ($\delta A \in \mathrm{gl}_2(\mathbb{R})$), $\beta^\alpha = 0 + \delta\beta^\alpha$, $\gamma^\alpha = 0 + \delta\gamma^\alpha$

- δA, $\delta\beta^\alpha = 0$, $\delta\gamma^\alpha = 0$ generates infinitesimal action:

  $(x^i; \xi_\alpha^i; y^i) \mapsto (x^i; \xi_\alpha^i + (\delta A)^\omega_\alpha \xi_\omega^i; y^i + \mathrm{Tr}(\delta A) y^i)$.

  $\Rightarrow$ $(1 \oplus 3)$ even vector fields over $(\Pi T)^2 M$

$$\mathrm{Deg} = \xi_1^i \partial_{\xi_1^i} + \xi_2^i \partial_{\xi_2^i} + 2y^i \partial_{y^i}, \quad L_+ = \xi_1^i \partial_{\xi_2^i},$$

$$L_3 = \tfrac{1}{2}\left\{\xi_1^i \partial_{\xi_1^i} - \xi_2^i \partial_{\xi_2^i}\right\}, \qquad L_- = \xi_2^i \partial_{\xi_1^i}. \tag{3}$$

---

[a] Talk presented at the XIth Regional Conference on Mathematical Physics, May 2004, Tehran, Iran.

[b] Later we recovered that English word *gormless* ≈ dull, foolish, lacking sense.

$\Rightarrow$ gorms are classified w.r.t. the action of Lie algebra $\mathrm{gl}_2(\mathbb{R})$ by two quantities: **degree** ***deg*** (eigenvalue of Deg field), and **spin**($\blacklozenge, \lozenge$) ($\blacklozenge$ is eigenvalue of total angular momentum $L^2 = L_- L_+ + L_3^2 + L_3$ operator and $\lozenge$ is L_3 field eigenvalue), where $\blacklozenge = 0, \frac{1}{2}, 1, \ldots, \dim M$ and $\lozenge = -\blacklozenge, \ldots, \blacklozenge$.

- $\delta A = 0$, $\delta\beta^\alpha$, $\delta\gamma^\alpha = 0$ generates infinitesimal action

  $$(x^i; \xi^i_\alpha; y^i) \mapsto (x^i + (\delta\beta)^\omega \xi^i_\omega; \xi^i_\alpha + (\delta\beta)^\omega \epsilon_{\omega\alpha} y^i; y^i).$$

  $\Rightarrow$ 2 odd vector fields on $(\Pi T)^2 M$

  $$d_\alpha = \xi^i_\alpha \partial_{x^i} + \epsilon_{\alpha\omega} y^i \partial_{\xi^i_\omega} \quad \alpha = 1, 2\,. \qquad (4)$$

  Basic properties of d fields are: $d_1^2 = 0 = d_2^2$ (nilpotentness) and $d_1 d_2 = -d_2 d_1$, d_α **increase degree** of the gorms, operator doublet $(d_1; d_2)$ forms spinor w.r.t. the action of $\mathrm{sl}_2(\mathbb{R})$, they can be regarded as a natural **gorm-analogs** of the **deRham differential**.

- $\delta A = 0$, $\delta\beta^\alpha = 0$, $\delta\gamma^\alpha$ generates infinitesimal action:

  $$(x^i; \xi^i_\alpha; y^i) \mapsto (x^i; \xi^i_\alpha; y^i + (\delta\gamma)^\omega \xi^i_\omega).$$

  $\Rightarrow$ 2 two odd vector fields on $(\Pi T)^2 M$

  $$R_\alpha = \xi^i_\alpha \partial_{y^i} \quad \alpha = 1, 2\,. \qquad (5)$$

  Basic properties of R fields are: $R_1^2 = 0 = R_2^2$ (nilpotentness) and $R_1 R_2 = -R_2 R_1$, R_α **decrease degree** of the gorms, operator doublet $(R_1; R_2)$ behaves like spinor w.r.t. the action of $\mathrm{sl}_2(\mathbb{R})$, they represent the new **degree decreasing(!) differentials**.

Cohomology of gorms w.r.t. d_α or R_α are naturally isomorphic to deRham cohomology of the manifold M.

An arbitrary element $\Phi \in \mathit{Diff}(M)$ is an invertible map from $M \to M$ and its action on $(\Pi T)^2 M$ is given by (1) and (2). To avoid some technicalities (but also loosing some generalities, for more details see Ref. [4]) we will only shortly conclude that arbitrary vector field $V = V^i(x)\partial_{x^i}$ on M (infinitesimal diffeomorphism) defines canonical sequence of vector fields on the supermanifold $(\Pi T)^2 M$

$$i^V = V^i(x)\partial_{y^i}, \quad j^V_\alpha = [d_\alpha, i^V], \quad \mathcal{L}^V = [d_1, [d_2, i^V]]. \qquad (6)$$

Even vector fields i^V and $\mathcal{L}^V$ on $(\Pi T)^2 M$ (due to their supercommutation relations) are natural analogs of the **interior product** and **Lie derivative** over gorms. Formula $[d_1, [d_2, i^V]]$, expressing $\mathcal{L}^V$ using differentials and interior product, is wittily named as **Cartan gormula**. The odd vector fields j^V_α represent some interstage mixture between interior product and Lie derivative.

Let us remark that setting $\theta^2 = 0$ in the definition of $(\Pi T)^2 M$ leads to the restriction on the subsupermanifold $(\Pi T)M \subset (\Pi T)^2 M$ with coordinates x^i and $\xi^i \equiv \xi^i_1$. Restricted actions of $\mathit{Diff}(\mathbb{R}^{0|2})$ and $\mathit{Diff}(M)$ to $(\Pi T)M$ defines vector fields, $\mathrm{Deg} = \xi^i \partial_{\xi^i}$, $d = \xi^i \partial_{x^i}$ (deRham differential), $i_V = V^i(x)\partial_{\xi^i}$ (interior product) and $\mathcal{L}_V = [d, i_V]$ (Lie derivative), which represent the standard Cartan calculus operations over forms $\Omega(M) \cong C^\infty((\Pi T)M)$.

2 Integration of Gorms and Euler Characteristic

Supermanifold $(\Pi T)^2 M$ admits natural $\mathit{Diff}(M)$-invariant (coordinate-independent) Berezin volume measure $\overline{\mathrm{d}x\mathrm{d}y\mathrm{d}\xi_2\mathrm{d}\xi_1}$. To **integrate a gorm** α **thus means**: expand α in ξ's, take the coefficient in front of $\xi^1_1 \ldots \xi^m_1 \xi^1_2 \ldots \xi^m_2$, and finally integrate[c] it over x's and y's. For example, let's take $M = \mathbb{R}$ and $\alpha(x, y; \xi_1, \xi_2) = e^{-(x^2+y^2)}\xi_1\xi_2$ $\Rightarrow \int \alpha = \int_{(\Pi T)^2 M} \overline{\mathrm{d}x\mathrm{d}y\mathrm{d}\xi_2\mathrm{d}\xi_1}\ \alpha(x, y; \xi_1, \xi_2) = \int_{M\times\mathbb{R}} \mathrm{d}x\mathrm{d}y\ e^{-(x^2+y^2)} = \pi$. The volume measure above is surprisingly also $\mathit{Diff}(\mathbb{R}^{0|2})$-invariant[d], what means that the (super)flow transformations of $(\Pi T)^2 M$ generated by Deg, L's, d's and R's vector fields have a Berezinian equal to unity. This statement is equivalent to the validity of gorm-analogs of the **Stokes theorem**: *for any vector field* $\mathcal{V}$, *which is linear combination of vector fields* (3)-(5) *and for an arbitrary integrable gorm* α, *works*: $\int \mathcal{V}(\alpha) = 0$.

[c] Since the y's integration is over non-compact fibre $\mathbb{R}^m$, only pseudodifferential gorms, which depend non-polynomially on y's with well behavior at infinity, have a reasonable (finite) integral.

[d] The similar asseveration, that natural volume measure over $(\Pi T)^n M$ is $\mathit{Diff}(\mathbb{R}^{0|n})$-invariant, works for all $n > 1$. Somehow pathologically behaves only the case when $n = 1$, where the full $\mathit{Diff}(\mathbb{R}^{0|1})$-invariance of the integral measure is partially broken by the field $\mathrm{Deg} = \xi^i \partial_{\xi^i}$.

In the last part of the paper, we will show the interesting link between gorms and topological properties of the manifold M. Suppose that M is a compact manifold with Riemannian metric tensor $g = g_{ij}(x)\mathrm{d}x^i \otimes \mathrm{d}x^j$, what is equivalent to the fact that g is quadratic function over TM w.r.t. y's, *i.e.* $g = \frac{1}{2}g_{ij}(x)y^iy^j$. Let us consider a general ***deg***= 4 and ***spin***= $(0,0)$ differential gorm γ on M such that: $d_\alpha\gamma = 0$ (closedness) and $g = \gamma|_{\xi_1=0=\xi_2}$ ("metricity"). It is easy to see that γ is uniquely determined by this conditions and the solution simply reads: $\gamma = \frac{1}{2}d_1d_2\left(R_2R_1(g)\right)$. Let's try compute a "metric gorm" γ integral $\int e^{-\gamma}$ (M is compact, g Riemannian $\Rightarrow e^{-\gamma}$ is integrable). Since we are limited by the three pages in this proceedings, we will present only result (for some useful hints and details see Ref. [4]):

$$\int e^{-\gamma} = \int_M \mathrm{Pfaf}(R) \approx \chi(M), \tag{7}$$

where $\mathrm{Pfaf}(R)$ is the *Pfaffian form* of the Riemann curvature tensor defined by the metric g and $\chi(M)$ is topological **Euler characteristic** of the manifold M. It is very easy to prove, by using the "gorms artillery", that $\chi(M)$ is topological invariant of the manifold. To see this, it is enough only infinitesimally change our starting input, namely metric: when $g \mapsto g + \delta g$, then $e^{-\gamma} \mapsto e^{-\gamma} + e^{-\gamma}d_1d_2(R_2R_1\delta g) = e^{-\gamma} + d_1\left\{e^{-\gamma}d_2(\cdots)\right\}$ (since γ is d_1 closed) and consequently $\chi(M) \approx \int e^{-\gamma} \mapsto \int e^{-\gamma} + \int d_1\left\{e^{-\gamma}d_2(\cdots)\right\} = \int e^{-\gamma} \approx \chi(M)$ (the prelast equation is justified by the validity of the Stokes theorem for $\mathcal{V} = d_1$).

Acknowledgments

Author thanks for the support to the VEGA project 1/0250/03, to Professor M. Chaichian for hospitality during the author's stage at the University of Helsinki, to the projects of the Academy of Finland No. 54023 and 104368, to the CIMO fellowship TM-03-1482 and to Professor F. Ardalan for the invitation to the XIth Regional Conference.

References

1. Y.I. Manin in *Gauge Fields Theory and Complex Geometry*, (Springer-Verlag, Berlin, 1988).
2. M. Kontsevich, *Lett. Math. Phys.* **66**, 157 (2003), arXiv: q-alg/9709040.
3. D. Kochan, *Czechoslov. Jour. Phys.* **54**, 177 (2004), arXiv: math.DG/0308232.
4. P. Ševera and D. Kochan, arXiv: math.DG/0307303.

NON-ABELIANIZABLE FIRST CLASS CONSTRAINTS

F. LORAN

Department of Physics, Isfahan University of Technology, Isfahan, Iran
E-mail: loran@cc.iut.ac.ir

We review the Faddeev-Popov method of quantization of first class constraint systems. We show that in the case of a finite set of non-Abelianizable first class constraints, the Faddeev-Popov determinant is vanishing for any choice of subsidiary constraints used as gauge fixing conditions.

1 Constraint System and Singular Lagrangian

[a] A singular Lagrangian $L = L(q_i, \dot{q}_i)$ is by definition a Lagrangian for which the corresponding Hessian matrix W_{ij} defined by the relation

$$W_{ij} = \frac{\partial^2 L}{\partial \dot{q}_i \partial \dot{q}_j}, \tag{1}$$

is not full rank. In terms of equations of motion, one can define a singular Lagrangian as a system for which some of the equations of motion are only some constraints on the coordinates q_i and $\dot{q}_i$'s. The equivalence of these two definitions can be verified by noting that the Euler-Lagrange equation for the coordinate q_i in terms of the Hessian Matrix W_{ij} is given by

$$W_{ij}\ddot{q}_i + V_i(q, \dot{q}) = 0, \tag{2}$$

where

$$V(q, \dot{q}) = \left(\frac{\partial^2 L}{\partial \dot{q}_i \partial q_j} \dot{q}_j - \frac{\partial L}{\partial q_i} \right). \tag{3}$$

Thus, if $\lambda^i_\alpha(q, \dot{q}_i)$ is a null vector of W_{ij}, there exist some constraints on coordinates given by the relation $\lambda^i_\alpha V_i = 0$, as far as $\lambda^i_\alpha W_{ij} = 0$.

One can reformulate constraint system in Hamiltonian formalism. Noting that p_i, the momentum conjugate to q_i is defined by the relation $p_i = \frac{\partial L}{\partial \dot{q}_i}$, one verifies that if $\det W_{ij} = 0$, then the physical phase space is a subset of the phase space (q_i, p_i) as far as there are some constraints on phase space coordinates $\phi_a(q, p) = 0$, obtained from the condition

$$\det \left| \frac{\partial p_i}{\partial \dot{q}_j} \right| = \det W_{ij} = 0.$$

ϕ_a's are called primary constraints. In general, the condition $\frac{d}{dt}\phi_a = 0$ generates some secondary constraints on phase-space coordinates. The Lagrangian constraints are part of these secondary constraints. As an example, consider the free $U(1)$ gauge field theory given by the Lagrangian density $\mathcal{L} = -\frac{1}{4} f_{\mu\nu} f^{\mu\nu}$. Defining the momenta π_μ conjugate to the gauge field a^μ, one verifies that

$$\pi_0 = \frac{\partial \mathcal{L}}{\partial \dot{a}^0} = f_{00}, \tag{4}$$

is vanishing identically. One calls the constraint $\pi_0 = 0$ a primary constraint. Requiring that $\dot{\pi}_0 = \{\pi_0, H\} = 0$, where H is the Hamiltonian, one obtains the Coulomb law $\partial_i \pi_i = 0$ as the secondary constraint.

2 First Class Constraints

Consider a constraint system with a set of constraints $\phi_i(q, p)$ satisfying the algebra

$$[\phi_i, \phi_j] = f_{ij}{}^k \phi_k, \tag{5}$$

where $[\ ,\ ]$ stands for the Poisson bracket. Such a set of constraints is called first class constraints. Dirac showed that first class constraints are generators of gauge transformation[1]. For example, in $U(1)$ gauge theory one can verify that the gauge transformation $a_\mu \to a_\mu + \partial_\mu \epsilon$ is generated by the generator of gauge transformation $G_\epsilon = \int (\dot{\epsilon}\phi_1 + \epsilon\phi_2)$ by the relation $\delta a^\mu = [a^\mu, G_\epsilon]$ where $\phi_1 = \pi_0$ and $\phi_2 = \partial_i \pi_i$.

3 Quantization of First Class Systems

Since a physical observable should be gauge invariant, one defines physical sates by the relation $\phi_i|\text{phys.}\rangle = 0$. In Faddeev-Popov method which is the generalization of Feynman path integral method to constraint systems, one defines the partition function by

$$Z = \int [d\mu] e^{-iS}, \tag{6}$$

[a] Talk presented at the XIth Regional Conference on Mathematical Physics, May 2004, Tehran, Iran.

where the Faddeev-Popov measure[2]

$$[d\mu] = [dp][dq](\det M)\delta(\phi_i)\delta(\omega_i).$$

ω_i's are some subsidiary constraints, called the gauge fixing conditions. These new constraints should be chosen such that the Faddeev-Popov determinant $\det M \neq 0$, where $M_{ij} = [\phi_i, \omega_j]$. An example for gauge fixing conditions is the Coulomb gauge $\nabla.\vec{a} = 0$ and the Lorenz gauge, $\partial_\mu a^\mu = 0$, known in $U(1)$ gauge field theory. In what follows we show that in the case of a finite set of non-Abelian first class constraints, the Faddeev-Popov determinant is vanishing for any choice of subsidiary constraints[3]. In other words, we show that there does not exist a set of subsidiary constraints ω_i's that eliminates the gauge transformation generated by non-Abelian first class constraints. The proof is as follows. First, note that

$$(\det M_{ij}) = \det\left(\frac{\partial\phi_i}{\partial z^\mu} J^{\mu\nu} \frac{\partial\omega_i}{\partial z^\nu}\right),$$

is vanishing for any chosen set of ω_i's if

$$\text{rank}\left(\frac{\partial\phi_i}{\partial z^\mu}\right) < N.$$

Here, N is the number of first class constraints, *i.e.*, $i = 1, \cdots, N$. z^μ's are phase space coordinates and $J_{\mu\nu}$ is the symplectic two-form defined by the relation $J^{\mu\nu} = [z^\mu, z^\nu]$. Second, note that if ϕ_i's are non-Abelian, *e.g.*, if they are elements of $so(N)$ Lie algebra, then $\text{rank}(\frac{\partial\phi_i}{\partial z^\mu}) < N$. Because if $\text{rank}(\frac{\partial\phi_i}{\partial z^\mu})$ was equal to N then using the theorem of implicit function, one could solve the constraints $\phi_i = 0$ in terms of a subset of coordinates z^i and could obtain an equivalent set of Abelian constraints[4] $\tilde{\phi}_i = z^i - z^i(z^a)$. By equivalence of two sets of constraints, one means that they give equivalent gauge groups and define equivalent physical phase space. Since a non-Abelian gauge group can not be equivalent to an Abelian gauge group, it is not possible even locally to solve non-Abelian constraints ϕ_i's and obtain equivalent Abelian constraints $\tilde{\phi}_i$'s. Therefore, $\text{rank}(\frac{\partial\phi_i}{\partial z^\mu}) < N$ and consequently, the Faddeev-Popov determinant is vanishing for any choice of gauge fixing conditions.

As an example, we show that the Faddeev-Popov determinant is vanishing for any chosen set of subsidiary constraints for $SO(3)$ gauge invariant model[3]: Consider the constraints of $SO(3)$ gauge model[5], $L_i = \epsilon_{ijk} x_j p_k$, $i = 1, 2, 3$. Assume three arbitrary subsidiary constraints ω_i's. Since L_i's are non-Abelianizable, the Faddeev-Popov determinant $\det(\{\omega_i, L_j\})$ is vanishing as can be verified as follows. Using the equality

$$\det\begin{pmatrix} a_{11} & a_{12} & a_{13} \\ a_{21} & a_{22} & a_{23} \\ a_{31} & a_{32} & a_{33} \end{pmatrix} = \epsilon_{ijk} a_{1i} a_{2j} a_{3k}, \qquad (7)$$

one finds that

$$\det(\{\omega_i, L_j\}) = \epsilon_{ijk}\{\omega_1, L_i\}\{\omega_2, L_j\}\{\omega_3, L_k\}$$
$$= -\epsilon_{ijk}\epsilon_{ia_1b_1}\epsilon_{ja_2b_2}\epsilon_{ka_3b_3} \prod_{c=1}^{3} \left(\frac{\partial\omega_c}{\partial x_{a_c}} x_{b_c} + \frac{\partial\omega_c}{\partial p_{a_c}} p_{b_c}\right). \qquad (8)$$

Two generic terms in the above sum can be distinguished:

$$P = \epsilon_{ijk}\epsilon_{iaa'}\epsilon_{jbb'}\epsilon_{kcc'} \left(\frac{\partial\omega_1}{\partial x_a} x_{a'}\right)\left(\frac{\partial\omega_2}{\partial x_b} x_{b'}\right)\left(\frac{\partial\omega_3}{\partial x_c} x_{c'}\right), \qquad (9)$$

and

$$Q = \epsilon_{ijk}\epsilon_{iaa'}\epsilon_{jbb'}\epsilon_{kcc'} \left(\frac{\partial\omega_1}{\partial x_a} x_{a'}\right)\left(\frac{\partial\omega_2}{\partial x_b} x_{b'}\right)\left(\frac{\partial\omega_3}{\partial p_c} p_{c'}\right). \qquad (10)$$

To calculate P, one realizes three generic terms:

$$P_1 = \left(\frac{\partial\omega_1}{\partial x}\right)\left(\frac{\partial\omega_2}{\partial x}\right)\left(\frac{\partial\omega_3}{\partial x}\right)\epsilon_{ijk}\epsilon_{i1a}\epsilon_{j1b}\epsilon_{k1c}(x_a x_b x_c),$$
$$P_2 = \left(\frac{\partial\omega_1}{\partial x}\right)\left(\frac{\partial\omega_2}{\partial x}\right)\left(\frac{\partial\omega_3}{\partial y}\right)\epsilon_{ijk}\epsilon_{i1a}\epsilon_{j1b}\epsilon_{k2c}(x_a x_b x_c),$$
$$P_3 = \left(\frac{\partial\omega_1}{\partial x}\right)\left(\frac{\partial\omega_2}{\partial y}\right)\left(\frac{\partial\omega_3}{\partial z}\right)\epsilon_{ijk}\epsilon_{i1a}\epsilon_{j2b}\epsilon_{k3c}(x_a x_b x_c). \qquad (11)$$

$P_1 = 0$ because here, $(i, j, k) \in \{2, 3\}$ and consequently ϵ_{ijk} is vanishing. $P_2 = 0$ because $\epsilon_{i1a}\epsilon_{j2_b} x_a x_b$ is symmetric under $a \leftrightarrow b$ and consequently under $i \leftrightarrow j$, though $\epsilon_{ijk} = -\epsilon_{jik}$. In addition, $P_3 = -yzx + zxy = 0$ [the first term corresponds to $a = 2$ and the second terms corresponds to $a = 3$ in (11)].

Q is the sum of four generic terms:

$$Q_1 = \frac{\partial\omega_1}{\partial x}\frac{\partial\omega_2}{\partial x}\frac{\partial\omega_3}{\partial p_x}\epsilon_{ijk}\epsilon_{i1a'}\epsilon_{j1b'}\epsilon_{k1c'} x_{a'} x_{b'} p_{c'},$$
$$Q_2 = \frac{\partial\omega_1}{\partial x}\frac{\partial\omega_2}{\partial x}\frac{\partial\omega_3}{\partial p_y}\epsilon_{ijk}\epsilon_{i1a'}\epsilon_{j1b'}\epsilon_{k2c'} x_{a'} x_{b'} p_{c'},$$
$$Q_3 = \frac{\partial\omega_1}{\partial x}\frac{\partial\omega_2}{\partial y}\frac{\partial\omega_3}{\partial p_x}\epsilon_{ijk}\epsilon_{i1a'}\epsilon_{j2b'}\epsilon_{k1c'} x_{a'} x_{b'} p_{c'},$$
$$Q_4 = \frac{\partial\omega_1}{\partial x}\frac{\partial\omega_2}{\partial y}\frac{\partial\omega_3}{\partial p_z}\epsilon_{ijk}\epsilon_{i1a'}\epsilon_{j2b'}\epsilon_{k3c'} x_{a'} x_{b'} p_{c'}. \qquad (12)$$

Q_1 and Q_2 are vanishing because under $i \leftrightarrow j$, $\epsilon_{i1a'}\epsilon_{j1b'}x_{a'}x_{b'}$ is symmetric but ϵ_{ijk} is antisymmetric. Using the identity, $\epsilon_{ijk}\epsilon_{j2b'} = -\delta_{i2}\delta_{kb'} + \delta_{ib'}\delta_{k2}$, one can show that Q_3 and Q_4 are some combination of L_i's. Therefore, Q is also vanishing on the constraint surface.

As far as the phase space of non-Abelian gauge theories like $SU(2)$ Yang-Mills theory becomes finite dimensional on compactified lattices, we anticipate that the above theorem can be considered as an explanation for the appearance of Gribov ambiguities in such theories.

References

1. P.A.M. Dirac, *Can. J. Math.* **2**, 129 (1950); *Proc. R. Soc. London Ser. A* **246**, 326 (1958); in *Lectures on Quantum Mechanics*, (Yeshiva University Press, New York, 1964).
2. L.D. Faddeev and V.N. Popov, *Phys. Lett.* B **25**, 30 (1967).
3. F. Loran, arXiv: hep-th/0303014.
4. M. Henneaux and C. Teitelboim in *Quantization of Gauge System*, (Princeton University Press, Princeton, New Jersey, 1992).
5. J. Govaerts and J.R. Klauder, *Annals Phys.* **274**, 251 (1999), arXiv: hep-th/9809119.

QUANTUM DEFORMATIONS OF RELATIVISTIC SYMMETRIES: SOME RECENT DEVELOPMENTS

J. LUKIERSKI
Institute for Theoretical Physics, University of Wrocław
Pl. Maxa Borna 9, 50-204 Wrocław, Poland
E-mail: lukier@ift.uni.wroc.pl

We first discuss different versions of noncommutative space-time and the corresponding appearance of quantum space-time groups. Further, we consider the relation between quantum deformations of relativistic symmetries and so-called doubly special relativity (DSR) theories.

1 Introduction

[a]Quantum deformations of Lie algebras and Lie group were motivated by quantum universe scattering method and introduced in 1980's as noncocommutative Hopf algebras[1-3]. Subsequently, the notion of quantum symmetries was tried for many symmetries occurring in physics, in particular for the basic relativistic symmetries, described by Poincaré algebra and Poincaré group, as well as anti-de-Sitter (AdS), de-Sitter (dS) and conformal symmetries.

Because the four-momenta generators, in contrast to the Lorentz rotations, are dimensionful, they introduce the notion of scaling into the space-time algebras. One should distinguish two types of quantum deformations

i) The first type includes a dimensionless deformation parameter q, which is invariant under rescaling of the four-momenta. The prototypes of such deformations are provided by Drinfeld-Jimbo (DJ) deformations[1,4] of AdS, dS or conformal algebra. One can show[5,6] that no DJ quantum deformation of Poincaré algebra, obtained by the extension of DJ deformation for the Lorentz subalgebra, exists.

The dimensionless deformation parameter of space-time symmetries appears less attractive from the point of view of physical applications. It is agreed that the quantum symmetries and noncommutative space-time coordinates should become relevant for very small distances (*e.g.* at Planck length $l_p \simeq 10^{-33}$cm). We conclude that the dimensional parameter in quantum algebra structure will characterize the distances at which the notions of classical geometry are not valid. Therefore it should be introduced.

ii) The second type of deformations of space-time symmetries involves a built-in elementary length, or elementary mass. Originally, such a deformation has been proposed in 1991 (see Ref. [7]), with the deformation parameter κ – the classical limits is provided by limit $\kappa \to \infty$. It was called the κ-deformation. Still it is not clear how to relate, by rigorous proof, the parameter κ with the Planck mass M_P ($M_P \simeq 10^{19}$GeV), but, due to the quantum gravity origin of noncommutativity of space-time coordinates, it is believed that they are linked very closely, and quite often are assumed to be identical.

2 Noncommutativity of Space-Time and Quantum Groups

The need of quantum space-time symmetries with dimensionful deformation parameter can be seen clearly from the noncommutativity of space-time coordinates. The general relations can be written as follows (see *e.g.* Ref. [8])

$$[\widehat{x}_\mu, \widehat{x}_\nu] = \frac{1}{\kappa^2} F(\kappa x) = \frac{1}{\kappa^2}\Theta^{(0)}_{\mu\nu} + \frac{1}{\kappa}\Theta^{(1)\nu}_{\mu\nu}\widehat{x}_\nu + \Theta^{(2)\rho\theta}_{\mu\nu}\widehat{x}_\rho\widehat{x}_\tau + \dots\,, \quad (1)$$

where we introduced the parameter κ in order to express the noncommutativity in terms of dimensionless coordinates $y_k = \kappa x_\mu$.

Let us consider the special cases with only one constant tensor $\Theta^{(k)\rho_1\dots\rho_k}_{\mu\nu} \neq 0$.

1) $\Theta^{(0)}_{\mu\nu} \neq 0, \quad \Theta^{(k)\rho_1\dots\rho_k}_{\mu\nu} = 0, \quad k = 1,2,3\cdots$.
This example was studied extensively, recently; the Poincaré symmetries are broken by a constant tensor but remain classical. Such a form of deformed space-time was obtained by Seiberg and Witten[9] by

[a]Talk presented at the XIth Regional Conference on Mathematical Physics, May 2004, Tehran, Iran.

considering the space-time manifold as described by the D-brane world volume in the presence of constant tensor field $B\mu\nu$. We recall that such a field is necessary for consistency of supergravity framework in $D=10$.

2) $\Theta_{\mu\nu}^{(1)\rho} \neq 0$, $\Theta_{\mu\nu}^{(k)\rho_1\cdots\rho_k} = 0$, $k = 0, 2, 3, 4\cdots$. In this case, we obtain the Lie-algebraic form of deformed space-time algebra. It appears that the κ-Minkowski space-time, obtained in the framework of standard κ-deformations of Poincaré symmetries[10−12] belongs to such a class of new theories. In 1996 the generalized κ-deformation of Poincaré symmetries was introduced[13], with the choice

$$\Theta_{\mu\nu}^{(1)\rho} = \frac{i}{\kappa}\left(a_\mu \delta_\nu{}^\rho - a_\nu \delta_\mu{}^\rho\right), \tag{2}$$

where a_μ denotes constant four-vector.
From (2), it follows that the noncommuting quantum direction in Minkowski space is described by the coordinate $\widehat{y} = a^\mu x_\mu$. It has been also shown that the classical r-matrix corresponding to (2) satisfies

- the modified Yang-Baxter (YB) equation, if $a_\mu^2 \neq 0$,
- the classical YB equation, if $a_\mu^2 = 0$.

The quantum deformations of relativistic symmetries with quantized light-cone direction ($a_\mu^2 = 0$) was firstly described by Ballesteros *et al.*[14] and was called null-plane quantum Poincaré symmetries. Further, it has been shown[15] that in such a case the quantization can be described by the twisting procedure[16,17]. If $\Theta_{\mu\nu}^{(1)\rho} \neq 0$, the translations $\widehat{v}_\mu$

$$\widehat{x}_\mu \longrightarrow \widehat{x}'_\mu = \widehat{x}_\mu + \widehat{v}_\kappa\,, \tag{3}$$

described by the coproduct of $\widehat{x}_\mu$, are also noncommutative

$$[\widehat{v}_\mu, \widehat{v}_\nu] = \Theta_{\mu\nu}^{(1)\rho}\widehat{v}_\rho\,. \tag{4}$$

The extension of noncommutative translations (4) to quantum Poincaré group is only possible for particular choices of $\Theta_{\mu\nu}^{(1)\rho}$, in particular for the one given by (2). The general classification of quantum Poincaré groups has been considered by Podleś and Woronowicz[18].

3) $\Theta_{\mu\nu}^{(2)\rho\tau} \neq 0$, $\Theta_{\mu\nu}^{(k)\rho_1\cdots\rho_k} = 0$, $k = 0, 1, 3, 4\cdots$. In this case the relation (1) does not contain any dimensionful parameter and it describes the quantum deformation of relativistic symmetries with dimensionless deformation parameter (*e.g* if the Lorentz sector is described by Drinfeld-Jimbo deformation). Such deformed space-time framework is described by braided quantum symmetries, because the quantum translations (3) do satisfy the relations

$$[\widehat{v}_\mu, \widehat{v}_\nu] = \Theta_{\mu\nu}^{(2)\rho\tau}\widehat{v}_\rho\widehat{v}_\tau\,, \tag{5}$$

but also one has to assume that

$$[\widehat{x}_\mu, \widehat{x}_\nu] \neq 0\,. \tag{6}$$

The first example of braided quantum Poincaré group has been presented by Majid[5].

3 Nonlinear Realizations of Relativistic Symmetries Versus Quantum Deformations

There are two sources of the modification of relativistic symmetries (see *e.g.* Ref. [19]).

i) One can change nonlinearly the basis of classical Poincaré algebra

$$\begin{aligned}
[M_{\mu\nu}^{(0)}, M_{\rho\tau}^{(0)}] &= i(\eta_{\mu\rho} M_{\nu\tau}^{(0)} + \ldots\,, \\
[M_{\mu\nu}^{(0)}, P_\rho^{(0)}] &= i(\eta_{\mu\rho} P_\nu^{(0)} - \eta_{\nu\rho} P_\mu^{(0)}\,, \\
[P_\mu^{(0)}, P_\nu^{(0)}] &= 0\,,
\end{aligned} \tag{7}$$

by introducing the deformation map – the invertible nonlinear functions of the generator. An important special class of deformation maps described only the change of four momentum basis

$$P_\mu^{(0)} \longrightarrow P_\mu = P_\mu(P^{(0)};\kappa) = F_\mu{}^\nu\left(\frac{P^{(0)}}{\kappa}\right) P_\nu^{(0)}\,, \tag{8}$$

while the Lorentz generators remain unchanged ($M_{\mu\nu} = M_{\mu\nu}^{(0)}$).
Introducing the inverse deformation map

$$P_\mu^{(0)} = P_\mu^{(0)}(P;\kappa) = \widetilde{F}_\mu{}^\nu\left(\frac{P}{\kappa}\right) P_\nu\,, \tag{9}$$

we see that the mass Casimir is modified

$$C_1 = P_\mu^{(0)}\, P^{(0)\mu} = P_\mu^{(0)}\left(\frac{P}{\kappa}\right) P_\mu^{(0)\mu}\left(\frac{P}{\kappa}\right) = m_0^2\,, \tag{10}$$

i.e. we obtained deformed nonlinear energy-momentum dispersion relation. Other consequence of the deformation map is the nonlinear modification of energy-momentum addition and conservation laws. The primitive coproduct for the generators $P_\mu^{(0)}$ is replaced by

$$\Delta(P_\mu) = P_\mu\left(P_\mu^{(0)} \otimes 1 + 1 \otimes P_\mu^{(0)}, \kappa\right)\,. \tag{11}$$

Denoting for two-particle system

$P_\mu(i)$ - the four-momenta of i-th particle ($i = 1, 2$),

$P_\mu(1,2)$ - the four-momenta of 2-particle system,

and using (10), (8) one can express the coproduct (11) as describing nonlinear energy-momentum addition law[20,21]

$$P_\mu(1,2) = P_\mu\left(P_\mu^{0)}(P(1);\kappa) + P_\mu^{0)}(P(1);\kappa);\kappa\right). \tag{12}$$

The composition law (12) is symmetric, what indicates that we are dealing with classical Lorentz symmetries nonlinearly realized in the four-momentum sector.

The choice of the deformation map which provides the deformed mass Casimir in bicrossproduct basis of κ-deformed Poincaré algebra[11,12]

$$C_1 = \left(2\kappa \sinh \frac{P_0}{2\kappa}\right)^2 - e^{-\frac{P_0}{\kappa}} \overrightarrow{P}^2 = M^2\,, \tag{13}$$

leads to Doubly Special Relativity theory (DSR) of Amelino-Camelia *et al.* (see *e.g.* Ref. [22]). Such a choice of deformed mass Casimir implies maximal value of three-momentum, if energy $\frac{E}{c} = P_0 \to \infty$. Indeed, in simple case $M^2 = 0$ one gets

$$\overrightarrow{P}^2 = \kappa^2\left(1 - e^{-\frac{P_0}{\kappa}}\right)^2 \underset{P_0\to\infty}{\longrightarrow} \kappa^2, \tag{14}$$

i.e., we obtain the operational definition of mass-like deformation parameter κ as maximal possible value of the three-momentum. It appears that selecting properly the deformation map (8) and corresponding nonlinear mass Casimirs (10), one can introduce three different variants of DSR theories[23].

ii) One can modify relativistic symmetries by introducing quantum deformations of Hopf algebra describing Poincaré symmetries.

The quantum deformations of relativistic symmetries can be easily distinguished from the classical relativistic symmetries in nonlinear disguise, if we observe that,

- quantum deformations imply nontrivial bialgebra structure, described by classical r-matrix ($I_i \equiv (M_{\mu\nu}^{(0)}, P_\mu^{(0)})$; $I_i \wedge I_j \equiv I_i \otimes I_j - I_j \otimes I_i$)

$$r = a^{ij}\, I_i^{(0)} \wedge I_j^{(0)}\,. \tag{15}$$

The classical r-matrix describes infinitesimal deformation of classical coproduct $\Delta^{(0)}$

$$\Delta(I_i^{(0)}) = \Delta^{(0)}(I_i^{(0)}) + \xi[r, \Delta^{(0)}(I_i^{(0)})] + \mathcal{O}(\xi^2)\,. \tag{16}$$

The formula (16) implies that the coproduct is not symmetric.

- The relativistic symmetries are quantum-deformed, the space-time coordinates are not commuting, contrary to the case of DSR framework.

We see therefore that there are simple criteria to distinguish between the modification due to change of classical basis and the one following from genuine quantum deformations [,].

The formula (16) can be "integrated" to arbitrary values of deformation parameter ξ, if we introduce the similarity transformation

$$\Delta_F(I_i^{(0)}) = F \circ \Delta^{(0)}(I_i^{(0)}) \circ F^{-1}\,, \tag{17}$$

where $F = F_i^{(1)} \otimes F_i^{(2)}$ is the twist function with the following linear term in the power expansion in ξ

$$F = 1 \otimes 1 + \xi\, a^{ij}\, I_i \otimes I_j + \mathcal{O}(\xi^2)\,. \tag{18}$$

The twist function F leads to associative coproducts (17), if it satisfies the equation

$$F_{12}\left(\Delta^{(0)} \otimes 1\right) F = F_{23}\left(1 \otimes \Delta^{(0)}\right) F\,. \tag{19}$$

Expanding (19) in powers of ξ, one gets from the bilinear terms the classical Yang-Baxter equation. Further, twist quantization modifies the converse (antipode) as follows

$$S_F(I_i^{(0)}) = u S^{(0)}(I_i^{(0)}) u^{-1}\,, \tag{20}$$

where $u = F_i^{(1)} \cdot S F_i^{(2)}$.

The twist quantization changes only the coproducts and coinverse – the classical Lie algebra relations and the counit remain unmodified.

4 Quantum Deformations of AdS and Conformal Algebras

Drinfeld twist quantization method can be applied to any deformation described infinitesimally by classical r-matrix satisfying CYBE. Recently, the classical r-matrices for $O(3,2)$ and $O(4,2)$ algebras, with generators belonging to the Borel subalgebra B_+, were explicitly written down and subsequently these bialgebras were quantized[24,15].

Let us consider as an example $O(3,2)$ algebra. If we introduce the Cartan-Weyl basis for $O(3,2) \simeq Sp(4)$ (see *e.g.* Ref. [7])

$$\underbrace{h_1, h_2}_{\substack{\text{Cartan}\\ \text{generators}}}\,,\ \underbrace{e_{\pm1}, e_{\pm2}}_{\substack{\text{simple root}\\ \text{generators}}}\,,\ \underbrace{e_{\pm3}, e_{\pm4}}_{\substack{\text{composite root}\\ \text{generators}}}\,, \tag{21}$$

the most general classical $O(3,2)$ r-matrix support in $B_+ \otimes B_+$ is the following

$$r = \alpha[(2h_1 + h_2) \wedge e_4 + 2e_1 \wedge e_2] + \xi h_2 \wedge e_2 + \rho e_2 \wedge e_4 \,. \qquad (22)$$

The corresponding twist function has been calculated in Ref. [24] and is the product of four factors: Jordanian twist, extended Jordanian twist, deformed Jordanian twist and Reshetikhin twist.

It should be pointed out that the generators of $O(3,2)$ can be physically assigned to $D = 3$ conformal or $D = 4$ AdS algebra. In first conformal case, one can show that the parameters α, ξ and ρ are dimensionful, and we arrive at the κ-deformation of $D = 3$ conformal algebra[24]. The assignment of $D = 4$ AdS generators leads to different conclusions: The deformation parameters can remain dimensionless because the role of dimensionful parameter is taken over by the AdS radius. The physical interpretation of the twist quantization of $O(4,2)$ algebra, as $D = 4$ conformal algebra, has been considered in Ref. [15], where all classical r-matrices with support in Borel subalgebra were quantized. In particular, in Ref. [15] the light-cone κ-deformation of Poincaré algebra (see (2) with $a_\mu^2 = 0$) has been extended to the particular κ-deformation of $D = 4$ conformal algebra. The alternative physical interpretation of twisted $O(4,2)$ as quantum $D = 5$ AdS algebra can be found in Ref. [25].

5 Final Remarks

The formalism of twist quantization has been recently extended to all classical superalgebras (see Ref. [26]), in particular to orthosymplectic superalgebras $OSp(n;2m)$. For $n = 1$ and $m = 2$, one obtains in this way new deformations of $D = 4$ AdS superalgebra. Subsequently, using suitable contraction method, one can obtain twist quantization of $D = 4$ Poincaré superalgebra.

References

1. V.G. Drinfeld in *Quantum Groups*, (Proc. Int. Congress of Mathematics, Berkeley, USA, 1986).
2. S.L. Woronowicz, *Comm. Math. Phys.* **111**, 613 (1987).
3. L.D. Faddeev, N.Yu. Reshetikhin and L.A. Takhtayan, *Leningrad Math. Journ.* **1**, 193 (1990).
4. L.Faddeev, N. Resetikhin and L. Takhtajan, *Alg. Anal.* **1**, 178 (1990).
5. S. Majid, *J. Math. Phys.* **34**, 2045 (1993).
6. P. Podleś and S.L. Woronowicz in *Proceedings of First Caribbean School on Mathematics and Theoretical Physics, Guadeloupe 1993*, ed. R. Coquereaux (World Scientific, Singapore, 1995).
7. J. Lukierski, A. Nowicki, H. Ruegg and V.N. Tolstoy, *Phys. Lett.* B **264**, 331 (1991).
8. P. Kosiński, J. Lukierski and P. Maślanka, *Czech. J. Phys.* **50**, 1283 (2000); *Phys. Atom. Nucl.* **64**, 2139 (2001).
9. N. Seiberg, E. Witten, *JHEP* **9909**, 032 (1999), arXiv: hep-th/9908142.
10. S. Zakrzewski, *J. Phys.* A **27**, 2075 (1994).
11. S. Majid and H. Ruegg, *Phys. Lett.* B **334**, 348 (1994).
12. J. Lukierski, H. Ruegg and W.J. Zakrzewski, *Ann. Phys.* **243**, 90 (1995), arXiv: hep-th/9312153.
13. P. Kosiński and P. Maślanka in *From Quamtum Field Theory to Quantum Groups*, ed. B. Jancewicz and J. Sobczyk (World Scientific, Singapore, 1996, p. 41); arXiv: q-alg/9512018.
14. A. Balesteros, F.J. Herranz, M.A. del Olmo and M. Santander, *Phys. Lett.* B **351**, 137 (1995).
15. V. Lyakhowski, J. Lukierski and M. Mozrzymas, *Phys. Lett.* B **538**, 375 (2002).
16. V. Drinfeld, *Dokl. Acad. Nauk SSSR* **273**, 531 (1983).
17. P.P. Kulish, V.D. and A.I. Mudrov, *J. Math. Phys.* **40**, 4569 (1999).
18. P. Podleś and S.L. Woronowicz, *Commun. Math. Phys.* **178**, 61 (1996), arXiv: hep-th/9412059.
19. J. Lukierski and A. Nowicki, *Czech. J. Phys.* **52**, 1261 (2002), arXiv: hep-th/0209017.
20. J. Lukierski and A. Nowicki, *Int. J. Mod. Phys.* A **18**, 7 (2003), arXiv: hep-th/0203065.
21. J. Judes and M. Visser, arXiv: qr-qc/0205067.
22. G. Amelino-Camelia, *Phys. Lett.* B **510**, 255 (2001); *Int. J. Mod. Phys.* D **11**, 35 (2002).
23. J. Lukierski and A. Nowicki, arXiv: hep-th/0210111.
24. V. Lyakhovski, J. Lukierski and M. Mozrzymas, *Mod. Phys. Lett.* A **18**, 753 (2003).
25. A. Ballesteros, N.R. Bruno and F.J. Herranz, *Phys. Lett.* B **574**, 273 (2003).
26. V. Tolstoy, arXiv: math.QA/0402433.

EXACTLY SOLVABLE FINITE DIFFERENCE MODELS OF LINEAR HARMONIC OSCILLATOR

S. M. NAGHIEV

Institute of Physics, Azerbaijan National Academy of Sciences
H. Javid Ave. 33, Baku AZ-1143, Azerbaijan
E-mail: azhep@physics.ab.az

R. M. IMANOV

Ganja State University, Faculty of Physics
A. Camil Str. 1, 374700, Ganja, Azerbaijan

The purpose of this talk is to review the results of investigations on the exactly solvable models of the linear harmonic oscillator, described by the finite-difference equations. Their wave functions are expressed through the orthogonal polynomials of a discrete variable. It is shown that, these models in the limit $h \to 0$, coincide with the well-known Hermite oscillators. Here, h is the step of difference differentiation.

1 Introduction

[a]An oscillator is called *harmonic* when its oscillation period is independent of its energy. In quantum theory, this statement leads to its characterization by a Hamiltonian operator whose energy spectrum is discrete, lower-bound, and equally spaced[1],

$$H\psi_n = E_n\psi_n, \qquad E_n \sim n + \text{constant}, \tag{1}$$

with $n = 0, 1, 2, \cdots$. Within the framework of Lie theory, this further indicates that only a few choices of operators and Hilbert spaces are available if the Hamiltonian operator is incorporated into some Lie algebra of low dimension.

If we relax the strict Schrödinger quantization rule, we find a family of harmonic oscillator models characterized by Hamiltonians that are *difference* operators (rather than *differential* operators). Their spectrum is also given by (1), with n either unbounded, or with an upper bound N. The wavefunctions are still continuously defined on intervals either unbounded or bounded, but the governing equation will relate their values only at discrete, equidistant points of he space; the Hilbert space of wavefunctions will then also have discrete measure. Thus "space" appears to be *discrete* rather than continuous.

2 Hermite, Charlier, Kravchuk, Meixner-Pollaczek and Meixner Oscillators

Here we collect the basic facts on the Hermite, Charlier, Kravchuk, Meixner-Pollaczek and Meixner oscillators. We introduce their Hamiltonian operators and wavefunctions, as well as raising and lowering operators for each oscillator model.

Hermite (quantum-mechanical) oscillator: The linear harmonic oscillator in nonrelativistic quantum mechanics is governed by the well-known Hamiltonian

$$H^{\mathrm{H}}(\xi) = \frac{\hbar\omega}{2}(\xi^2 - \partial_\xi^2) = \hbar\omega\,[a^+a + 1/2]\,, \tag{2}$$

where $\xi = \sqrt{m\omega/\hbar}\,x$ is a dimensionless coordinate. We indicate $\partial_\xi = d/d\xi$, and the creation and annihilation operators are defined as usual

$$a^+ = \frac{1}{\sqrt{2}}(\xi - \partial_\xi)\,, \quad a = \frac{1}{\sqrt{2}}(\xi + \partial_\xi)\,, \qquad [a,\, a^+] = 1. \tag{3}$$

Eigenfunctions of the Hamiltonian (2) are expressed in terms of the *Hermite* polynomials $H_n(\xi)$, $n = 0, 1, 2, \cdots$.

Charlier oscillator: A difference (or discrete) analogue of the linear harmonic oscillator (2), can be built on the half-line in terms of the *Charlier* polynomials[2] $C_n(x; \mu)$, for any fixed $\mu > 0$ and $n = 0, 1, 2, \cdots$.

---

[a]Talk presented by S.M.N. at the XIth Regional Conference on Mathematical Physics, May 2004, Tehran, Iran.

The Hamiltonian for the Charlier oscillator model is a difference operator[2]

$$H^{\mathrm{C}}(\xi) = \\ = \hbar\omega\Big[2\mu + \tfrac{1}{2} + \xi/h_1 - \sqrt{\mu(\mu+1+\xi/h_1)}\, e^{h_1\partial_\xi} \\ -\sqrt{\mu(\mu+\xi/h_1)}\, e^{-h_1\partial_\xi}\Big], \quad (4)$$

where by definition

$$e^{\pm y\partial_x} f(x) = f(x \pm y), \quad (5)$$

is a shift operator by y with the step $h_1 = 1/\sqrt{2\mu}$.

As in the nonrelativistic case (2), it is possible to factorize[2] the Hamiltonian (4)

$$H^{\mathrm{C}}(\xi) = \hbar\omega\,(b^+ b + \tfrac{1}{2}), \quad (6)$$

by means of the difference operators

$$b = \sqrt{\mu+1+\xi/h_1}\, e^{h_1\partial_\xi} - \sqrt{\mu}, \\ b^+ = \sqrt{\mu+\xi/h_1}\, e^{-h_1\partial_\xi} - \sqrt{\mu}. \quad (7)$$

These operators satisfy the Heisenberg commutation relation

$$[b, b^+] = 1. \quad (8)$$

Kravchuk oscillators: Another difference analogue of the harmonic oscillator[3] can be built on the finite interval $[0, N]$, where N is some positive integer, in terms of the Kravchuk polynomials $K_n(x; p, N)$.[4] This is a family of polynomials, parametrized by $0 < p < 1$, of degree $n = 0, 1, 2, \cdots$ in the variable x.

The corresponding Kravchuk oscillator Hamiltonian[3] is a difference operator with step $h_2 = \sqrt{2Npq}$,

$$H^{\mathrm{K}}(\xi) = \hbar\omega\,\big(2p(1-p)N + \tfrac{1}{2} + (\tfrac{1}{2} - p)\,\xi/h_2 \\ -\sqrt{p(1-p)}\,\left[\alpha(\xi)\, e^{h_2\partial_\xi} + \alpha(\xi - h_2)\, e^{-h_2\partial_\xi}\right]\big), \quad (9)$$

$$\alpha(\xi) = \sqrt{(qN - \xi/h_2)(pN + 1 + \xi/h_2)}. \quad (10)$$

Now, it has been shown in [4] that the difference operators

$$A(\xi) = (1-p)\,\alpha(\xi)\, e^{h_2\partial_\xi} - p e^{-h_2\partial_\xi}\,\alpha(\xi) \\ + \sqrt{p(1-p)}\,[(2p-1)N + 2\xi/h_2]\,, \quad (11)$$

$$A^+(\xi) = (1-p) e^{-h_2\partial_\xi}\,\alpha(\xi) - p\,\alpha(\xi)\, e^{h_2\partial_\xi} \\ + \sqrt{p(1-p)}\,[(2p-1)N + 2\xi/h_2]\,, \quad (12)$$

together with the operator

$$A_0(\xi) = \frac{1}{\hbar\omega}\left[H^{\mathrm{K}}(\xi) - \tfrac{1}{2}(N+1)\right], \quad (13)$$

close under commutation as the algebra $so(3)$ of the rotation group,

$$[A_0, A] = -A, \quad [A_0, A^+] = A^+, \quad [A^+, A] = 2A_0. \quad (14)$$

Meixner oscillators: We now organize the properties of the Meixner polynomials according to the scheme followed in the previous subsections. Known orthogonality relations for the Meixner polynomials $M_n(\xi; \beta, \gamma)$ with $\beta > 0$ and $0 < \gamma < 1$, allow us to construct another difference analogue of the harmonic oscillator[4].

Meixner oscillator has a Hamiltonian operator

$$H(\xi) = \frac{1+\gamma}{1-\gamma}\,(\xi + \tfrac{1}{2}\beta) \\ -\frac{\sqrt{\gamma}}{1-\gamma}\left(\mu(\xi) e^{\partial_\xi} + \mu(\xi-1) e^{-\partial_\xi}\right), \quad (15)$$

$$\mu(\xi) = \sqrt{(\xi+1)(\xi+\beta)}. \quad (16)$$

We now build the difference operators

$$K_+ = \frac{\gamma}{1-\gamma}\,\mu(\xi) e^{\partial_\xi} \\ + \frac{1}{1-\gamma}\, e^{-\partial_\xi}\mu(\xi) - \frac{2\sqrt{\gamma}}{1-\gamma}(\xi + \tfrac{1}{2}\beta)\,, \quad (17)$$

$$K_- = \frac{1}{1-\gamma}\,\mu(\xi) e^{\partial_\xi} \\ + \frac{\gamma}{1-\gamma}\, e^{-\partial_\xi}\mu(\xi) - \frac{2\sqrt{\gamma}}{1-\gamma}(\xi + \tfrac{1}{2}\beta)\,. \quad (18)$$

Together with $K_0 = H$, they now form the closed Lie algebra sp$(2, \Re)$,

$$[K_0, K_\pm] = \pm K_\pm\,, \qquad [K_-, K_+] = 2K_0\,. \quad (19)$$

Meixner-Pollaczek (relativistic) oscillators: There is a family of Meixner-Pollaczek polynomials $P_n^\lambda(x; \phi)$ which satisfies the orthogonality relation

$$\int_{-\infty}^{\infty} P_n^\lambda(\xi; \phi)\, P_{n'}^\lambda(\xi; \phi)\, \rho^{\mathrm{P}}(\xi)\, d\xi = \delta_{n,n'}\,\frac{\Gamma(2\lambda+n)}{n!}, \quad (20)$$

with respect to a continuous measure with the weight

$$\rho^{\mathrm{P}}(\xi) = \frac{1}{2\pi}(2\sin\phi)^{2\lambda}|\Gamma(\lambda + i\xi)|^2 \exp[(2\phi - \pi)\xi]. \quad (21)$$

The reason why we mention these polynomials here is the following. In Ref. [5], it was shown that the Meixner polynomials $M_n(\xi; \beta, \gamma)$ and the Meixner-Pollaczek polynomials $P_n^\lambda(x; \phi)$ are in fact interrelated by

$$P_n^\lambda(\xi; \phi) = \frac{e^{-in\phi}}{n!}(2\lambda)_n M_n(i\xi - \lambda; 2\lambda, e^{-2i\phi}). \quad (22)$$

In the relativistic model of the linear harmonic oscillator, proposed in Ref. [6], the wavefunctions in configuration space are expressed in terms of the Meixner-Pollaczek polynomials. In other words, the relation (22) gives the connection between the relativistic harmonic oscillator and the Meixner oscillator, discussed in this paper.

References

1. M. Moshinsky and Yu.F. Smirnov in *The Harmonic Oscillator in Modern Physics*, Contemporary Concepts in Physics, Vol. 9 (Harwood Academic, New York, 1996).
2. N.M. Atakishiyev and S.K. Suslov in *Modern Group Analysis: Methods and Applications*, (Elm, Baku, 1989), pp. 17–20 (in Russian).
3. N.M. Atakishiyev and S.K. Suslov, *Theor. Math. Phys.* **85**, 1055 (1991).
4. R. Koekoek and R.F. Swarttouw, *Report 98-17*, Delft University of Technology (1998).
5. N.M. Atakishiyev and S.K. Suslov, *J. Phys.* A: Math. Gen. **18**, 1583 (1985).
6. A.D. Donkov, V.G. Kadyshevsky, M.D. Mateev and R.M. Mir-Kasimov, *Teor. Mat. Fiz.* **8**, 61 (1971); N.M. Atakishiyev, R.M. Mir-Kasimov and S.M. Nagiyev, *Theor. Math. Phys.* **44**, 592 (1981); N.M. Atakishiyev, *Theor. Math. Phys.* **58**, 166 (1984).

THERMODYNAMIC BETHE ANSATZ (TBA)

W. NAHM
Dublin Institute for Advanced Studies
10 Burlington Road, Dublin 4, Ireland
Email: wnahm@stp.dias.ie

Integrable perturbations of conformally invariant Quantum Field Theories in two-dimensions can be investigated by the Bethe Ansatz. Usually one considers its thermodynamic approximation, but the conformally invariant limit can be studied more precisely and efficiently by the Bethe Ansatz itself. This is demonstrated for the calculation of the partition function of reflectionless theories. If certain states decouple, the Bethe Anzats may yield the exact spectrum of the Hamilton operator in the conformal limit. Finally, some known analytic properties of the solutions of the thermodynamic Bethe Ansatz are derived in detail.

1 Introduction

[a]From a mathematical point of view, Quantum Field Theory (QFT) is still a very difficult subject, and very few examples are under complete control. In part this is due to the fact that physicists concentrated their efforts on the cases of immediate practical importance, like QED, QCD and more generally the standard model. Mathematically, these are very difficult and will remain so for a while, but studies of simpler Quantum Field Theories can serve as a bridge between phenomenology and mathematical understanding. It helps that some of the simpler theories have experimental applications, too, for example to the study of phase transitions in surface systems. More importantly, they allow many insights into the structure of general QFTs. Last but not least they lead to important developments in pure mathematics, from an understanding of monstrous moonshine via the representation theory of loop groups to mirror symmetry and beyond. Some of these applications came under the heading of string theory, though they require nothing else but conformally invariant quantum field theory in two dimensions (CFT). Nonperturbative string theory may of course provide even deeper insights and yields many QFTs as low energy limits. But, here, we will concentrate on the Quantum Field Theories themselves. Since they are local, it is possible to compactify space, and this will be one of our major tools.

Free QFTs are well understood. If they have no massless excitations, their Hilbert space is a Fock space and the vector space of all their local fields has a basis of normal ordered products of the free fields and their derivatives. Fock spaces are graded by particle numbers. The N-particle subspace is contained in the N-fold tensor product of 1-particle spaces. In Minkowski space, the latter form irreducible representations of the Lorentz group and can be described by group theory alone. Alternatively, one may describe them as spaces of positive energy solutions of complexified linear equations of motion. The latter description still works when one compactifies space, since one still has time translation invariance and a corresponding energy operator. We always will compactify in a way which preserves spatial translation symmetry, such that we have a momentum operator, too. Thus, we need periodicity, possibly twisted by a phase. In particular, for fermions both periodicity (Ramond sector) or anti-periodicity (Neveu-Schwarz sector) are natural. In any case the complete system should be periodic, such that the total momentum is quantized in the usual way. For Neveu-Schwarz fermions this means that we need an even number of particles.

Apart from free QFTs, the conformally invariant theories (CFTs) in $1+1$ dimensions form the simplest class. Here, we always consider compactified space. In this case one has a state-field correspondence, such that the Hilbert space of states and the vector space of fields are isomorphic, up to the usual issues of completion. Partition functions are modular. For rational CFTs one knows almost as much as for free QFTs, though the n-point functions are more complicated and yield some generalization of hypergeometric functions. In principle, non-rational

[a]Invited plenary talk presented at the XIth Regional Conference on Mathematical Physics, May 2004, Tehran, Iran.

theories can be constructed by marginal perturbations around rational ones, but much less is known about them, apart from a few quasi-rational cases.

After CFT, the next step should concern massive integrable theories in $1+1$ dimensions. If these theories are considered on the real line, their Hilbert space and the scattering matrix S are known. The Hilbert space has Fock space descriptions in terms of incoming and outgoing particles, with a mass spectrum given by algebraic numbers, when one of the masses m is taken as unit. The S-matrix relates the two Fock space descriptions by a unitary matrix. In the cases to be considered below it is an elementary function of the particle momenta.

When we compactify space on a circle of circumference l, the only dimensionless parameter is ml. Scaling of m and l for constant ml yields isomorphic theories. In the limit $ml \to \infty$, one obtains the S-matrix theory just mentioned. In all known examples the limit $ml \to 0$ exists, too, and yields a conformally invariant theory. The relation between the two limits can be thought of as a duality, similar to the duality between M-theory in 11 dimensions and IIA string theory in 10 dimensions.

As usual for dualities, information can be transferred from one limit to the other. In the CFT the vector space of all local fields and their grading by conformal spin and conformal dimension (also called scaling dimension) is known. The conformal spin takes discrete values, so it does not change with ml. For $ml > 0$, one no longer has scale invariance, thus the grading by the scaling dimension is lost. Changes of ml are described by superrenormalizable perturbations, however, so fields just mix with fields of lower scaling dimension. In other words, the filtration of the vector space of local fields by the scaling dimension is preserved.

So far the most successful method to transfer information from the S-matrix to the CFT and from the CFT to the massive QFT is the Bethe Ansatz[1]. Usually a mean-field approximation to its equations was used, the thermodynamic Bethe Ansatz (TBA). Its equations are transcendental, but have a hidden mathematical structure of great elegance. Some of the underlying structure is clouded by the approximation, however. We shall see that the Bethe Ansatz itself allows a deeper and simpler understanding of the relations between the CFT and the massive theory than its thermodynamic approximation. This is symbolized by the title of the present talk (which differs slightly from its announcement). For more context see Ref. [3].

Our results bring a neglected question into sharper focus. To what extent is the Bethe Ansatz true? Is it just an efficient approximation in some parameter range, or is it exact when interpreted correctly? We cannot answer these questions, but the results of the next section support the validity of the Ansatz and allow simple tests.

The last section deals with the thermodynamic Bethe Ansatz. It is essentially a comment on Ref. [2], which should be helpful for further research, since that paper is rather brief.

2 The Bethe Ansatz and its Conformal Limit

When ml is large and the momenta remain finite, the description of QFT states in terms of particles is still valid as long as the particles remain well separated. Particles may have different types $a = 1, \ldots, r$ with corresponding masses m_a. In integrable theories one has conserved quantities with infinitely many conformal spins, which for separated particles become functions of the particle momenta. When particles scatter, the set of these momenta and in particular the particle number is conserved. In general, the particle type associated to one momentum may change by the scattering, but by Lorentz invariance the change must involve particle types with the same mass. We only will consider cases for which no change of particle type occurs at all. Such theories are called reflectionless integrable QFTs. Other integrable theories may be treated by similar but more complicated methods.

We exploit Lorentz invariance by writing energy and momentum of a particle of type a as

$$(E, p) = m_a(\cosh\theta, \sinh\theta).$$

Lorentz transformations act as constant shifts on the rapidities θ.

Now consider a state in a reflectionless integrable QFT with two particles of types a, b and momenta p_1 and p_2. Let $|x_1 - x_2|$ be large compared to the particle sizes. The probability density for finding the first particle at x_1 and the second at x_2 is given by $\exp(ip_1x_1 + ip_2x_2)$, up to a phase factor, which we normalize to 1 for $x_1 \ll x_2$. For $x_1 \gg x_2$, we can write it in the form $\exp i\delta_{ab}(\theta_1 - \theta_2)$, due to Lorentz invariance. Let $\theta_1 > \theta_2$. Then $x_1 \ll x_2$ applies to large negative times and $x_1 \gg x_2$ to large positive ones. In other words, in this range the phase factor agrees with the scattering matrix, which transforms the free particle description at $t = -\infty$ to the one at $t = +\infty$. For $\theta_1 < \theta_2$ the phase factor is the inverse of the scattering matrix. When $a = b$, the correct permutation symmetry of the wave function must be taken into account, but this can be done at the end of all other calculations and only imposes restrictions at $p_1 = p_2$. Independently of the wave function symmetry, we call particles bosons when $p_1 = p_2$ is allowed and fermions otherwise.

The derivative of the phase shift δ is measurable. Take, for example, a standard Gaussian wave package as incoming state with energy E. The average time delay for a particle to reach a point on the other side of the scatterer is given by $d\delta/dE$. When the derivative is positive, we have a positive time delay, as expected for a repulsive interaction. The phase shift δ itself is not measurable without ambiguity, but there are constraints when we compactify our space. In this case the regions $x_1 \gg x_2$ and $x_1 \ll x_2$ are glued together by periodicity. At high rapidity difference, one expects that masses are unimportant, so we approach the conformal case where there is no scattering. One may have superselection rules, which slightly complicate the calculations, but let us focus on the simpler situation where we may put in any particle of sufficiently large rapidity without affecting the others. Thus, its S-matrix elements must be 1 and $\delta_{ab}(\pm\infty)$ must be multiples of 2π. We fix them by putting $\delta_{ab}(-\infty) = 0$ for all $a, b = 1, \ldots, r$. With this convention $\delta_{ab}(\theta)$ becomes a well-defined function on the real axis, since it is the integral of δ'_{ab} between $-\infty$ and θ.

When the rapidity difference is $+\infty$, we can write

$$\delta_{ab}(+\infty) = -2\pi\tilde{A}_{ab},$$

with $\tilde{A}_{ab} \in \mathbb{Z}$. The minus sign will be convenient later, since we will consider attractive interactions, for which $\tilde{A}_{ab} > 0$. Fractional $\tilde{A}_{ab}$ may occur, but induces superselection sectors.

Now, we consider a system of N^a particles of types $a = 1, \ldots, r$, with momenta p_i^a, $i = 1, \ldots, N^a$. When all particles are far apart, we can describe them by a wave function given by $\exp(\sum_{a=1}^{r}\sum_{i=1}^{N^a} ip_i^a x_i^a)$, apart from a phase factor which depends on the ordering of the x_i^a. When x_i^a moves from the left of x_j^b to the right, with no other change in the ordering, we expect an extra phase

$$\exp(i\delta_{ab}(\theta_i^a - \theta_j^b)).$$

When we compactify space with period l, the momenta are constrained by the periodicity of the wave functions. This yields the Bethe Ansatz equations

$$p_i^a l + \sum_{b=1}^{r}\sum_{j=1}^{N^b} \delta_{ab}(\theta_i^a - \theta_j^b) - \delta_{aa}(0) = 2\pi n_i^a,$$

for $i = 1, \cdots, N$ and with integral n_i^a for untwisted periodicity conditions. When the phase shifts are redefined by multiples of 2π, the n_i^a change by integers, but with our convention, their values are fixed. For free particles the phase shifts vanish and the momenta are quantized by $p_i^a = 2\pi n_i^a/l$. For free bosons, all choices of the n_i^a are allowed and, for free fermions, the only restriction is $n_i^a \neq n_j^a$.

Periodic boundary conditions preserve translational invariance, so the total momentum

$$P = \sum_{a=1}^{r}\sum_{i=1}^{N^a} p_i^a,$$

is still quantized in units of $2\pi/l$. Let us apply this condition to a two-particle state with varying l. With $\theta = \theta_1 - \theta_2$, one finds that $\delta_{ab}(\theta) + \delta_{ba}(-\theta)$ is independent of θ. Because of $\delta_{ab}(-\infty) = 0$, we have

$$\delta_{ab}(\theta) + \delta_{ba}(-\theta) = -2\pi\tilde{A}_{ab}.$$

Note that $\tilde{A}_{ab} = \tilde{A}_{ba}$, so we can regard $\tilde{A}$ as a symmetric $r \times r$ matrix. For $a = b$, we have $\delta_{00}(0) = -\pi\tilde{A}_{aa}$. For a state with N_a particles of type a, we obtain

$$\frac{Pl}{2\pi} = \sum_{a=1}^{r}\sum_{i=1}^{N^a} n_i^a + \frac{1}{2}\sum_{a,b=1}^{r} Nr^a\tilde{A}_{ab}N^b - \frac{1}{2}\sum_{a=1}^{r} N^a\tilde{A}_{aa}.$$

We will assume that the Bethe Ansatz equations determine the θ_i^a uniquely for any choice of the n_i^a. This is easy to show for repulsive interactions but wrong for strong attraction, so it must be investigated in detail for the integrable QFTs under consideration. We also assume the closely related property that $n_i^a > n_j^a$ implies $\theta_i^a > \theta_j^a$. Once the θ_i are determined, the total energy of the state becomes

$$E = E_0 + \sum_{a=1}^{r}\sum_{i=1}^{N^a} m_a \cosh\theta_i^a.$$

Here E_0 is the ground state energy which one has in the absence of particles. Apart from E_0, the partition function of the theory is known, once one knows the allowed choices of the n_i^a.

For finite l, there always will be some overlap of the particles, so *a priori* one can only expect that the Bethe Ansatz is an approximation. We will see, however, that it has good chances to be exact for the theories under consideration. A severe check is provided by the limit $ml \to 0$, where the particle overlap becomes maximal. In this limit, the theory becomes conformally invariant and has a known mass spectrum, thus the correctness of the Bethe Ansatz is easy to check.

First, let us note that the rapidity differences between right-movers with $\theta > 0$ and left-movers with $\theta < 0$ are of order $|\log(ml)|$, so their scattering becomes trivial and the Bethe Ansatz reduces to separate systems of equations for right- and left-movers. Let us consider a system of right-movers only. In this case $\cosh\theta_i^a \simeq \sinh\theta_i^a$, for all particles, such that the limit of $(E - E_0)l$ is equal to Pl, which was calculated above. For fermionic behavior, the state of lowest energy at fixed N^a has $n_{i+1}^a = n_i^a + 1$ for $i = 1, \cdots, N^a - 1$. At large N^a one expects that these states have an N^a-independent $n_1^a = \tilde{b}_a$. Indeed, adding an extra particle at the high rapidity end will not change the low rapidities, and by the convention $\delta_{ab}(+\infty) = 0$, we have made sure that the low n_i^a do not change either. Thus, the energy of the lowest state with particle numbers N^a is $2\pi Q(N)/l$, where

$$Q(N) = 2\pi E_0/l + \sum_{a,b=1}^{r} N^a \tilde{A}_{ab} N^b/2 + \sum_{a=1}^{r} \left(\binom{N^a}{2} + (\tilde{b}_a - \tilde{A}_{aa}/2)N^a \right).$$

Using a notation for $Q(0)$ which comes from CFT, we rewrite this expression as

$$Q(N) = h - c/24 + \sum_{a,b=1}^{r} N^a A_{ab} N^b/2 + \sum_{a=1}^{r} b_a N^a.$$

In particular, $A = \tilde{A} + I_r$, where I_r is the $r \times r$ unit matrix.

Note added: Compare the substantially more complicated TBA derivation of this result which just appeared in Ref. [6].

Now we can calculate the partition function for right-movers. We first do it in the case of a single species of massless free fermions in the Neveu-Schwarz sector. A state of N right-moving particles is labeled by half-integral $(n_1, \ldots, n_N)$ with $0 < n_1 < \ldots < n_N$. We write $n_1 = 1/2 + s_1$, $n_2 = n_1 + 1 + s_2$, $n_3 = n_2 + 1 + s_3$ etc. The total energy is given by

$$\sum_{i=1}^{N} n_i = N^2/2 + \sum_{i=1}^{N} (N + 1 - i) s_i.$$

Thus, the N-particle contribution to the partition function has the form

$$q^{N^2/2}/(q)_N,$$

where the q-deformed factorial $(q)_N$ is defined by $(q)_N = (1-q)(1-q^2)\cdots(1-q^N)$. In general the analogous calculation yields

$$Z = \sum_{N \in \mathbb{N}^r} \frac{q^{Q(N)}}{(q)_{N^1} \cdots (q)_{N^r}},$$

as partition function of the right-movers. Note, however, that this calculation needs not to be valid for small values of the N^a, where n_1^a may be different from $\tilde{b}_a$. Since one cannot expect that any modification of a finite number of terms in the preceding formula for Z would lead to a modular function, some states must decouple in the CFT limit. Assume in particular that $N^a = 1$ and $N^b = 0$ for all $b \neq a$. The lowest n_1^a contributing to the CFT is $A_{aa} + b_a$. All states with $0 < n_1^a < A_{aa} + b_a$ must decouple. This will be a severe test for the exact validity of the Bethe Ansatz equations.

The ground state energy can be determined by the condition that the partition function is a modular function. In the free fermion case this yields $h = 0$ and $c = 1/2$. For the general case, see Ref. [4]. With $q = \exp(2\pi i \tau)$ and $\tilde{q} = \exp(-2\pi i/\tau)$, modular behavior requires for $\tau \to 0$

$$Z \sim \tilde{q}^{-c/24},$$

where c is central charge for unitary CFTs, or more generally, the effective central charge. We can evaluate the partition function by a saddle point approximation. When N^a increases by one, the summand in our formula for Z changes by

$$\frac{q^{\sum_{b=1}^{r} A_{ab} N^b}}{1 - q^{N^a}}.$$

Since $q \to 1$, factors like q^{b_a} can be neglected. At the stationary point we need

$$q^{\sum_{b=1}^{r} A_{ab} N^b} = 1 - q^{N^a},$$

or equivalently

$$x_a + \prod_{b=1}^{r} x_b^{A_{ab}} = 1,$$

where $x_a = q^{N^a}$. This is an algebraic equation which is well worth studying for all reflectionless integrable QFTs. In particular, it seems that its Galois group is always Abelian, so the equation can be solved in terms of roots of unity.

When we approximate $(q)_N$ by an integral, we find

$$\log(q)_N \simeq \frac{\frac{\pi^2}{6} - Li_2(q^N)}{\log q},$$

where

$$Li_2(x) = \sum_{n=1}^{\infty} \frac{x^n}{n^2}.$$

Using the Rogers dilogarithm

$$L(x) = Li_2(x) + \log x \log(1-x)/2,$$

and $L(x) + L(1-x) = \pi^2/6$, we find

$$\frac{\pi^2}{6}c = \sum_{a=1}^{r} L(1 - x_a).$$

Usually this result is derived from the thermodynamic approximation of the Bethe Ansatz to be discussed below, but here, we see that it follows much simpler and much more naturally from the conformal limit of the Bethe Ansatz itself. In the thermodynamic description one uses $Y_a = x_a/(1-x_a)$, and the algebraic relations take the form

$$\prod_{b=1}^{r} Y_b^{A_{ab}} = \prod_{b=1}^{r} (1+Y_b)^{\tilde{A}_{ab}}.$$

These relations deform to interesting equations for functions $Y_a(\theta)$, as we shall see below in a simple example.

The simplest non-free example is the perturbation of the $(2,5)$ minimal model by the field of scaling dimension $-2/5$. Here, one has $r = 1$ and $\tilde{A} = 1$, thus $A = 2$. The algebraic equation $x + x^2 = 1$ is solved by the golden ratio $x = (\sqrt{5}-1)/2 = \exp(2\pi i/5) + \exp(-2\pi i/5)$. One has $Y^2 = 1 + Y$ and $Y = 1 + x$. The effective central charge is $6L(1-x)/\pi^2 = 2/5$. In the two superselection sectors of the model, one finds $Q(N) = N^2 - 1/60$ and $Q(N) = N^2 + N + 11/60$, respectively, in particular $b = 0$ and $b = 1$. This yields the sum sides of the famous Rogers-Ramanujan identities

$$\sum_{N=0}^{\infty} \frac{q^{N^2}}{(q)_N} = \prod_{\substack{n=1 \\ n=1,4 \mod 5}}^{\infty} (1-q^n)^{-1},$$

$$\sum_{N=0}^{\infty} \frac{q^{N^2+N}}{(q)_N} = \prod_{\substack{n=1 \\ n=2,3 \mod 5}}^{\infty} (1-q^n)^{-1}.$$

In this case the product sides have interpretations in terms of Verma modules for the Virasoro algebra, but the existence of such a product decomposition is a much more general phenomenon and needs a better understanding.

Another striking example is the perturbation of the free fermion CFT with $c = 1/2$ by the field of scaling dimension $1/8$. Here, $r = 1$ and $A/2$ is the inverse of $C(E_8)$, the Cartan matrix of E_8. This yields the conjectural identity

$$\sum_{N \in \mathbb{N}^8} \frac{q^{Q(N)}}{(q)_{N^1} \cdots (q)_{N^8}} = \sum_{m=0}^{\infty} \frac{q^{2m^2}}{(q)_{2m}},$$

with $Q(N) = \sum_{a,b=1}^{8} N^a C(E_8)^{-1}_{ab} N^b$.[5] Note the restriction to an even number of Neveu-Schwarz fermions on the right hand side.

3 The Thermodynamic Bethe Ansatz

So far we have considered the CFT limit of the integrable theories. Massive theories are often considered at finite temperature $1/\beta$ in the limit $ml \to \infty$, with $m\beta$ fixed. In this case, the partition function is dominated by contributions from large momenta and large values of n. Since one has $ml \exp(\theta) \sim 2\pi n$, thus $\Delta\theta \sim \Delta n/n$, one gets arbitrarily large occupation numbers in each finite interval $\Delta\theta$ and can use a mean-field approximation. We order the rapidities of a typical state by $\theta_i^a < \theta_{i+1}^a$ (for fermions of type a) and analogously for n_i^a. For such states, we introduce averaged functions n^a with $n_i^a \simeq n^a(\theta_i^a)$. We count the number of occupied states by functions c^a with $c^a(\theta_i^a) = i$. For large l we expect that these functions have smooth limits. Accordingly, the Bethe Ansatz equations take the form

$$m_a l \sinh(\theta) + \sum_{b=1}^{r} \int \delta_{ab}(\theta - \theta') dc^b(\theta') = 2\pi n^a(\theta).$$

With $dn^a/d\theta = l\rho^a$ and $dc^a/d\theta = l\chi^a$, we obtain

$$\cosh(\theta) + \sum_{b=1}^{r} \int \delta'_{ab}(\theta - \theta') \chi^b(\theta') d\theta' = 2\pi \rho^a(\theta).$$

The energy of such a typical high temperature state is given by

$$E = l\sum_{a=1}^{r} m_a \int \cosh(\theta)\chi^a(\theta)d\theta.$$

The entropy of states with given ρ^a, χ^a is given by

$$\sum \binom{l\rho(\theta)\Delta\theta}{l\chi(\theta)\Delta\theta}.$$

By Stirlings formula one finds

$$S = l\sum_{a=1}^{r}\int \Big(\chi^a(\theta)\log\frac{\rho^a(\theta)}{\rho^a(\theta)-\chi^a(\theta)} + \chi^a(\theta)\log\frac{\rho^a(\theta)-\chi^a(\theta)}{\chi^a(\theta)}\Big)\,d\theta.$$

For a typical finite temperature state the variation of the free energy $E - S/\beta$ vanishes. We vary with respect to the χ^a and obtain the corresponding variations of the ρ^a by the Bethe Ansatz equations, in the form

$$2\pi\ \delta\rho^a(\theta') = \sum_{b=1}^{r}\int \phi_{ba}(\theta-\theta')\ \delta\chi^b(\theta)\ d\theta.$$

Here $\phi_{ab}(\theta) = \delta'_{ab}(\theta)$ and we have used $\phi_{ab}(\theta) = \phi_{ba}(-\theta)$. Minimizing the free energy yields the TBA equation

$$\epsilon^a(\theta) = \beta m_a\cosh(\theta) - \frac{1}{2\pi}\sum_{b=1}^{r}\int \phi_{ba}(\theta-\theta')L^b(\theta')d\theta',$$

where we have put

$$\epsilon^a(\theta) = \log\frac{\rho^a(\theta)-\chi^a(\theta)}{\chi^a(\theta)},$$

and

$$L^a(\theta) = \log\frac{\rho^a(\theta)}{\rho^a(\theta)-\chi^a(\theta)}.$$

In particular,

$$e^{L^a(\theta)} = 1 + e^{-\epsilon^a(\theta)}.$$

For a given mass spectrum and given ϕ_{ab}, the thermodynamic Bethe Ansatz can be considered as an integral equation for the $\epsilon^a(\theta)$. It turns out that their exponents $Y_a(\theta) = \exp(\epsilon^a(\theta))$ have remarkable properties, in particular, they are entire periodic functions of θ, with a period related to the roots of unity mentioned above[2]. We rederive these conclusions of Zamolodchikov, which in part were stated without proofs.

The function $L^a(\theta)$ rapidly go to zero for $\beta m_a\cosh(\theta) \gg 1$, but for $\beta \to 0$ and fixed θ they approach constants L^a, and analogously for ϵ^a. Thus, $L^a(\theta)$ has a central plateau. Since

$$\frac{1}{2\pi}\int_{-\infty}^{+\infty}\phi_{ba}(\theta)d\theta = \tilde{A}_{ba},$$

we find $\epsilon^a = \sum_{b=1}^{r}\tilde{A}_{ba}L^b$, which translates into

$$\exp(-L^a) + \exp(-\sum_{b=1}^{r}A_{ba}L^b) = 1.$$

For $x_a = \exp(-L^a)$, this is exactly the algebraic equation we had obtained above in the CFT context. Note that $Y_a = \exp(\epsilon^a)$.

The decay of $L^a(\theta)$ from L^a to zero has a limiting shape for small β, which can be obtained from the solution of

$$\epsilon^a(\theta) = (m_a/m_1)\exp(\theta) - \frac{1}{2\pi}\sum_{b=1}^{r}\int_{-\infty}^{+\infty}\phi_{ba}(\theta-\theta')L^b(\theta')d\theta'.$$

The plateau now extends to $-\infty$. Here, we just consider the simplest case, namely, the integrable perturbation of the $(2,5)$ minimal model, though generalizations are easy. Recall that $r = 1$, $\tilde{A} = 1$. Both ϵ and L are equal to the logarithm of the golden ratio. The exact S matrix yields

$$\phi(\theta) = \frac{\sinh(\theta/2 + i\pi/3)\sinh(\theta/2 + i\pi/6)}{\sinh(\theta/2 - i\pi/3)\sinh(\theta/2 - i\pi/6)}.$$

The thermodynamic Bethe Ansatz equation allows an easy analytic continuation of $\epsilon(\theta)$ away from the real axis. We first present the calculations in the high temperature limit, though they apply to any temperature. For $-\pi/3 < \Im\theta < \pi/3$ we still have

$$\epsilon(\theta) = \exp(\theta) - \frac{1}{2\pi}\int_{-\infty}^{+\infty}\phi(\theta-\theta')L(\theta')d\theta',$$

with an integration along the real θ' axis. In particular, $\epsilon(\theta)$ is regular in this domain and at its boundaries, and approaches the logarithm of the golden ratio when its real part tends to $-\infty$. For $\theta = i\pi/3$ we obtain a pole of $\phi(\theta)$, such that the integration path has to be shifted. When we move it back to the real axis we get a contribution from the pole. Since its residue is i, we obtain

$$\epsilon(\theta) = \exp(\theta) - \frac{1}{2\pi}\int_{-\infty}^{+\infty} \phi(\theta - \theta')L(\theta')d\theta' + L(\theta - i\pi/3),$$

for $\pi/3 < \Im\theta < 2\pi/3$. On the other hand, one has

$$\phi(\theta + i\pi/3) + \phi(\theta - i\pi/3) = \phi(\theta).$$

This yields

$$\epsilon(\theta + i\pi/3) + \epsilon(\theta - i\pi/3) = \epsilon(\theta) + L(\theta),$$

and by exponentiation

$$Y(\theta + i\pi/3)Y(\theta - i\pi/3) = 1 + Y(\theta).$$

Surprisingly, iteration yields $Y(\theta + 5i\pi/3) = Y(\theta)$.

Since $\epsilon(\theta)$ is regular for $-\pi/3 \leq \Im\theta \leq \pi/3$, its exponent $Y(\theta)$ is regular and different from zero in this range. Using $Y(\theta + i\pi/3) = (1 + Y(\theta))/Y(\theta - i\pi/3)$ we see that $Y(\theta)$ is regular for $\pi/3 \leq |\Im\theta| \leq \pi$, and by periodicity everywhere. When the real part of θ approaches negative infinity, $Y(\theta)$ tends to the golden ration for arbitrary $\Im\theta$. Altogether, $Y(\theta)$ is an entire function of $\exp(6\theta/5)$.

The property

$$\exp(s(\theta + i\pi/3)) + \exp(s(\theta - i\pi/3)) = \exp(s\theta),$$

is true for $s = 1, 5 \mod 6$. These exponents correspond to the conformal spins of the conserved QFT fields. When one replaces $\exp\theta$ in the high temperature version of the TBA with any linear combinations of such $\exp(s\theta)$, the analysis given above does not change, except for the fact that negative s yields an essential singularity at $\exp(6\theta/5) = 0$. This is the situation for the TBA equation at finite β, which contains the term $\beta m \cosh\theta$. It is quite possible that the corresponding functions $Y(\theta)$ have an elementary description. In the high temperature limit, it seems that their only zeros lie at $\Im\theta = 5i\pi/6$, up to periodicity, so one has an entire function of order $5/6$ with zeros along a single ray. Here is another example were mathematicians might help and another case of the close links between seemingly different domains of mathematics which are exposed by almost all studies of quantum field theory.

References

1. A.B. Zamolodchikov, *Nucl. Phys.* B **342**, 695 (1990).
2. A.B. Zamolodchikov, *Phys. Lett.* B **253**, 391 (1991).
3. W. Nahm in *Les Houches Lecture Notes* (2003), arXiv: hep-th/0404120.
4. M. Terhoeven, *Mod. Phys. Lett.* A **9**, 133 (1994).
5. R. Kedem, T.R. Klassen, B.M. McCoy and E. Melzer, *Phys. Lett.* B **307**, 68 (1993).
6. N. Mann and J. Polchinski in *From Fields to Strings: Circumnavigating Theoretical Physics*, arXiv: hep-th/0408162.

CONNECTION BETWEEN $\mathcal{N}=4$ SUPERCONFORMAL ALGEBRA WITH $D(2/1;\alpha)$ IN ZERO MODE

J. SADEGHI

Institute for Studies in Theoretical Physics and Mathematics (IPM), School of Physics
P.O. Box 19395-5531, Tehran Iran
and
Department of Physics, Faculty of Sciences, Mazandaran University
P.O. Box 47415-416, Babolsar, Iran
E-mail: pouriya@ipm.ac.ir

Recently, conformal algebras with extended supersymmetry have attracted a lot of attention. These algebras have a basic importance in analysis of algebraic structure of varieties of string theories and are also relevant to the description of critical properties of some statistical models. The character formula for some of unitary representation in case of $\mathcal{N}=4$ have been derived long time ago, and shown that there are two representations which are massive and massless for R and NS sector. The partition function from $\mathcal{N}=4$ algebra has a connection with $sl(2|1;C)$ subalgebra of $D(2|1;\alpha)$. So in this talk, we investigate the representation of $\mathcal{N}=4$ and $D(2|1;\alpha)$ and also we show that they have deep relation to zero mode. In order to relate the representation algebras, we must first obtain the generators of $D(2|1;\alpha)$ in general mode, and then compare with $\mathcal{N}=4$ superconformal algebra in zero mode. We also discuss the character formula in these two algebras.

1 Introduction

[a]The elegance of conformal field theory is due to new infinite dimensional lie algebras, called Kac-Moody algebras. They were discovered by two mathematian Kac and Moody in 1967. Kac-Moody algebras is a generalization of ordinary algebra (finite-dimensional), such that its generators obey

$$[T^i_m, T^j_n] = if^{ijl}T^l_{m+n} + Km\delta^{ij}_{m,-n}. \tag{1}$$

The algebras look very much like an ordinary Lie algebra, except for the infinite integer index m on each generator and the constant k, which is called the level, and also the Virasoro algebra

$$[L_m, L_n] = (n-m)L_{m+n} + \frac{c}{12}n(n^2-1)\delta_{m+n,0}. \tag{2}$$

The relation between these two algebras becomes,

$$[L_m, T^i_n] = -nT^i_{m+n}, \tag{3}$$

and depends on the representation of the chosen algebra. We can get a relation between the level k of the Kac-Moody algebra and the central term of the Virasoro algebra. It is interesting to note that this commutation relation not only gives relation to algebra but also we can construct a representation of the Virasoro algebra in terms of the Kac-Moody algebra. We also know that, there are several applications $sl(2)$, $sl(2|1)$ and $\mathcal{N}=4$ algebra. We start first with the $SL(2;R)$ WZNW model. In the case where WZNW models are based on compact Lie groups such as $SU(2)$, the unitary representation are finite-dimensional and correlation functions can be written as a finite sum of conformal blocks by bootstrap approach[1]. The non-compact group give a much more complicated situation and the general solution is not known. There has been much work done on analyzing this. We know that the group $SL(2;R)$, which is one of the simplest non-compact Lie groups, is important in many key areas, *e.g.* in two-dimensional gravity. The second application is the Quantum Hall Plateau transition, where the $SL(2;R)$ WZNW model has recently been proposed as a low energy effective field theory[2]. Finally, the string theory on AdS_3, which is the $SL(2;R)$ group manifold, is described by a $SL(2;R)$ WZNW theory[3]. Our results on $SL(2|1;C)$ will be also useful for the correspondence between the theory of $\mathcal{N}=2$ non-critical strings and the $SL(2|1;R)/SL(2|1;R)$ gauged WZNW model.

1.1 *Superalgebra* $D(2|1;\alpha)$

We would like to begin the discussion on superalgebra $D(2|1;\alpha)$ with $C/\{0,-1,\infty\}$. This superalgebra can be split into two parts, one part is fermion the other is boson. It is required that we have two super-

[a]Talk presented at the XIth Regional Conference on Mathematical Physics, May 2004, Tehran, Iran.

space which consist of L_0 (boson) and L_1 (fermion). If a, b and c belong to L_0 and L_1 superspaces, we would have the following relations[4]

$$[a,[b,c]]+[b,[c,a]]+[c,[a,b]]=0, \qquad \{a,b,c\}\in L_0, \tag{4}$$

and

$$[a,\{b,c\}]+\{b,[c,a]\}-\{c,[a,b]\}=0, \qquad [a,b,c]\in L_1. \tag{5}$$

We also know that the superalgebra $D(2|1;\alpha)$ is rank 3, so we have 17 generators, including 9 boson and 8 fermion generators. For this reason, we need to define odd and even roots for fermion and boson, respectively. The roots α_1 (odd) and α_2, α_3 (even) make regular $sl(2|1)$ subalgebras, which are in the direction of α_2, α_3 and $\alpha_\theta = 2\alpha_1+\alpha_2+\alpha_3$. We therefore have

$$sl(2|1)_{\alpha_2}, sl(2|1)_{\alpha_3}, sl(2|1)_{\alpha_\theta}. \tag{6}$$

In this case, the bosonic generators are $T^{\pm,\pm}$, $T^{\pm 3}$, T^{θ^3} and $T^{\pm\theta}$ and fermionic generators are $J'^{\pm}$, $J^{\mp}$, $\bar{J}'^{\pm}$ and $\bar{J}^{\pm}$. The coxeter number h^ν for $D(2|1;\alpha)$ is zero, and its superdimension, which is the number of bosonic generators minus the number of fermionic generators, *i.e.* we have $sdim = 9-8 = 1$. The central charge of the Virasoro algebra satisfied by the Sugawara energy-momentum tensor of the affine algebra is therefore one for any value of the level k since,

$$c = \frac{ksdim}{k+h^\nu} = 1.$$

Let us write the a_j^p with $p,j=1,2,3$ (see Ref. [6])

$$[a_j^p, a_{j'}^{p'}] = \delta^{pp'}\sum_{j''=1}^{3}\epsilon_{jj'j''}a_{j''}^p, \tag{7}$$

and the elements of $Q_{rst}(r,s=1,2)$ being chosen so that

$$[a_j', Q_{rst}] = \sum_{r'=1}^{2}(a_j)_{r'r}Q_{r'st}, \tag{8}$$

where $a_i = \frac{i}{2}\sigma_i$. So using (7), (8) and

$$a_1^{(1)} \pm +ia_2^{(1)} = \frac{1}{\sqrt{\gamma}}T^{\pm,\pm}, a_3^1 = \frac{1}{\sqrt{\gamma}}T^{+3}, a_i^3 = iT_i^\theta, \tag{9}$$

as well as

$$\begin{aligned} Q_{111} &= J'^+, & Q_{112} &= \bar{J}^-, & Q_{122} &= J^+,\\ Q_{222} &= J'^-, & Q_{212} &= \bar{J}'^+, & Q_{221} &= \bar{J}^+,\\ Q_{211} &= J^-, & Q_{121} &= \bar{J}'^-, \end{aligned} \tag{10}$$

and also the relation[4] $\{Q_{rst}, Q_{r's't}\}$, we can obtain the commutation relation of $D(2|1;\alpha)$. These results can be compared to $\mathcal{N}=4$ superconformal algebra in zero mode (ref), which leads us to connect the characters formula in these algebras.

1.2 $N = 4$ *Superconformal Algebra*

The $\mathcal{N}=4$ superconformal (SCA) algebra are labeled by two quantum number h and l, which are eigenvalue and isospin, respectively[5]. The $\mathcal{N}=4$ SCA , A_γ parameterized by the central charge, c, of the Virasoro algebra and one additional parameter $\gamma \geq 0$, or equivalently, by the levels k^+ and k^- of the two commuting Kac-Moody algebras. The generators of $\mathcal{N}=4$ SCA are the Virasoro operator of $L(z)$, the four supercharges of $\frac{3}{2}$ $G_\pm(z)$, $G_{\pm k}(z)$ and the six operators of one-dimension $T^{\pm,+}(z)$, $T^{\pm,-}(z)$ and $T^{\pm 3}(z)$, one-dimension current $U(z)$ and four-dimension $\frac{1}{2}$ operators $Q_\pm(z)$ and $Q_{\pm k}(z)$. This $\mathcal{N}=4$ algebra A_γ contains, among other substructures, two commuting $su(2)$ Kac-Moody algebras and a finite dimensional sub-superalgebra $D(2|;\frac{\gamma}{1-\gamma})$ where $\gamma\in R$. These parameters are related through $c=\frac{6k^+k^-}{k}$, where $k=k^++k^-$, $\gamma=\frac{k^-}{k}$, $1-\gamma=\frac{k^+}{k}$ and $k=\frac{c}{6\gamma(1-\gamma)}$.

1.3 *Relation Between Two Algebras*

$D(2/1;\alpha k_2)_{k_1}$ decomposes to

$$SL(2/1)_{K_1}\otimes U(1) = SL(2)_{k_1}\oplus SL(2)_{k_2}\oplus SL(2)_1,$$

where $k_1 = \frac{p}{u}-1$ and $k_2=\frac{u}{p}-1$. The untwisted $sl(2/1;C)$ admissible characters are given by[6]

$$X_{r,s,u,p}^{sl(2/1;C)}(z,\xi,q) = z^{\frac{r-1}{2}-\frac{s-1}{2}\frac{p}{u}}\xi^{\frac{r-1}{2}-\frac{s+1}{2}\frac{p}{u}}q^{r-1-s\frac{p}{u}} \times \frac{\theta_{1,0}(z^{\frac{1}{2}}\xi^{\frac{1}{2}},q)\theta_{1,0}(z^{\frac{1}{2}}\xi^{-\frac{1}{2}},q)}{\theta_{1,1}(z,q)\eta(q)^3}\Psi_{r,s,u,p}(z,\xi,q), \tag{11}$$

where

$$\Psi_{r,s,u,p}(z,\xi,q) = \sum_{m\in Z} q^{m^2up-mu(r-1)} \times \left\{\frac{q^{mp(s-1)}z^{-mp}}{1+z^{-\frac{1}{2}}\xi^{-\frac{1}{2}}q^{mu-1}} - \frac{q^{(s-1)(r-1)}z^{1-r}q^{-mp(s-1)}z^{mp}}{1+z^{\frac{1}{2}}\xi^{-\frac{1}{2}}q^{mu-s}}\right\}, \tag{12}$$

and $1-p \leq r \leq p$, $1 \leq s \leq u$. Setting $p=1$, and therefore $r=1$, in the above relations provides u^2 characters $X_{1,s,u,1;\theta}$ with $1 \leq s \leq u$ and $0 \leq \theta \leq u-1$ (θ is spectral flow quantity). The restricted range for the integer variable θ in this case is the consequence of the fact that the $sl(2|1)$ characters are quasi-periodic for a shift $\xi \longrightarrow \xi q^{2\alpha u}$, $\alpha \in Z$. More precisely,

$$X_{1,s,u,1;\theta}(q,z,\xi q^{2\alpha u}) = q^{-\alpha^2 u(u-1)}\xi^{-\alpha(u-1)} \times X_{1,s,u,1;\theta}(q,z,\xi). \quad (13)$$

because the character of a twisted module $N;\,\theta$ is expressed through the untwisted character X_N as

$$X_{1,s,u;\theta}(q,z,\xi q^{2\alpha u}) = \xi^{k\alpha u} q^{k\alpha^2 u^2} X_{1,s,u,1;\theta}(q,z,\xi). \quad (14)$$

The equation (14) is merely the consequence of the spectral flow, while the equation (13) is most easily established when rewriting the characters in product form, through a residue analysis, for instance. In the range $1 \leq s \leq u$ and $u-s \leq u-1$, one obtains

$$X_{1,s,u,1;\theta}(q,z,\xi) = q^{-\frac{(\theta-u)^2)}{u}-\frac{(s+1)(\theta-u)}{u}-\frac{s}{u}} \times z^{-\frac{s-1}{2u}}\xi^{1-\frac{2\theta+s+1}{2u}} \times \frac{q^{-\frac{u}{8}}\eta(q)^3}{q^u z^{\frac{1}{2}}\xi^{\frac{1}{2}}q^{-(u-\theta-1)})} \times \frac{\theta_{1,1}(q^u, zq^{-(s-1)})}{\theta_{1,0}(q^u, z^{\frac{1}{2}}\xi^{-\frac{1}{2}}q^{-(u-\theta-s)})}. \quad (15)$$

Comparing to the product formulas given in Ref. [6] yields the following identification of the $\frac{1}{2}u(u+1)$, so-called class IV Ramond characters

$$X_{1,s,u,1;\theta}(q,z,\xi) = X^{R,IV}_{s-1,u-\theta-1}(q,z,\xi). \quad (16)$$

In the case of $\mathcal{N}=4$, there exist two classes of unitary representations, massive and massless. In the massive case for Ramond sector[5], we have the following condition

$$h > \frac{k}{4}, l = \frac{1}{2}, 1, \frac{3}{2}, \cdots, \frac{k}{2}; R,$$

and for the massless case we also have

$$h > \frac{k}{4}, l = 0, \frac{1}{2}, 1, \frac{3}{2}, \cdots, \frac{k}{2}-\frac{1}{2}; R.$$

From Ref. [5], we get the following massless character

$$q^{-\frac{k}{4}}X_0^R(k,l;\sigma,\nu) = F^R(\sigma,\nu,\tau) \times \sum_{m\in Z}\left\{\frac{z^{(k+1)m+l}q^{(k+1)m^2+2lm}}{(1+z^{\frac{-1}{2}}\xi^{\frac{1}{2}}q^{-m})(1+z^{\frac{-1}{2}}\xi^{\frac{-1}{2}}q^{-m})} - \frac{z^{-(k+1)m-l}q^{(k+1)m^2+2lm}}{(1+z^{\frac{1}{2}}\xi^{\frac{1}{2}}q^{-m})(1+z^{\frac{1}{2}}\xi^{\frac{-1}{2}}q^{-m})}\right\}. \quad (17)$$

Here, we have changed their notation and also took the $k \in +Z$ as level of the $\mathcal{N}=4$ algebra. Now multiplying the numerator and denominator of second term of the sum by $q^{2m}z^{-1}$, while replacing m with $-m$ in the first term and setting $l=0$ for the above sum. With $k+1=p$, this is identical to sum of the integrable characters equation (11) (with $u=1$, $s=1$) and since the characters have F^R in common with Ψ, we get

$$X^{IV sl(2|1)} = q^{-\frac{k}{4}}X^{N=4}(k, l=0; 2\pi\sigma, 2\pi\nu). \quad (18)$$

This relation is preserved under spectral flow and one easily shows that (18) is also satisfied with $l=\frac{k}{2}$. At this stage, we can see that the $sl(2|1;C)$ characters are very closely connected to $\mathcal{N}=4$ superconformal characters.

References

1. V.G. Knizhnik and A.B. Zamolodechikov, *Nucl. Phys.* B **247**, 83 (1984) .
2. M.J. Bhassen *et al.*, *Nucl. Phys.* B **580**, 688 (2000).
3. J. Maldacena, *Adv. Theor. Math. Phys.* **2**, 231 (1998).
4. J.F. Cornwell in *Group Theory in Physics Volume III, Supersymmetries and Infinite Dimensional Algebras*, (Acedemic Press, San Diego, Fourth Printing, 1989).
5. T. Eguchi and A. Taormina, *Phys. Lett.* B **210**, 125 (1988).
6. P. Bowcock, B.L. Feigin, A.M. Semikhatov and A. Taormina, *Commun. Math. Phys.* **214**, 495 (2000).

GLIMPSES OF A NEW PHYSICS

B. G. SIDHARTH
International Institute for Applicable Mathematics & Information Sciences
B.M. Birla Science Centre
Adarsh Nagar, Hyderabad - 500 063, India
E-mail: hyd1_birlasc@sancharnet.in

Using an irreducible quantum of area, we deduce a cosmology which correctly predicts a dark energy driven accelerating universe and is also consistent with observation apart from deducing supposedly inexplicable, accidental relations.

1 Introduction

[a]In spite of several fruitless decades of work, the two great intellectual pillars of twentieth century Physics, General Relativity and Quantum Theory have remained irreconcilable, though each has proved to be successful in its own domain. Leaving other considerations aside, the unsatisfactory character of the short distance behavior of the Standard Model itself points to Physics beyond the Standard Model as noted by 't Hooft. Over the past few decades some progress towards the Quantum Theory of Gravity has been made by introducing a new concept, be it in Quantum Superstring Theory or other Quantum Gravity approaches, as also the one to be described below. In these schemes, there is a minimum spacetime cut-off at the Planck or more generally the Compton scale.

2 The Minimum Cut-Off

To fix physical ideas, let us start with the Kerr-Newman metric, applied to the electron[1]. This purely classical description, amazingly enough, yields the purely Quantum Mechanical $g = 2$ factor of the electron, though there is now a naked singularity, in that the horizon becomes complex.

On the other hand, the position coordinate of the electron from the Dirac theory is also given by a complex coordinate. Moreover, the imaginary parts in either case have the same order, that of the electron Compton wavelength. As Dirac argued, meaningful Physics is recovered only on averaging over intervals of the order of the Compton scale, in which case the imaginary part of the coordinate disappears. When minimum spacetime intervals are introduced, we immediately have a (Lorentz invariant) noncommutative geometry, which in a simple form is[1]

$$[dx^\mu, dx^\nu] = \Theta^{\mu\nu}\beta, \beta \sim 0(l^2). \tag{1}$$

The crux is this "Quantum of Area" l^2, which has emerged as an all important irreducible unit in the latest theories[2]. Let us now consider the implications in the usual non-Abelian gauge theory.

As is well known we consider a generalization of the usual phase function λ to include fields with internal degrees of freedom. For example, λ would be replaced by A_μ to give the Gauge Field, q being the "charge"

$$F_{\mu\nu} = \partial_\mu A_\nu - \partial_\nu A_\mu - \imath q[A_\mu, A_\nu]. \tag{2}$$

To generate massive gauge bosons, in analogy with superconductivity theory, we consider a new phase adjusted Gauge Field after the symmetry is broken

$$W_\mu = A_\mu - \frac{1}{q}\partial_\mu \phi. \tag{3}$$

W_μ generates the mass via a Higgs mechanism.

Let us now consider in the Gauge Field transformation, an additional phase term, $f(x)$ using (1). Then there is a contribution to the second order in coordinate differentials, in addition to (2).

$$\frac{1}{2}\left[\partial_\mu B_\nu - \partial_\nu B_\mu\right][dx^\mu, dx^\nu] + (\text{symmetric term}), \tag{4}$$

where in (4), $B_\mu \equiv \partial_\mu f$. Effectively, A_μ is replaced by

$$A_\mu \to A_\mu + B_\mu = A_\mu + \partial_\mu f. \tag{5}$$

Comparing (5) with (3), we can immediately see that the effect of noncommutativity is precisely that of providing a new symmetry breaking term. Specializing to a spherically symmetric field for simplicity,

[a]Invited plenary talk presented at the XIth Regional Conference on Mathematical Physics, May 2004, Tehran, Iran.

instead of the usual Maxwell equations in the gauge field context, we have,

$$\vec{E} \to \vec{E} - \vec{\nabla} f = \vec{\nabla} Q - \vec{\nabla} f. \quad (6)$$

Finally, we have for a point Gauge charge, from (6),

$$Q = \int_v \frac{(\rho + \lambda(r))}{r}. \quad (7)$$

In (7) $\lambda(r)$ represents the effect of the noncommutativity and is an order of l^2 effect, that is, it falls off rapidly. As such, it represents a field mediated by massive particles. It may be remarked that a similar argument using equations like (1) and (4) can be used to argue that one could obtain a reconciliation of electromagnetism and gravitation[3].

3 Cosmology

We consider a Cosmology in which given the well-known $N \sim 10^{80}$ elementary particles, typically pions in the universe, it follows that the pions can be thought of as being created from a Quantum vacuum, in a phase transition by $n \sim 10^{40}$ Planck particles on the lines of the Prigogine Cosmology (see Ref. [4,5]). In fact, let us think of the units Quanta of area in the Quantum vacuum as being oriented randomly. Focussing our attention on the normals, we have a situation similar to the Ising Model. Furthermore, considering the amplitudes of these elementary elements of the Quantum vacuum, we have[1] a nonlinear Schrödinger equation

$$\imath\hbar \frac{\partial\psi}{\partial t}\frac{-\hbar^2}{2m'}\frac{\partial^2\psi}{\partial x^2} + \int \psi^*(x')\psi(x)\psi(x')U(x')dx'. \quad (8)$$

Equation (8) is the complete analogue of the Landau-Ginsburg equation, with correlation length given by

$$\xi = (\frac{\gamma}{\propto})^{\frac{1}{2}}(\gamma \equiv \hbar^2/2m). \quad (9)$$

It can be seen from (9) that this is just the Compton length. Equation (8) describes a Landau-Ginsburg like phase transition, and the normals get oriented. We note that in critical point phenomena, we have the reduced order parameter $\bar{Q}$ and the reduced correlation length $\bar{\xi}$. Near the critical point we have

$$\bar{Q}^\nu = \bar{\xi}^\beta. \quad (10)$$

In (10) typically $\nu \approx 2\beta$. As $\sqrt{Q} \sim \frac{1}{\sqrt{n}}$ given n particles, and in view of the fractal two dimensionality of the Quantum path

$$\bar{Q} \sim \frac{1}{\sqrt{n}}, \bar{\xi} = (l_P/l)^2, \, l = \sqrt{n} l_P. \quad (11)$$

In the above phase transition in which the Planck areas and elementary particles are created, given n particles at any stage, $\sqrt{n}$ particles are fluctuationally created in the minimum spacetime cut-off intervals l, τ.
To proceed, it was shown in the above references [4,5] that

$$m = \sqrt{n} m_P, \, M = \sqrt{n'} m_P. \quad (12)$$

M and m denote the mass of the universe and the pion, m_P the Planck mass and $n' \sim 10^{120}$, with similar equations for the $N \sim 10^{80}$ pions. Equating the gravitational potential energy of the pion in a three-dimensional isotropic sphere of pions of radius R, the radius of the universe, with the rest energy of the pion, we can deduce the well-known relation

$$R \approx \frac{GM}{c^2}, \quad (13)$$

where M can be obtained from (12), or, for the pions, $M = Nm$. We now use the fluctuation in the particle number $\sim \sqrt{N}$ as above,

$$\frac{dn}{dt} = \frac{\sqrt{n}}{\tau_P},$$

with a similar equation for N. Integrating T being the age of the universe

$$T = (\hbar/mpc^2)\sqrt{n} = \frac{\hbar}{mc^2}\sqrt{N}, \quad (14)$$

$$\frac{dR}{dt} \approx HR, \quad (15)$$

where H in (15) is the Hubble constant, and using (12) and (13), H is given by

$$H = Gm_P/(c^2\tau_P R) = \frac{Gm^3c}{\hbar^2}. \quad (16)$$

Equation (13) and (14) show that in this formulation, the correct mass, radius and age of the universe can be deduced given N as the sole cosmological parameter. Equation (16) is the so-called mysterious Weinberg formula. There is also a small cosmological constant Λ consistent with observation,

$$\Lambda \leq \mathcal{O}(H^2).$$

So all puzzling large number relations are explained here.

To proceed, we observe that because of the fluctuation of $\sim \sqrt{N}$, there is an excess electrical potential energy of the electron, which, in fact, we have identified as its inertial energy. That is[1],

$$\sqrt{N}e^2/R \approx mc^2, \text{whence}\, e^2/Gm^2 \sim \sqrt{N} \approx 10^{40}. \quad (17)$$

We also get the well-known Eddington formula, and a relation for G

$$R = \sqrt{N}l \,\text{or}\, R = \sqrt{n'}l_P, \; \frac{Gm}{lc^2} = \frac{1}{\sqrt{N}} \propto T^{-1}. \quad (18)$$

The above model predicts a dark energy driven ever expanding and accelerating universe, whose density keeps decreasing. The work of Perlmutter and others as also observations from the Wilkinson Microwave Anisotropy Probe and the Sloan Digital Sky Survey have confirmed the view.

4 Discussion

As noted, it was shown that the pion and the universe itself could be thought of as being made up of Planck oscillators[5]

$$\omega = \omega_P l_P / l, \quad (19)$$

where the subscript P denotes the Planck scale and l would be for the pion, its Compton scale and for the universe itself it would denote R. For the pion, (19) gives the pion mass m, which shows that the pion is the lowest energy state of $\sim 10^{40}$ Planck oscillators. For the universe with $n' \sim 10^{120}$ Planck oscillators, l on the r.h.s. of (19) would be the radius R and then the expression on the l.h.s. would yield the lowest energy state in this case. The highest energy state of n' oscillators would then be, $n'\omega$ which on using (19) yields the correct mass of the universe. Moreover, the inverse dependence on distance, of the energy, in (19) indicates that this could also be characterized as the potential of an inverse square interaction, for example, the gravitational interaction. In that case, (13) on using (19) gives back (18). What this means, is that without taking recourse to gravitation in the first instance and using the fact that the energy is that of the underpinning of a normal mode of Planck oscillators, it then follows that gravitation shows up as a manifestation of this energy, distributed over N particles of the universe. Thus, gravitation is now reduced to the status of a statistical measure of residual energies as in Zakharov's formulation using the Planck scale as is confirmed by (17).

In the Bekenstein Black Hole Radiation Formula if we introduce for G, it is time dependent version (18) then we get for the radiation time, M being the mass of the Schwarzschild Black Hole, this time

$$T = \Theta M, \quad \Theta \equiv \left\{ \frac{\pi\gamma^2(30.8)^3}{\hbar c^4} \right\}^{\frac{1}{3}}, \gamma \equiv \frac{l\hbar}{m^2}. \quad (20)$$

One can easily verify that (20) gives the Planck time for a Planck mass and $T \sim 10^{17}$ seconds, the age of the universe, for the universe itself.

Another way of interpreting the quantum of area is that the area of a pion is n times the elemental Planck area and similarly the area of the universe is N times the elementary particle area. This can be interpreted as due to the two dimensionality of the Brownian Quantum path in (11), thought of as the random walk equation.

Acknowledgments

It is a pleasure to thank IPM Tehran for the invitation and hospitality.

References

1. B.G. Sidharth in *Chaotic Universe: From the Planck to the Hubble Scale*, (Nova Science Publishers, Inc., New York, 2001).
2. J. Baez, *Nature* **421**, 702 (2003).
3. B.G. Sidharth, to appear in *Annales de la Fondation Louis de Broglie.*
4. B.G. Sidharth, *Found. Phys. Lett.* **15**, 577 (2002).
5. B.G. Sidharth, *Found. Phys. Lett.* **17**, 503 (2004).

HIDDEN PROPERTY OF EXTENDED JORDANIAN TWISTS FOR LIE SUPERALGEBRAS

V. N. TOLSTOY
Institute of Nuclear Physics, Moscow State University
119992 Moscow, Russia
E-mail: tolstoy@nucl-th.sinp.msu.ru

A general construction of chains of extended Jordanian twists for Lie superalgebras is presented. To this end, we use automorphisms which give trivial coproducts for subalgebras, provided the subalgebras are kernels of cobrackets for the corresponding classical r-matrices. It turns out that these automorphisms are connected with the extended Jordanian twists.

1 Introduction

[a]Jordanian type deformations of finite-dimensional simple Lie algebras belong to an important class of triangular (non-standard) deformations. They are described by chains of extended Jordanian twists. Full chains of such twists for all complex simple Lie algebras of the classical series A_n, B_n, C_n and D_n were constructed in Refs. [1-4]. It was shown that the chains for the case of the series B_n, C_n and D_n contain automorphisms which trivialize coproducts for subalgebras of the algebras. These automorphisms were constructed by fitting.

In this paper, we give a general construction of chains of extended Jordanian twists for finite-dimensional complex simple Lie superalgebras. To this end, we use automorphisms which give trivial coproducts for elements of subalgebras provided the subalgebras are kernels of cobrackets for the corresponding r-matrices. We show that these automorphisms are connected with the extended Jordanian twists.

2 Classical r-Matrices of Jordanian Type

Let $\mathfrak{g}$ be any finite-dimensional complex simple Lie superalgebra, then $\mathfrak{g} = \mathfrak{n}_- \oplus \mathfrak{h} \oplus \mathfrak{n}_+$, where $\mathfrak{n}_\pm$ are maximal nilpotent subalgebras and $\mathfrak{h}$ is a Cartan subalgebra. The subalgebra $\mathfrak{n}_+$ ($\mathfrak{n}_-$) is generated by the positive (negative) root vectors e_β ($e_{-\beta}$) for all $\beta \in \Delta_+(\mathfrak{g})$. The symbol $\mathfrak{b}_+$ will denote the Borel subalgebra of $\mathfrak{g}$, $\mathfrak{b}_+ := \mathfrak{h} \oplus \mathfrak{n}_+$. Let θ be a maximal root of $\mathfrak{g}$, and let a Cartan element $h_\theta \in \mathfrak{h}$ and a root vector $e_\theta \in \mathfrak{n}_+$ satisfies the relation

$$[h_\theta,\, e_\theta] = e_\theta. \tag{1}$$

The elements h_θ and e_θ are homogeneous, *i.e.*

$$\deg(h_\theta) = 0,\ \deg(e_\theta) = 0, \text{ or } 1\ . \tag{2}$$

Moreover, let homogeneous elements $e_{\gamma_{\pm i}}$ indexed by the symbols i and $-i$, $i \in I = \{1, 2, \cdots, N\}$ satisfy the relations

$$\begin{aligned}
&[h_\theta,\, e_{\gamma_i}] = (1 - t_{\gamma_i})\, e_{\gamma_i},\\
&[h_\theta,\, e_{\gamma_{-i}}] = t_{\gamma_i}\, e_{\gamma_{-i}} \quad (t_{\gamma_i} \in \mathbb{C}),\\
&[e_{\gamma_{\pm i}},\, e_\theta] = 0 \quad (i \in I),\\
&[e_{\gamma_k},\, e_{\gamma_l}] = \delta_{k,-l}\, e_\theta \quad (k > l \in I \textstyle\bigcup (-I)),\\
&\deg(e_\theta) = \deg(e_{\gamma_i}) + \deg(e_{\gamma_{-i}}), \qquad (\mathrm{mod}\, 2).
\end{aligned} \tag{3}$$

For the Lie superalgebra $\mathfrak{g}$ the brackets $[\cdot,\,\cdot]$ always denote the supercommutator

$$[x, y] := xy - (-1)^{\deg(x)\deg(y)} yx, \tag{4}$$

for any homogeneous elements x and y.

Consider the even skew-symmetric two-tensor

$$\begin{aligned}
r_{\theta,N}(\xi) = \xi \Big(&h_\theta \wedge e_\theta + \\
&+ \sum_{i=1}^{N} (-1)^{\deg(e_{\gamma_i})\deg(e_{\gamma_{-i}})} e_{\gamma_i} \wedge e_{\gamma_{-i}} \Big),
\end{aligned} \tag{5}$$

where

$$\begin{aligned}
\deg(\xi) &= \deg(e_\theta)\\
&= \deg(e_{\gamma_i}) + \deg(e_{\gamma_{-i}}), \qquad (\mathrm{mod}\, 2),
\end{aligned} \tag{6}$$

and we assume that the operation "$\wedge$" in (5) is graded

$$\begin{aligned}
e_{\gamma_i} \wedge e_{\gamma_{-i}} := {}& e_{\gamma_i} \otimes e_{\gamma_{-i}}\\
&- (-1)^{\deg(e_{\gamma_i})\deg(e_{\gamma_{-i}})} e_{\gamma_{-i}} \otimes e_{\gamma_i}.
\end{aligned} \tag{7}$$

---

[a]Talk presented at the XIth Regional Conference on Mathematical Physics, May 2004, Tehran, Iran.

It is not hard to check that the element (5) satisfies the classical Yang-Baxter equation (CYBE), and it is called the extended Jordanian r-matrix of N-order. Let N be maximal order, *i.e.*, we assume that another elements $e_{\gamma_{\pm j}} \in \mathfrak{n}_+$, $j > N$, which satisfy the relations (3) do not exist. Such element, (5), is called the extended Jordanian r-matrix of maximal order[5].

Consider a maximal subalgebra $\mathfrak{b}'_+ \in \mathfrak{b}_+$ which co-commutes with the maximal extended Jordanian r-matrix (5), $\mathfrak{b}'_+ := \operatorname{Ker}\delta \in \mathfrak{b}_+$

$$\begin{aligned}\xi\delta(x) &:= [x \otimes 1 + 1 \otimes x,\, r_{\theta,N}(\xi)] \\ &= [\Delta(x),\, r_{\theta,N}(\xi)] = 0 \quad (\forall x \in \mathfrak{b}'_+).\end{aligned} \tag{8}$$

Let $r_{\theta_1,N_1}(\xi_1) \in \mathfrak{b}'_+ \otimes \mathfrak{b}'_+$ is also an extended Jordanian r-matrix of the form (5) with a maximal root $\theta_1 \in \mathfrak{h}'$ and maximal order N_1. Then the sum

$$r_{\theta,N;\,\theta_1,N_1}(\xi,\xi_1) := r_{\theta,N}(\xi) + r_{\theta_1,N_1}(\xi_1), \tag{9}$$

is also a classical r-matrix.

Further, we consider a maximal subalgebra $\mathfrak{b}''_+ \in \mathfrak{b}'_+$ which co-commutes with the maximal extended Jordanian r-matrix $r_{\theta_1,N_1}(\xi_1)$ and we construct an extended Jordanian r-matrix of maximal order, $r_{\theta_2,N_2}(\xi_2)$. Continuing this process as result, we obtain a canonical chain of subalgebras

$$\mathfrak{b}_+ \supset \mathfrak{b}'_+ \supset \mathfrak{b}''_+ \cdots \supset \mathfrak{b}^{(k)}_+, \tag{10}$$

and the resulting r-matrix

$$\begin{aligned}&r_{\theta,N;\,\theta_1,N_1;\ldots;\theta_k,N_k}(\xi,\xi_1,\cdots,\xi_k) = \\ &= r_{\theta,N}(\xi) + r_{\theta_1,N_1}(\xi_1) + \cdots + r_{\theta_k,N_k}(\xi_k).\end{aligned} \tag{11}$$

If the chain (10) is maximal, *i.e.* it is constructed in correspondence to the maximal orders $N, N_1, \ldots N_k$, then the r-matrix (11) is called the maximal classical r-matrix of Jordanian type for the Lie superalgebra $\mathfrak{g}$.

The maximal classical r-matrices of Jordanian type for all classical Lie superalgebras $\mathfrak{sl}(m|n)$ and $\mathfrak{osp}(M|2n)$ were given in explicit form in Ref. [5].

3 Chains of Extended Jordanian Twists

The twisting two-tensor $F_{\theta,N}(\xi)$ corresponding to the r-matrix (5) has the form

$$F_{\theta,N}(\xi) = \mathcal{F}_N(\xi)F_J, \tag{12}$$

where the two-tensor F_J is the Jordanian twist and $\mathcal{F}_N$ is the extension of the Jordanian twist (see Ref. [5]). These two-tensors are given by the formulas

$$F_J = \exp(2h_\theta \otimes \sigma_\theta)\,, \tag{13}$$

$$\begin{aligned}\mathcal{F}_N(\xi) &= \Big(\prod_{i=1}^{N'} \exp(\xi(-1)^{\deg e_{\gamma_i}\deg e_{\gamma_{-i}}} \times \\ &\qquad \times e_{\gamma_i} \otimes e_{\gamma_{-i}} \exp(-2t_{\gamma_i}\sigma_\theta))\Big)\mathcal{F}_S \\ &= \exp\Big(\xi \sum_{i=1}^{N'}(-1)^{\deg e_{\gamma_i}\deg e_{\gamma_{-i}}} \times \\ &\qquad \times e_{\gamma_i} \otimes e_{\gamma_{-i}} \exp(-2t_{\gamma_i}\sigma_\theta)\Big)\mathcal{F}_S,\end{aligned} \tag{14}$$

where

$$\begin{aligned}\mathcal{F}_S &= \Big(1 - \xi\frac{e_{\theta/2}}{\exp\sigma_\theta + 1} \otimes \frac{e_{\theta/2}}{\exp\sigma_\theta + 1}\Big) \times \\ &\quad \times \sqrt{\frac{(\exp\sigma_\theta + 1) \otimes (\exp\sigma_\theta + 1)}{2(\exp\sigma_\theta \otimes \exp\sigma_\theta + 1)}}\,,\end{aligned} \tag{15}$$

if $\theta/2$ is a root (*i.e.* $\theta/2 = \gamma_i = \gamma_{-i}$ for some i), $e^2_{\theta/2} = e_\theta$, $N' = N - 1$, and

$$\mathcal{F}_S = 1\,, \tag{16}$$

if $\theta/2$ is not any root, $N' = N$. Moreover

$$\begin{aligned}\deg(\xi) &= \deg(e_\theta) \\ &= \deg(e_{\gamma_i}) + \deg(e_{\gamma_{-i}}), \qquad (\mathrm{mod}\, 2)\,,\end{aligned} \tag{17}$$

and

$$\sigma_\theta := \frac{1}{2}\ln(1 + \xi e_\theta)\,. \tag{18}$$

It should be noted that if the root vector e_θ is odd, then $\sigma_\theta = \frac{1}{2}\xi e_\theta$.

Theorem (hidden property) *(i) Let $\mathfrak{g}'$ be a subalgebra of $\mathfrak{g}$, which co-commutes with the r-matrix (5), $\delta_\xi(\mathfrak{g}') = 0$ [see (8)]. If an invertible element w_ξ of some extension of $U(\mathfrak{g})$ satisfies the equations*

$$\begin{aligned}&[\Delta(x), (w_\xi^{-1} \otimes w_\xi^{-1})\Delta_\xi(w_\xi)F_{\theta,N}(\xi)] = 0, \\ &w_\xi \equiv 1 \bmod \xi, \quad \varepsilon(w_\xi) = 1, \quad (\forall x \in \mathfrak{g}'),\end{aligned} \tag{19}$$

then the automorphism $w_\xi x w_\xi^{-1}$ trivializes the twisted coproduct $\Delta_\xi(\,\cdot\,) := F_{\theta,N}(\xi)\Delta(\,\cdot\,)F^{-1}_{\theta,N}(\xi)$ in $\mathfrak{g}'$

$$\begin{aligned}&\Delta_\xi(w_\xi x w_\xi^{-1}) \\ &\quad := w_\xi x w_\xi^{-1} \otimes 1 + 1 \otimes w_\xi x w_\xi^{-1}, \quad x \in \mathfrak{g}'.\end{aligned} \tag{20}$$

(ii) The element $w_\xi \equiv \sqrt{u(\mathcal{F}_N(\xi))}$ *satisfies the equations (19), where* $u(\mathcal{F}_N(\xi))$ *is the Hopf "folding" of the two-tensor (14)*

$$u(\mathcal{F}_N(\xi)) = ((S_\xi \otimes \mathrm{Id})\mathcal{F}_N(\xi)) \circ 1 \tag{21}$$

$$= ((\mathrm{Id} \otimes S_J)\mathcal{F}_N(\xi)) \circ 1 \tag{22}$$

$$= \Big(\prod_{i=1}^{N'}\sum_{n=0}^{\infty} \frac{(-\xi)^n}{n!} \times (-1)^{n \deg e_{\gamma_i} \deg e_{\gamma_{-i}}} e^n_{\gamma_i} e^n_{\gamma_{-i}})\Big) u_S \tag{23}$$

$$= \exp\left(\frac{-2\xi\sigma_\theta}{\exp(2\sigma_\theta)-1} \times \sum_{i=1}^{N'} (-1)^{\deg e_{\gamma_i} \deg e_{\gamma_i}} e_{\gamma_i} e_{\gamma_{-i}}\right) u_S. \tag{24}$$

Here, u_S *is the folding of the super-tensor* $\mathcal{F}_S$*:* $u_S = \exp(\frac{1}{2}\sigma)$ *if* $\theta/2$ *is a root, and* $u_S = 1$ *if* $\theta/2$ *is not any root;* S_J *is the antipode after the Jordanian twist (13); the operation "*$\circ$*" means* $(a \otimes b) \circ x = axb$*. The inverse element* $u^{-1}(\mathcal{F}_N(\xi))$ *is given the following explicit formula*

$$u^{-1}(\mathcal{F}_N(\xi)) = ((\mathrm{Id} \otimes S_\xi)\mathcal{F}_N^{-1}(\xi)) \circ 1 \tag{25}$$

$$= ((S_J \otimes \mathrm{Id})\mathcal{F}_N^{-1}(\xi)) \circ 1 \tag{26}$$

$$= \Big(\prod_{i=1}^{N'}\sum_{n=0}^{\infty} \frac{(-\xi)^n \exp(-2n\sigma)}{n!} \times (-1)^{n \deg e_{\gamma_i} \deg e_{\gamma_{-i}}} e^n_{\gamma_i} e^n_{\gamma_{-i}})\Big) u_S^{-1} \tag{27}$$

$$= \exp\Big(\frac{2\xi\sigma_\theta}{\exp(2\sigma_\theta)-1} \times \sum_{i=1}^{N'} (-1)^{\deg e_{\gamma_i} \deg e_{\gamma_i}} e_{\gamma_i} e_{\gamma_{-i}}\Big) u_S^{-1}. \tag{28}$$

Moreover

$$\sqrt{u(\mathcal{F}_N(\xi))} = \Big(\prod_{i=1}^{N'}\sum_{n=0}^{\infty} \frac{(-\xi)^n}{n!(\exp(\sigma_\theta)+1)^n} \times (-1)^{n \deg e_{\gamma_i} \deg e_{\gamma_{-i}}} e^n_{\gamma_i} e^n_{\gamma_{-i}}\Big) \sqrt{u_S} \tag{29}$$

$$= \exp\Big(\frac{-\xi\sigma_\theta}{\exp(2\sigma_\theta)-1} \times \sum_{i=1}^{N'} (-1)^{\deg e_{\gamma_i} \deg e_{\gamma_i}} e_{\gamma_i} e_{\gamma_{-i}}\Big) \sqrt{u_S}, \tag{30}$$

and

$$\sqrt{u^{-1}(\mathcal{F}_N(\xi))} = \Big(\prod_{i=1}^{N'}\sum_{n=0}^{\infty} \frac{\xi^n \exp(-n\sigma)}{n!(\exp(\sigma_\theta)+1)^n} \times (-1)^{n \deg e_{\gamma_i} \deg e_{\gamma_{-i}}} e^n_{\gamma_i} e^n_{\gamma_{-i}}\Big) \sqrt{u_S^{-1}} \tag{31}$$

$$= \exp\Big(\frac{\xi\sigma_\theta}{\exp(2\sigma_\theta)-1} \times \sum_{i=1}^{N'} (-1)^{\deg e_{\gamma_i} \deg e_{\gamma_i}} e_{\gamma_i} e_{\gamma_{-i}}\Big) \sqrt{u_S^{-1}}. \tag{32}$$

With the help of the elements $w_\xi = \sqrt{u(\mathcal{F}_N(\xi))}$, $w_{\xi_1} = \sqrt{u(\mathcal{F}_{N_1}(\xi_1))}, \cdots, w_{\xi_k} = \sqrt{u(\mathcal{F}_{N_k}(\xi_k))}$, the total twist chain corresponding to the r-matrix (11) can be presented as follows

$$\begin{aligned} &F_{\theta,N;\theta_1,N_1;\ldots;\theta_k,N_k}(\xi,\xi_1,\ldots,\xi_k) = \\ &= F_{\theta_k,N_k}(\xi,\xi_1,\ldots,\xi_{k-1};\xi_k)\cdots F_{\theta_2,N_2}(\xi,\xi_1;\xi_2) \\ &\times F_{\theta_1,N_1}(\xi;\xi_1)F_{\theta,N}(\xi)\ , \end{aligned} \tag{33}$$

where

$$\begin{aligned} &F_{\theta_i,N_i}(\xi,\xi_1,\cdots,\xi_{i-1};\xi_i) := (w_{\xi_{i-1}} \otimes w_{\xi_{i-1}})\cdots \\ &\times (w_{\xi_1} \otimes w_{\xi_1})(w_\xi \otimes w_\xi)F_{\theta_i,N_i}(\xi_i)(w_\xi^{-1} \otimes w_\xi^{-1}) \\ &\times (w_{\xi_1}^{-1} \otimes w_{\xi_1}^{-1})\cdots(w_{\xi_{i-1}}^{-1} \otimes w_{\xi_{i-1}}^{-1}), \\ &(i=1,\ldots,k). \end{aligned} \tag{34}$$

Acknowledgments

This work was supported by Russian Foundation for Fundamental Research, grant No. RFBR-02-01-00668, and CRDF RMI-2334-MD-02.

References

1. P.P. Kulish, V.D. Lyakhovsky and A.I. Mudrov, *Journ. Math. Phys.* **40**, 4569 (1999).
2. P.P. Kulish, V.D. Lyakhovsky and M.A. del Olmo, *Journ. Phys. A: Math. Gen.* **32**, 8671 (1999).
3. V.D. Lyakhovsky, S. Stolin and P.P. Kulish, *J. Math. Phys. Gen.* **42**, 5006 (2000).
4. D.N. Ananikyan, P.P. Kulish and V.D. Lyakhovsky, *St. Petersburg Math. J.* **14**, 385 (2003).
5. V.N. Tolstoy, arXiv: math.QA/0402433.

CHAPTER 6: NONCOMMUTATIVE FIELD THEORY & STRING THEORY

VANISHING VACUUM ENERGY IN NONSUPERSYMMETRIC ORIENTIFOLDS

C. ANGELANTONJ

Institut für Physik, Humboldt Universität zu Berlin
Newtonstr. 15, D-12489 Berlin, Germany
E-mail: angelant@physik.hu-berlin.de

We review a novel source for supersymmetry breaking in orientifold models, that gives a vanishing contribution to the vacuum energy at genus zero and three-half. We also argue that all the corresponding perturbative contributions to the vacuum energy from higher-genus Riemann surfaces vanish identically.

1 Introduction

[a]One of the outstanding problems in String Theory is to understand why the cosmological constant is extremely small and possibly zero after supersymmetry is broken. In fact, whenever supersymmetry is broken at a scale $M_{\rm susy}$ the four-dimensional vacuum energy Λ gets a one-loop contribution of order $M^4_{\rm susy}$, and the fields in a supermultiplet have masses shifted by $\delta m^2 = M^2_{\rm susy}$. From this simple statement it is thus evident that whenever the same energy scale sets both the cosmological constant and the gaugino mass, one unavoidably runs into clashes with the experimental data. As a result, it would be interesting if two independent mechanisms be employed, the first setting the magnitude of the vacuum energy (possibly to zero, or to a very small value compatible with actual astronomical observations) and the second fixing the gaugino mass around 1 TeV.

Several ways to break supersymmetry in orientifold models have been recently devised. They follow mainly under the following classification: intersecting branes or, in the T-dual language magnetized backgrounds, Scherk-Schwarz compactifications and Brane Supersymmetry Breaking. Each mechanism has some interesting properties but at the same time suffers from the difficulty to get reasonable values for Λ and $m_{\rm gaugino}$. For example, intersecting brane constructions have shown to be a natural setting to realize the Standard Model pattern of gauge symmetries and matter within String Theory[1]. However, aside from omnipresent of tachyonic instabilities, it seems to be quite difficult to give masses to the superpartners of the $SU(3) \times SU(2) \times U(1)$ gauge bosons, and moreover a non-vanishing vacuum energy is generated already at the disk level[2] by the intersection angles (or by the strength of the background magnetic fields). This is not the case instead for Scherk-Schwarz deformations[3], where the mass of fermionic superpartners is altered by their different boundary conditions along the compact coordinates, while a cosmological constant is generated at one-loop and its magnitude is set by the size of the compact directions[4]: $\Lambda \sim R^{-4}$. As a result very large extra dimensions[5], consistently allowed in perturbative type I constructions, might make the vacuum energy compatible with the astronomical bounds. When D-branes are introduced, one has to distinguish between branes wrapping the Scherk-Schwarz directions and branes orthogonal to them[6]. In the former case supersymmetry is broken, the gauginos get a mass of order R^{-1} and thus are too light if the astronomical bounds for Λ are met. In the latter case instead, the D-brane excitations are not affected since they do not depend on the transverse coordinates, and the open-string spectrum stays classically supersymmetric. Finally, Brane Supersymmetry Breaking[7] is a pure stringy mechanism affecting only the open-string excitations. The supersymmetric bulk is then coupled to nonsupersymmetric branes where the mass splitting is set by the string scale itself, that in perturbative type I vacua can be lowered to few TeV. In its simplest realization, however, Brane Supersymmetry Breaking yields a non-vanishing vacuum energy already at the disk level, with $\Lambda \sim (n^\circ_{\rm B} - n^\circ_{\rm F})(\text{few TeV})^4$, where $n^\circ_{\rm B,F}$ counts the number of bosonic and fermionic degrees of freedom in the open-string sector. As a result, Λ is several orders of magnitude bigger than the observed value, unless the D-branes have Fermi-Bose degenerate excitations. Indeed, a class of nonsupersymmetric type I vacua with suppressed cosmological con-

[a]Talk presented at the XIth Regional Conference on Mathematical Physics, May 2004, Tehran, Iran.

stant and almost exact Fermi-Bose degenerate open-string spectrum was presented in Ref. [8], based on suitable combinations of Scherk-Schwarz deformations and Brane Supersymmetry Breaking.

In the following, we review another class of nonsupersymmetric vacua with vanishing perturbative vacuum energy[9], where a classically supersymmetric closed-string spectrum is coupled to a nonsupersymmetric open-string spectrum with exact Fermi-Bose degeneracy.

2 A Simple Six-Dimensional Example

Let us consider the compactification of the type IIB superstring on a rigid torus at the $SO(8)$ enhanced symmetry point

$$\mathcal{T} = |V_8 - S_8|^2 \left(|O_8|^2 + |V_8|^2 + |S_8|^2 + |C_8|^2\right) . \quad (1)$$

A world-sheet parity projection[10,11] of the spectrum in (1) amounts to introducing the Klein-bottle amplitude

$$\mathcal{K} = \tfrac{1}{2}(V_8 - S_8)\,(O_8 + V_8 + S_8 + C_8)\,, \quad (2)$$

so that the Ω-invariant six-dimensional massless excitations comprise an $\mathcal{N} = (1,1)$ supergravity multiplet coupled to four vector multiplets. In the transverse channel

$$\tilde{\mathcal{K}} = \frac{2^4}{2}(V_8 - S_8)\,O_8, \quad (3)$$

develops nonvanishing NS-NS and R-R tadpoles that require the introduction of D-branes to compensate the tension and charge of O-planes.

In the absence of Wilson lines, *i.e.* for N coincident D-branes, the transverse-channel annulus amplitude reads

$$\tilde{\mathcal{A}} = \frac{2^{-4}}{2}\,N^2\,(V_8 - S_8)\,(O_8 + V_8 + S_8 + C_8)\,, \quad (4)$$

and together with (3) implies the transverse-channel Möbius amplitude

$$\tilde{\mathcal{M}} = -N\,(\hat{V}_8 - \hat{S}_8)\,\hat{O}_8\,. \quad (5)$$

However, this is not the only possible choice for $\tilde{\mathcal{M}}$. The standard hated characters for the internal lattice contribution decompose with respect to $\mathrm{SO}(4) \times \mathrm{SO}(4)$ according to

$$\hat{O}_8 = \hat{O}_4\hat{O}_4 - \hat{V}_4\hat{V}_4\,, \quad (6)$$

but following Ref. [12], one may introduce discrete Wilson lines to modify the $\mathrm{SO}(4) \times \mathrm{SO}(4)$ decomposition according to

$$\hat{O}'_8 = \hat{O}_4\hat{O}_4 + \hat{V}_4\hat{V}_4\,, \quad (7)$$

and write the alternative Möbius amplitude

$$\tilde{\mathcal{M}}' = -N\,(\hat{V}_8 - \hat{S}_8)\,\hat{O}'_8\,. \quad (8)$$

Although this modification preserves tadpole cancellation $N = 16$ in the transverse-channel, it does affect the open-string spectrum, since the corresponding P transformation is also modified. In fact, since for the $SO(4)$ characters P interchanges $\hat{O}_4$ and $\hat{V}_4$, one finds that

$$P: \quad \hat{O}_8 \to -\hat{O}_8\,, \quad \text{but} \quad \hat{O}'_8 \to +\hat{O}'_8\,. \quad (9)$$

Therefore, the loop-channel annulus amplitude

$$\mathcal{A} = \tfrac{1}{2}\,N^2\,(V_8 - S_8)\,O_8, \quad (10)$$

has two consistent (supersymmetric) projections

$$\mathcal{M} = +\tfrac{1}{2}\,N\,(\hat{V}_8 - \hat{S}_8)\,\hat{O}_8\,, \quad (11)$$

and

$$\mathcal{M}' = -\tfrac{1}{2}\,N\,(\hat{V}_8 - \hat{S}_8)\,\hat{O}'_8. \quad (12)$$

The former yields a $USp(16)$ gauge group while the latter yields an $SO(16)$ gauge group.

Notice, however, that one could have well decided to modify the internal lattice only in the NS or only in R sector of the D-brane, or asymmetrically on two stacks of D-branes labeled by Chan-Paton multiplicities N and M. The corresponding direct-channel annulus and Möbius amplitudes then read

$$\begin{aligned}\mathcal{A} = \tfrac{1}{2}\,(V_8 - S_8)\,&\left[(N^2 + M^2)\,(O_4O_4 + V_4V_4)\right.\\ &\left.+2NM\,(O_4C_4 + V_4S_4)\right]\,, \end{aligned} \quad (13)$$

and

$$\begin{aligned}\mathcal{M} = \tfrac{1}{2}\,\hat{V}_8 &\left(N\,\hat{O}_8 - M\,\hat{O}'_8\right)\\ &-\tfrac{1}{2}\,\hat{S}_8\left(-N\,\hat{O}'_8 + M\,\hat{O}_8\right)\,, \end{aligned} \quad (14)$$

with $N + M = 16$ fixed by tadpole conditions.

It is then obvious that taking two identical stacks of branes, $N = M = 8$, yields an exact Fermi-Bose degenerate open-string spectrum, whose massless excitations comprise six-dimensional gauge bosons and four scalars in the adjoint of the Chan-Paton group

$$G_{\mathrm{CP}} = \mathrm{USp}(8) \otimes \mathrm{SO}(8), \quad (15)$$

and non-chiral fermions in the representations $(27 \oplus 1, 1) \oplus (1, 35 \oplus 1)$. As a result no contributions to the vacuum energy are generated at one-loop while supersymmetry is broken at the string scale on the D-branes.

Of course, Fermi-Bose degeneracy is not enough to guarantee that the vacuum energy stays small perturbatively, since higher-order corrections might well spoil this result. We shall now provide arguments that actually this is not the case: at any order in perturbation theory, no contributions to the vacuum energy are generated, if the branes are separated in equal sets, *i.e.* if $N = M$. This is obvious for closed Riemann surfaces, both oriented and unoriented, of arbitrary genus, since the closed-string sector is not affected by the deformation and therefore has the same properties as in the supersymmetric type I string case.

When boundaries are present, one has to be more careful, since the nonsupersymmetric deformation might well induce nonvanishing contributions to the vacuum energy. As shown in Ref. [9], however, these contributions are always multiplied by a numerical coefficient proportional to $(M - N)$, that vanishes for our choice of brane displacement. Due to lack of space, we cannot enter in more details, and refer the interested reader to the original paper [9], where convincing evidence for the vanishing of higher-order contributions is given.

Acknowledgments

I would like to thank M. Cardella for a stimulating collaboration and the organizers of the XIth Regional Conference on Mathematical Physics for their kind invitation and warm hospitality.

References

1. R. Blumenhagen, B. Kors, D. Lüst and T. Ott, *Nucl. Phys.* B **616**, 3 (2001), arXiv: hep-th/0107138; L.E. Ibanez, F. Marchesano and R. Rabadan, *JHEP* **0111**, 002 (2001), arXiv: hep-th/0105155.
2. C. Angelantonj, I. Antoniadis, E. Dudas and A. Sagnotti, *Phys. Lett.* B **489**, 233 (2000), arXiv: hep-th/0007090.
3. J. Scherk and J. H. Schwarz, *Nucl. Phys.* B **153**, 61 (1979); R. Rohm, *Nucl. Phys.* B **237**, 553 (1984); C. Kounnas and M. Porrati, *Nucl. Phys.* B **310**, 355 (1988); S. Ferrara, C. Kounnas, M. Porrati and F. Zwirner, *Nucl. Phys.* B **318**, 75 (1989).
4. H. Itoyama and T. R. Taylor, *Phys. Lett.* B **186**, 129 (1987).
5. I. Antoniadis, *Phys. Lett.* B **246**, 377 (1990); N. Arkani-Hamed, S. Dimopoulos and G. Dvali, *Phys. Lett.* B **429**, 263 (1998), arXiv: hep-ph/9803315; I. Antoniadis, N. Arkani-Hamed, S. Dimopoulos and G. R. Dvali, *Phys. Lett.* B **436**, 257 (1998), arXiv: hep-ph/9804398.
6. I. Antoniadis, E. Dudas and A. Sagnotti, *Nucl. Phys.* B **544**, 469 (1999), arXiv: hep-th/9807011.
7. S. Sugimoto, *Prog. Theor. Phys.* **102**, 685 (1999), arXiv: hep-th/9905159; I. Antoniadis, E. Dudas and A. Sagnotti, *Phys. Lett.* B **464**, 38 (1999), arXiv: hep-th/9908023.
8. C. Angelantonj and I. Antoniadis, *Nucl. Phys.* B **676**, 129 (2004), arXiv: hep-th/0307254.
9. C. Angelantonj and M. Cardella, arXiv: hep-th/0403107.
10. A. Sagnotti in *Cargese Summer Institute on Non-Perturbative Methods in Field Theory*, (Cargese, France, 1987), arXiv: hep-th/0208020.
11. C. Angelantonj and A. Sagnotti, *Phys. Rept.* **371**, 1 (2002); *Erratum-ibid.* **376**, 339 (2003), arXiv: hep-th/0204089.
12. M. Bianchi and A. Sagnotti, *Phys. Lett.* B **247**, 517 (1990), *Nucl. Phys.* B **361**, 519 (1991).

EXACT WILSONIAN EFFECTIVE SUPERPOTENTIAL FOR NONCOMMUTATIVE $\mathcal{N}=1$ SUPERSYMMETRIC $U(1)$

F. ARDALAN and N. SADOOGHI

Department of Physics, Sharif University of Technology
P.O. Box 11365-9161, Tehran, Iran
and
Institute for Theoretical Physics and Mathematics (IPM), School of Physics
P.O. Box 19395-5531, Tehran, Iran
Emails: ardalan@ipm.ir, sadooghi@sharif.edu

Using the covariant and invariant Konishi anomalies for noncommutative $\mathcal{N}=1$ supersymmetric $U(1)$ gauge theory, we calculate the exact Wilsonian low energy effective superpotential in the limit of large and small $\sqrt{\Theta}|p|$, with Θ the noncommutativity parameter. In the small $\sqrt{\Theta}|p|$ limit the noncommutative superpotential depends on a gauge invariant superfield which includes supersymmetric Wilson line.

1 Introduction

[a]The exact form of the Wilsonian low energy effective superpotential of ordinary $\mathcal{N}=1$ supersymmetric Quantum Field Theories can be determines using kinematical constraints such as holomorphy and various symmetries, and also certain dynamical information about the asymptotic behavior of the theory[1,2].

In this article, we will determine, using the Konishi anomalies of the theory, the exact Wilsonian effective superpotential for noncommutative $\mathcal{N}=1$ supersymmetric $U(1)$ gauge theory with particular emphasis on the role of nonplanar diagrams.

To do this we use the results of the planar and nonplanar anomalies of noncommutative nonsupersymmetric $U(1)$ gauge theories calculated previously in [3]. The important observation there was that a noncommutative gauge theory with matter fields in the fundamental representation consists of a gauge invariant, $j_{\mu,5}^{inv.}\equiv\bar{\psi}_\alpha(\gamma_\mu\gamma_5)^{\alpha\beta}\star\psi_\beta$, and a gauge covariant axial vector current, $J_{\mu,5}^{cov.}\equiv\psi_\beta\star\bar{\psi}_\alpha(\gamma_\mu\gamma_5)^{\alpha\beta}$. As it turns out, the anomaly corresponding to the covariant current arises from planar diagrams, and is given by a $\star$-deformation of the ordinary ABJ anomaly. The anomaly from invariant current, however, arises from infinitely many nonplanar diagrams. It involves a nonlocal open Wilson line and can be written in terms of generalized $\star$-products. It exhibits further the noncommutative UV/IR mixing[4].

This article is organized as follows: In Sec. 2, we will calculate the covariant (planar) and invariant (nonplanar) Konishi anomalies for $\mathcal{N}=1$ supersymmetric noncommutative $U(1)$, that will be used to determine the exact Wilsonian effective superpotential for small and large $\sqrt{\Theta}|p|$ limit, separately in Sec. 3.

2 Planar and Nonplanar Konishi Anomalies

Let us consider the partition function of noncommutative $\mathcal{N}=1$ supersymmetric $U(1)$

$$\mathcal{Z}=\int\mathcal{D}\Phi\,\mathcal{D}\bar{\Phi}\;e^{-iS_{matter}},\tag{1}$$

with the matter field action

$$S_{matter}=\int d^8z\;\bar{\Phi}(z)\star e^{V(z)}\star\Phi(z).\tag{2}$$

Here, ϕ and V are the standard chiral matter and gauge field supermultiplets and the $\star$-product is defined by

$$f(x)\star g(x)\equiv$$
$$\equiv f(x+\xi)\exp\left(\frac{i\Theta_{\mu\nu}}{2}\frac{\partial}{\partial\xi_\mu}\frac{\partial}{\partial\zeta_\nu}\right)g(x+\zeta)\Big|_{\xi=\zeta=0},$$

where Θ is the noncommutativity parameter. To calculate the *covariant* Konishi anomaly corresponding to the covariant current $J_{cov.}\equiv\Phi\star\bar{\Phi}\star e^V$ using the path integral method of Fujikawa, we use the invariance of the partition function (1) under *fundamental* local transformation $\delta_A\Phi(z)=iA(z)\star\Phi(z)$, with $A(z)$ an arbitrary function. Varying the matter field

[a]Talk presented by N.S. at the XIth Regional Conference on Mathematical Physics, May 2004, Tehran, Iran.

action with respect to this transformation and comparing the result with the corresponding Jacobian, we arrive at

$$\bar{D}^2 J_{cov.}(z) = i \sum_n \bar{D}^2 \left(\Phi_n(z) \star \bar{\Phi}_n(z)\right), \qquad (3)$$

which is to be regularized using the Gaussian damping factor e^{L/M^2} with $L \equiv \frac{1}{16}\bar{D}^2 e^{-V} D^2 e^V$ and the Pauli-Villars regulator mass M. After some standard algebra[5], the covariant Konishi equation for nonvanishing tree-level superpotential W_{tree} reads

$$-\frac{\bar{D}^2}{4} J_{cov.} = \Phi(z) \star \frac{\partial W_{tree}}{\partial \Phi(z)} + S_{cov.}(z), \qquad (4)$$

with the covariant Konishi anomaly

$$S_{cov.}(z) = -\frac{1}{32\pi^2} W_\alpha(z) \star W^\alpha(z), \qquad (5)$$

where $W_\alpha \equiv \bar{D}^2 e^{-V} D_\alpha e^V$. This result is the expected $\star$-deformation of the Konishi anomaly of ordinary $\mathcal{N} = 1$ supersymmetric $U(1)$.

To calculate the *invariant* Konishi anomaly corresponding to the invariant current $J_{inv.} \equiv \bar{\Phi} \star e^V \star \Phi$, we consider the invariance of the partition function (1) under the *antifundamental* local transformation $\delta_A \Phi(z) = i\Phi(z) \star A(z)$ and get

$$\bar{D}^2 J_{inv.}(z) = i \sum_n \bar{D}^2 \left(\bar{\Phi}(z) \star \Phi(z)\right). \qquad (6)$$

Here, in contrast to the previous case, since the l.h.s. of this equation in invariant under $\star$-gauge transformation, the r.h.s. must be regulated using a gauge invariant version of the Gaussian damping factor e^{L/M^2}. As in [6], we smear the operator L along an open Wilson line[7] with the length proportional to $\sqrt{\Theta}$. Expanding then the supersymmetric Wilson line in the orders of external gauge field and after some straightforward calculation, we arrive at

$$-\frac{\bar{D}^2}{4} J_{inv.} = S_{inv.},$$

with

$$S_{inv.} = -\frac{1}{32\pi^2} \int \frac{d^4k_1}{(2\pi)^4} \frac{d^4k_2}{(2\pi)^4}\, e^{-\frac{M^2}{4} q \circ q} \times e^{-ik_1 x}\, W_\alpha(k_1, \theta)\, \frac{\sin(k_1 \times k_2)}{(k_1 \times k_2)} W_\alpha(k_2, \theta)\, e^{-ik_2 x} + \cdots, \qquad (7)$$

with $q = k_1 + k_2$, $k_{1 \times k_2} \equiv \frac{\Theta_{\mu\nu}}{2} k_{1\mu} k_{2\nu}$, and $q \circ q \equiv q_\mu \Theta^{\mu\nu} \Theta_{\nu\rho} q^\rho$. Here, two different limits $\frac{q \circ q}{4} \gg \frac{1}{M^2}$ and $\frac{q \circ q}{4} \ll \frac{1}{M^2}$ lead to two different results

$$S_{inv.} = \begin{cases} 0, & \frac{q \circ q}{4} \gg \frac{1}{M^2}, \\ -\frac{1}{32\pi^2} W_\alpha \star' W^\alpha + \cdots, & \frac{q \circ q}{4} \ll \frac{1}{M^2}, \end{cases} \qquad (8)$$

with the generalized $\star$-product

$$f(x) \star' g(x) \equiv \equiv f(x+\xi) \frac{\sin\left(\frac{\Theta_{\mu\nu}}{2} \frac{\partial}{\partial \xi_\mu} \frac{\partial}{\partial \zeta_\nu}\right)}{\left(\frac{\Theta_{\mu\nu}}{2} \frac{\partial}{\partial \xi_\mu} \frac{\partial}{\partial \zeta_\nu}\right)} g(x+\zeta)\Big|_{\xi=\zeta=0}.$$

According to this result, the invariant Konishi anomaly is therefore affected by UV/IR mixing. Using this results the invariant Konsihi equation including the effects of the tree level superpotential reads

$$-\frac{\bar{D}^2}{4} J_{inv.} = \frac{\partial W_{tree}}{\partial \Phi(z)} \star \Phi(z) + S_{inv.}(z), \qquad (9)$$

with $S_{inv.}(z)$ given in (8).

3 Exact Effective Superpotential

In this section we calculate the low energy effective superpotential of noncommutative $\mathcal{N} = 1$ supersymmetric $U(1)$ as a function of generalized noncommutative meson matrix $T_{ij} \equiv \tilde{Q}_i \star Q_j$, $i, j = 1, \cdots, N_f$ and gaugino bilinears $S \propto W_\alpha \star W^\alpha$ and $S' \propto W_\alpha \star' W^\alpha + \cdots$, which turn out to be the relevant low energy degrees of freedom in the large and small $\sqrt{\Theta}|p|$ limit, respectively. In S' the extra terms denote the contribution of the open Wilson line. To do this, we use the one-loop β-function of the theory, the anomaly corresponding to $U_A(1)$ axial and $U_R(1)$ R-symmetry of the theory and the fact that in supersymmetric theories, due to nonrenormalization theorem which is also valid in the noncommutative case, the full effective superpotential is given by the sum of nonrenormalized tree level superpotential and the Veneziano-Yankielowicz[1] (VY) dynamical superpotential $W_{dyn.}$

$$W_{eff.} = W_{tree}(T; m, \Lambda) + W_{dyn.}(T, S; \Lambda_{N_f}). \qquad (10)$$

The one-loop β-function of noncommutative $\mathcal{N} = 1$ supersymmetric $U(1)$ with N_f flavors in the fundamental representation is given by $\beta(g) = -\frac{g^3}{16\pi^2}\beta_{N_f}$ with the coefficient[3,8]

$$\beta_{N_f} = \begin{cases} +2(3 - N_f), & |p| \gg \frac{1}{\sqrt{\Theta}}, \\ -2(3 - N_f), & |p| \ll \frac{1}{\sqrt{\Theta}}. \end{cases} \qquad (11)$$

It exhibits therefore the UV/IR mixing phenomenon, which affects also the anomalies corresponding to the

$U_A(1)$ axial and $U_R(1)$ R-symmetries. Further, according to the results in [6] the variation of the supersymmetric Lagrangian with respect to these global transformation reads

$$\delta_A \mathcal{L} = \begin{cases} 0, & \frac{p\circ p}{4} \gg \frac{1}{M^2}, \\ 2N_f \alpha\, \mathcal{A}', & \frac{p\circ p}{4} \ll \frac{1}{M^2}, \end{cases} \tag{12}$$

and

$$\delta_R \mathcal{L} = \begin{cases} 2\alpha R(\lambda)\mathcal{A}, & \frac{p\circ p}{4} \gg \frac{1}{M^2}, \\ 2N_f \alpha R(\psi)\mathcal{A}', & \frac{p\circ p}{4} \ll \frac{1}{M^2}, \end{cases} \tag{13}$$

respectively. Here, $\mathcal{A}$ and $\mathcal{A}'$ are the planar and nonplanar noncommutative ABJ anomalies. They are defined by $\mathcal{A} \equiv -\frac{1}{32\pi^2} F_{\mu\nu} \star \tilde{F}^{\mu\nu}$ and $\mathcal{A}' \equiv -\frac{1}{32\pi^2} F_{\mu\nu} \star' \tilde{F}^{\mu\nu} + \cdots$, where the ellipses denote the contribution of the open Wilson line. Further $R(\lambda)$ and $R(\psi)$ are the R-charges of the chiral gaugino λ in the adjoint representation and chiral fermion ψ in the fundamental representation.

According to (12) and (13), the relevant degrees of freedom in the *small* $\sqrt{\Theta}|p|$ limit are the meson field T and $S' \equiv -S_{inv.}$, with the *invariant* Konishi anomaly from (8). Varying the effective action with respect to $U_A(1)$ and $U_R(1)$ transformations, we arrive at two differential equations

$$T\frac{\partial W_{dyn.}}{\partial T} = -N_f S',$$

$$-W_{dyn.} + S' \frac{\partial W_{dyn.}}{\partial S'} + \frac{2}{3}T\,\frac{\partial W_{dyn.}}{\partial T} = \frac{N_f}{3}S',$$

which determine the VY dynamical superpotential in *small* $\sqrt{\Theta}|p|$ limit

$$W_{dyn.}(T, S'; \Lambda_{N_f}, \Lambda_\Theta) =$$
$$= S' \left(\log\left(\frac{S'^{N_f} \Lambda_\Theta^{(N_f+6)}}{\Lambda_{N_f}^{2(3+N_f)}\ \det T} \right) - N_f \right). \tag{14}$$

Here, in addition to the RG scale Λ_{N_f}, a new mass scale $\Lambda_\Theta \equiv \frac{1}{\sqrt{\Theta}}$ is introduced. The dependence on Λ_{N_f} and Λ_Θ in (14) are determined using the one-loop β-function (11) and the results from (12) and (13), so that the argument of logarithm in (14) is dimensionless and has vanishing R- and axial charges. Further integrating out the gaugino field S' from (14), the effective Affleck-Dine-Seiberg[2] (ADS) superpotential of the meson field reads

$$W_{eff.}^{ADS} = m \text{ tr } T + \lambda \text{ tr } T^2$$
$$-N_f \left(\frac{\Lambda_{N_f}^{-2(3+N_f)} \Lambda_\Theta^{(N_f+6)}}{\det T} \right)^{-\frac{1}{N_f}}, \tag{15}$$

where after (10), we have added the tree level superpotential $W_{tree} = m \text{ tr } T + \lambda \text{ tr } T \star T$ to the pure dynamical ADS superpotential arising from (14).

Similarly, the relevant degrees of freedom in the *large* $\sqrt{\Theta}|p|$ limit is, apart from the meson field T, the gaugino field $S \equiv -S_{cov.}$, with $S_{cov.}$ the *covariant* Konishi anomaly from (5). Using the information from (12) and (13) the differential equations leading to the VY dynamical superpotentials are

$$T\frac{\partial W_{dyn.}}{\partial T} = 0,$$

$$-W_{dyn.} + S\ \frac{\partial W_{dyn.}}{\partial S} + \frac{2}{3}T\ \frac{\partial W_{dyn.}}{\partial T} = -S,$$

which leads to VY dynamical superpotential in *large* $\sqrt{\Theta}|p|$ limit

$$W_{dyn.}(S; \Lambda_{N_f}, \Lambda_\Theta) =$$
$$= -S \left(\log\left(\frac{S\Lambda_\Theta^{3-2N_f}}{\Lambda_{N_f}^{2(3-N_f)}} \right) - 1 \right). \tag{16}$$

Here, the results of the one-loop β-function (11) as well as anomalies (12) and (13) are again used to determine the dependence of the dynamical superpotential (16) on the two relevant mass scales Λ_{N_f} and Λ_Θ. Integrating out the gaugino field the ADS superpotential of the meson fields reads

$$W_{eff.}^{ADS} = m \text{ tr } T + \lambda \text{ tr } T^2 + \Lambda_{N_f}^{6-2N_f} \Lambda_\Theta^{2N_f-3}. \tag{17}$$

We must emphasize that although the dependence of the effective superpotential (14) on the superfield S' is similar to the superpotential of ordinary SQCD[2], the dependence of S' on the Wilson line is highly nontrivial.

References

1. T.R. Taylor, G. Veneziano and S. Yankielowicz, *Nucl. Phys.* B **218**, 493 (1983).
2. K. Intriligator and N. Seiberg, *Nucl. Phys. Proc. Suppl.* 45 **BC**, 1 (1996), arXiv: hep-th/9509006.
3. F. Ardalan and N. Sadooghi, *Int. J. Mod. Phys.* A **16**, 3151 (2001); *ibid.* A **17**, 123 (2002).
4. S. Minwalla, M. van Raamsdonk and N. Seiberg, *JHEP* **0002**, 020 (2000).
5. K.-i. Konishi and K.-i.Shizuya, *Nuovo Cim.* A **90**, 111 (1985).

6. F. Ardalan and N. Sadooghi, to appear in *Int. J. Mod. Phys.* A (2005), arXiv: hep-th/0307155.
7. S. Marculescu and I. Mezinescu, *Nucl. Phys.* B **181**, 127 (1981); R.I. Mkrytchyan, *Nucl. Phys.* B **198**, 295 (1982); M. Awada and F. Mansouri, *Phys. Lett.* B **384**, 111 (1996).
8. V.V. Khoze and G, Travaglini, *JHEP* **0101**, 026 (2001); L. Alvarez-Gaumé and M.A. Vazquez-Mozo, *Nucl. Phys.* B **668**, 293 (2003).

ASPECTS OF NONCOMMUTATIVE GAUGE THEORIES AND THEIR COMMUTATIVE EQUIVALENTS

R. BANERJEE

S. N. Bose National Centre for Basic Sciences, JD Block, Sector 3
Salt Lake, Kolkata 700098, India
E-mail: rabin@bose.res.in

We discuss some exact Seiberg-Witten type maps for noncommutative electrodynamics. Their implications for anomalies in different (noncommutative and commutative) descriptions is also analyzed.

Exact SW Map for NC Electrodynamics

[a]Ever since Seiberg and Witten[1] gave a map (SW map) connecting a noncommutative (NC) gauge theory with its commutative equivalent, it has found startling applications as well as connections to other branches of physics[2]. This map ensures the stability of gauge transformations in the commutative and NC descriptions. The original map was given for the potentials and the field strengths, but it was only valid up to the leading order in the NC parameter. In this paper, we review some exact results on SW-type maps involving various objects like the action, currents, anomalies, etc.

The action for NC electrodynamics is given by

$$\widehat{S}_{\rm NC} = -\frac{1}{4}\int \mathrm{d}^4x\, \widehat{F}_{\mu\nu}\star\widehat{F}^{\mu\nu}, \tag{1}$$

where $\widehat{F}_{\mu\nu} = \partial_\mu\widehat{A}_\nu - \partial_\nu\widehat{A}_\mu - \mathrm{i}\,\widehat{A}_\mu\star\widehat{A}_\nu + \mathrm{i}\,\widehat{A}_\nu\star\widehat{A}_\mu$ and

$$(A\star B)(x) = \exp\left(\frac{\mathrm{i}}{2}\theta^{\alpha\beta}\partial^x_\alpha\partial^y_\beta\right)A(x)B(y)\Big|_{x=y}. \tag{2}$$

To express the action in terms of its commutative equivalents, recall that the low-energy effective action on a single Dp-brane is given by the DBI action[3],

$$S(g_s,g,A,B) = \frac{2\pi}{g_s(2\pi k)^{\frac{p+1}{2}}}\int \mathrm{d}^{p+1}x\,\sqrt{-\det\left(g+k(F+B)\right)}, \tag{3}$$

where $F_{\mu\nu} = \partial_\mu A_\nu - \partial_\nu A_\mu$ and the expression involves the usual closed string variables (g_s, g).

As shown in Ref. [1], there is a general description interpolating between the commutative description with closed string parameters and the NC description with open string parameters. The effective action in these variables is given by

$$\widehat{S}(G_s,G,\widehat{A},\theta) = \frac{2\pi}{G_s(2\pi k)^{\frac{p+1}{2}}}\int \mathrm{d}^{p+1}x\,\sqrt{-\det\left(G+k(\widehat{F}+\Phi)\right)}, \tag{4}$$

where the connection between the open and closed string parameters is

$$\frac{1}{G+k\Phi}+\frac{\theta}{k} = \frac{1}{g+kB}, \qquad G_s = g_s\sqrt{\frac{\det(G+k\Phi)}{\det(g+kB)}}. \tag{5}$$

For $\Phi = B$ leading to $G=g$, $G_s = g_s$, $\theta = 0$, the action $\widehat{S}$ reproduces the commutative description (3), while $\Phi = 0$ yields the familiar NC description. Furthermore, it was shown in Ref. [1] that, in the slowly-varying-field approximation, the DBI actions are independent of the choice of Φ, so that

$$\widehat{S}(G_s,G,\widehat{A},\theta) = S(g_s,g,A,B) + O(\partial F). \tag{6}$$

Using the identities (5) in the explicit structure for $\widehat{S}$, S and then exploiting (6) yields

$$\int \mathrm{d}^{p+1}x\,\sqrt{-\det\left(G+k(\widehat{F}+\Phi)\right)} = \int \mathrm{d}^{p+1}x\,\sqrt{\det(1+F\theta)} \times\sqrt{-\det\left(G+k(\Phi+\boldsymbol{F})\right)} + O(\partial F), \tag{7}$$

where

$$\boldsymbol{F} = \frac{1}{1+F\theta}F,$$

and a matrix notation $AB = A_{\mu\alpha}B^{\alpha\mu}$ or $(AB)_{\mu\nu} = A_{\mu\alpha}B^{\alpha}{}_{\nu}$ is being used. In the zero-slope limit $k\to 0$,

[a]Invited talk presented at the XIth Regional Conference on Mathematical Physics, May 2004, Tehran, Iran.

(7) for $\Phi = 0$ and $p = 3$ defines the exact non-linear action of θ-expanded NC electrodynamics (1)

$$-\frac{1}{4}\int d^4x\, \widehat{F}_{\mu\nu} \star \widehat{F}^{\mu\nu}$$
$$+\frac{1}{4}\int d^4x \sqrt{\det(1+F\theta)}\left(\frac{1}{1+F\theta}F\frac{1}{1+F\theta}F\right). \quad (8)$$

If we introduce an effective non-symmetric "metric" induced by dynamical gauge fields such that

$$g_{\mu\nu} = \eta_{\mu\nu} + (F\theta)_{\mu\nu}, \quad g^{\mu\nu} = \left(\frac{1}{\eta + F\theta}\right)^{\mu\nu},$$

then the NC Maxwell action (8) looks like the ordinary Maxwell action coupled to the induced metric $g_{\mu\nu}$

$$-\frac{1}{4}\int d^4x\, \widehat{F}_{\mu\nu} \star \widehat{F}^{\mu\nu}$$
$$= -\frac{1}{4}\int d^4x \sqrt{-\det g}\, g^{\mu\alpha} g^{\beta\nu} F_{\mu\nu} F_{\alpha\beta}. \quad (9)$$

This result was earlier obtained in Refs. [4, 5]. Up to $O(\theta)$, an explicit expansion reveals

$$-\frac{1}{4}\int d^4x\, \widehat{F}_{\mu\nu} \star \widehat{F}^{\mu\nu} = -\frac{1}{4}\int d^4x \Big[F_{\mu\nu}F^{\mu\nu}$$
$$+\theta^{\alpha\beta}\left(2F_{\mu\alpha}F_{\nu\beta} - \frac{1}{2}F_{\alpha\beta}F_{\mu\nu}\right)F^{\mu\nu}\Big] + O(\theta^2). \quad (10)$$

This agrees with the result obtained by directly using the SW map[1]

$$\widehat{F}_{\mu\nu} = F_{\mu\nu} + \theta^{\alpha\beta}(F_{\mu\alpha}F_{\nu\beta} - A_\alpha\partial_\beta F_{\mu\nu}) + O(\theta^2), \quad (11)$$

in the NC action (1).

Exact SW Map for Sources

Let us next consider the inclusion of the sources to (1)

$$\widehat{S}(\widehat{A},\widehat{\psi}) = -\frac{1}{4}\int d^4x\, \widehat{F}_{\mu\nu} \star \widehat{F}^{\mu\nu} + \widehat{S}_{\rm M}(\widehat{\psi},\widehat{A})$$
$$= \widehat{S}_{\rm ph}(\widehat{A}) + \widehat{S}_{\rm M}(\widehat{\psi},\widehat{A}). \quad (12)$$

Then the equation of motion for $\widehat{A}_\mu$ is

$$\frac{\delta \widehat{S}_{\rm ph}}{\delta \widehat{A}_\mu} = \widehat{\rm D}_\nu \star \widehat{F}^{\nu\mu} = -\widehat{J}^\mu,$$

where

$$\widehat{J}^\mu := \left.\frac{\delta \widehat{S}_{\rm M}}{\delta \widehat{A}_\mu}\right|_{\widehat{\psi}}.$$

It is clear that $\widehat{J}^\mu$ transforms covariantly and is also covariantly conserved ($\widehat{\rm D}_\mu \star \widehat{J}^\mu = 0$).

Now the action (12) is rewritten, using the SW map (8) in particular, to obtain a U(1) gauge-invariant action defined on commutative space

$$\widehat{S}\left(\widehat{A},\widehat{\psi}\right)\Big|_{\rm SW\ map} := S_{\rm ph}(A) + S_{\rm M}(\psi, A),$$

where $S_{\rm ph}(A)$ equals the r.h.s. side of (8). The equation of motion is

$$\frac{\delta S_{\rm ph}}{\delta A_\mu} = -J^\mu,$$

where

$$J^\mu := \left.\frac{\delta S_{\rm M}}{\delta A_\mu}\right|_{\psi}.$$

This J_μ is gauge invariant and satisfies the ordinary conservation law ($\partial_\mu J^\mu = 0$).

It is now feasible to derive a relation between $\widehat{J_\mu}$ and J_μ by noticing that

$$\widehat{J}^\mu(x) = \left.\frac{\delta \widehat{S}_{\rm M}}{\delta \widehat{A}_\mu}\right|_{\widehat{\psi}}$$
$$= \int d^4y \left[\left.\frac{\delta S_{\rm M}}{\delta A_\nu(y)}\right|_{\psi} \frac{\delta A_\nu(y)}{\delta \widehat{A}_\mu(x)} + \left.\frac{\delta S_{\rm M}}{\delta \psi_\alpha(y)}\right|_{A} \frac{\delta \psi_\alpha(y)}{\delta \widehat{A}_\mu(x)}\right]$$
$$= \int d^4y\, J^\nu(y)\frac{\delta A_\nu(y)}{\delta \widehat{A}_\mu(x)}, \quad (13)$$

where the equation of motion for ψ_α, *i.e.*,

$$\delta S_{\rm M}/\delta\psi_\alpha = 0,$$

has been used. This is the exact SW map for the sources. It was earlier derived in Ref. [6]. Knowing the original SW map among the potentials, it is possible to get explicit expressions for the above map. Up to $O(\theta)$ we find

$$\widehat{J}^\mu = J^\mu + (\theta F J)^\mu + \partial_\alpha\left(\theta^{\alpha\beta}A_\beta J^\mu\right) + O(\theta^2). \quad (14)$$

For higher order corrections, we refer to Ref. [7]. It is easy to check that this map ensures the stability of gauge transformations. Thus inserting the ordinary gauge transformations ($\delta J_\mu = 0$, $\delta F_{\mu\nu} = 0, \delta A_\mu = \partial_\mu\lambda$), yields

$$\delta\widehat{J}_\mu = \theta^{\alpha\beta}\partial_\alpha J_\mu \partial_\beta\lambda = \theta^{\alpha\beta}\partial_\alpha \widehat{J}_\mu \partial_\beta\widehat{\lambda} + O(\theta^2),$$

which reproduces the desired covariant transformation (up to $O(\theta)$) for $\widehat{J}_\mu$.

1 A Map for Anomalies

The relation (14) may be used to provide a map for anomalies. Taking the covariant derivative on both sides of (14) yields

$$\widehat{\mathrm{D}}^\mu \star \widehat{J}_\mu = \partial^\mu J_\mu + \theta^{\alpha\beta}\partial_\alpha\left(A_\beta \partial^\mu J_\mu\right), \qquad (15)$$

where

$$\widehat{\mathrm{D}}^\mu \star \widehat{J}_\mu = \partial^\mu \widehat{J}_\mu + \theta^{\alpha\beta}\partial_\alpha A^\mu \partial_\beta J_\mu + O(\theta^2).$$

This relation is obviously compatible for conserved currents ($\partial_\mu J^\mu = 0$) in the commutative description and covariantly conserved currents ($\widehat{\mathrm{D}}_\mu \star \widehat{J}^\mu = 0$) in the NC description. In fact, it serves as a non-trivial consistency check on our formalism. What happens if there is an anomaly so that the currents are no longer conserved? This situation arises if we consider axial/chiral currents instead of vector currents and take loop effects into account. We prove that the map (15) is still valid providing a connection between the ABJ anomaly $\mathscr{A} = \partial_\mu J^{\mu 5} = (1/16\pi^2)\varepsilon_{\mu\nu\lambda\rho}F^{\mu\nu}F^{\lambda\rho}$ in the usual case and the planar (covariant) anomaly $\widehat{\mathscr{A}} = \widehat{\mathrm{D}}_\mu \star \widehat{J}^{\mu 5} = (1/16\pi^2)\varepsilon_{\mu\nu\lambda\rho}\widehat{F}^{\mu\nu} \star \widehat{F}^{\lambda\rho}$ in the NC case, so that

$$\widehat{\mathscr{A}} = \mathscr{A} + \theta^{\alpha\beta}\partial_\alpha(A_\beta \mathscr{A}), \qquad (16)$$

as follows from a simple extension of (15). Inserting $\mathscr{A}$ and using the identity[8]

$$\theta^{\alpha\beta}F_{\alpha\beta}\varepsilon_{\mu\nu\lambda\rho}F^{\mu\nu}F^{\lambda\rho} = -4\varepsilon_{\mu\nu\lambda\rho}(F\theta F)^{\mu\nu}F^{\lambda\rho},$$

immediately yields the SW-transformed planar anomaly ($\widehat{\mathscr{A}}(\widehat{F}) = \widehat{\mathscr{A}}(F, A)$) obtained by using the map (11) in $\widehat{\mathscr{A}}$. This completes the analysis up to $O(\theta)$. For higher orders it was shown in [7] that the anomaly map obtained by this technique holds only in the slowly-varying-field approximation; *i.e.*, where the $\star$-product appearing in $\widehat{\mathscr{A}}$ can be ignored.

Discussions

We conclude our paper by discussing some other possibilities. For example, the exact SW map for scalar field is found to be[5]

$$\frac{1}{2}\int \mathrm{d}^4x\,(\widehat{\mathrm{D}}^\mu \star \widehat{\phi}) \star (\widehat{\mathrm{D}}_\mu \star \widehat{\phi})$$
$$= \frac{1}{2}\int \mathrm{d}^4x\,\sqrt{\det(1+F\theta)}\left(\frac{1}{1+F\theta}\frac{1}{1+\theta F}\right)^{\mu\nu} \times \partial_\mu\phi\partial_\nu\phi. \qquad (17)$$

The extension for non-abelian $\mathrm{U}(N)$ groups is rather non-trivial. The analogue of (8) is difficult since the corresponding string based arguments are highly ambiguous. However, there is no such problem for the sources and be get[7]

$$\widehat{J}^{\mu,a}(x) = \int \mathrm{d}^4y\, J^{\nu,b}(y)\frac{\delta A_\nu^b(y)}{\delta \widehat{A}_\mu^a(x)},$$

which is the expected modification to (13).

Acknowledgements

This work is based on collaborations with K. Kumar, C. Lee and H. S. Yang. I thank them all, as also the organizers of the workshop for a very pleasant atmosphere, academic or otherwise.

References

1. N. Seiberg and E. Witten, *JHEP* **9909**, 032 (1999), arXiv: hep-th/9908142.
2. R. Jackiw, arXiv: physics/0209108.
3. E.S. Fradkin and A.A. Tseytlin, *Phys. Lett.* B **163**, 123 (1985); A. Abouelsaood, C.G. Callan, Jr., C.R. Nappi and S.A. Yost, *Nucl. Phys.* B **280**, 599 (1987).
4. H.S. Yang, arXiv: hep-th/0402002.
5. R. Banerjee and H.S. Yang, arXiv: hep-th/0404064.
6. R. Banerjee, C. Lee and H.S. Yang, to appear in *Phys. Rev.* D, arXiv: hep-th/0312103.
7. R. Banerjee and K. Kumar, arXiv: hep-th/0404110.
8. R. Banerjee and S. Ghosh, *Phys. Lett.* B **533**, 162 (2002), arXiv: hep-th/0110177, R. Banerjee, *Int. J. Mod. Phys.* A **19**, 613 (2004), arXiv: hep-th/0301174.

THE CONTINUUM LIMIT OF THE NONCOMMUTATIVE $\lambda\phi^4$ MODEL

W. BIETENHOLZ and F. HOFHEINZ
Institut für Physik, Humboldt Universität zu Berlin
Newtonstr. 15, D-12489 Berlin, Germany
E-mails: bietenho@physik.hu-berlin.de, hofheinz@physik.hu-berlin.de

J. NISHIMURA
High Energy Accelerator Research Organization (KEK)
1-1 Oho, Tsukuba 305-0801, Japan
E-mail: jnishi@post.kek.jp

We present a numerical study of the $\lambda\phi^4$ model in three Euclidean dimensions, where the two spatial coordinates are noncommutative (NC). We first show the explicit phase diagram of this model on a lattice. The ordered regime splits into a phase of uniform order and a "striped phase". Then, we discuss the dispersion relation, which allows us to introduce a dimensionful lattice spacing. Thus, we can study a double scaling limit to zero lattice spacing and infinite volume, which keeps the noncommutativity parameter constant. The dispersion relation in the disordered phase stabilizes in this limit, which represents a nonperturbative renormalization. From its shape, we infer that the striped phase persists in the continuum, and we observe UV/IR mixing as a nonperturbative effect.

1 The Noncommutative Plane

[a]We consider a NC plane given by Hermitian coordinate operators $\hat{x}_\mu$, which obeys

$$[\hat{x}_\mu, \hat{x}_\nu] = i\theta\epsilon_{\mu\nu} \ . \tag{1}$$

We impose a (fuzzy) lattice structure on this plane by means of the operator identity

$$\exp\left(i\frac{2\pi}{a}\hat{x}_\mu\right) = \hat{1\!\!1} \ . \tag{2}$$

The periodicity of the momentum components k_μ implies

$$\theta k_\mu / 2a \in \mathbb{Z} \ . \tag{3}$$

Hence the lattice is automatically periodic, say, over the lattice volume $N \times N$. Then the noncommutativity parameter corresponds to

$$\theta = Na^2/\pi \ . \tag{4}$$

We see that the limits to the continuum ($a \to 0$) and to the infinite volume ($Na \to \infty$) should be taken simultaneously, if we want to keep θ finite. In particular, we are interested in the double scaling limit $a \to 0$, $N \to \infty$, which keeps $\theta = const.$

2 The 3d NC $\lambda\phi^4$ Model

In the $\star$-product formulation, the action of the NC $\lambda\phi^4$ model in Euclidean space takes the form

$$S[\phi] = \int d^dx \left[\frac{1}{2}\partial_\mu\phi\partial_\mu\phi + \frac{m^2}{2}\phi^2 + \frac{\lambda}{4}\phi\star\phi\star\phi\star\phi\right] .$$

Since the $\star$-product does not affect the bilinear terms, λ determines the strength of NC effects. We consider this model in $d = 3$ with a commutative Euclidean time and a NC plane. Its formulation on a N^3 lattice can be mapped onto a matrix model[1] of the form

$$S[\bar{\phi}] = N\mathrm{Tr}\sum_{t=1}^{N}\left\{\frac{1}{2}\sum_{\mu=1}^{2}\left[\Gamma_\mu\bar{\phi}(t)\Gamma_\mu^\dagger - \bar{\phi}(t)\right]^2 + \frac{1}{2}\left[\bar{\phi}(t+1) - \bar{\phi}(t)\right]^2 + \frac{m^2}{2}\bar{\phi}^2(t) + \frac{\lambda}{4}\bar{\phi}^4(t)\right\} , \tag{5}$$

where each time site $t = 1, \cdots, N$ accommodates a Hermitian $N \times N$ matrix $\bar{\phi}(t)$. The "twist eaters" Γ_μ provide a shift by one lattice unit in a spatial direction, if they obey the 't Hooft-Weyl algebra

$$\Gamma_\mu\Gamma_\nu = Z_{\nu\mu}\Gamma_\nu\Gamma_\mu \ . \tag{6}$$

The twist $Z_{\mu\nu} = Z^*_{\nu\mu}$ is a phase factor; in our formulation it reads $Z_{12} = \exp(i\pi(N+1)/N)$.

[a]Talk presented by W.B. at the XIth Regional Conference on Mathematical Physics, May 2004, Tehran, Iran.

3 The Phase Diagram

In contrast to the $\star$-product formulation, the matrix formulation of (5) is suitable for Monte-Carlo simulations. Our numerical results are described in Ref. [2], and in several proceeding contributions as well as a Ph.D. thesis[3]. Similar techniques were applied to arrive at nonperturbative results for 2d NC field theories, in particular for the $\lambda\phi^4$ model on a NC plane[4,2] and on a fuzzy sphere[5], and for NC QED_2.[6]

In the 3d model described above, we first explored the phase diagram. We found it to be stable for $N \gtrsim 25$ in the plane spanned by the axes $N^2 m^2$ and $N^2\lambda$, see Fig. 1.

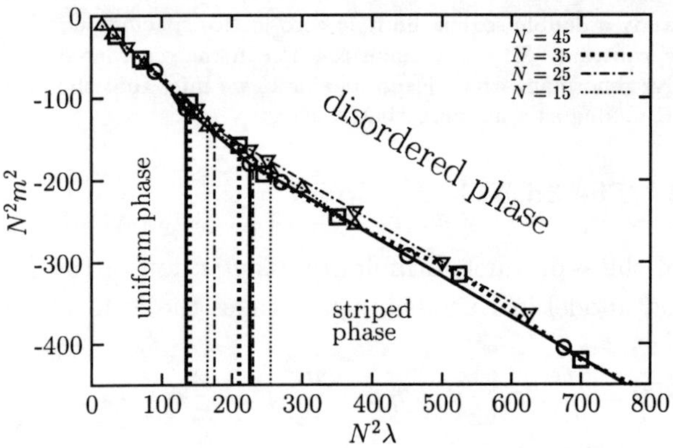

Figure 1. The phase diagram of the 3d NC $\lambda\phi^4$ model, obtained from numerical simulations on a N^3 lattice.

Strongly negative m^2 leads to ordering. At weak λ, the order is uniform as in the commutative world, whereas at stronger λ (corresponding to an amplified θ) stripe patterns dominate. This picture agrees with analytic conjectures[7]. The occurrence of a striped phase is a qualitative difference from the commutative $\lambda\phi^4$ model, though similar effects are known for instance in the (commutative) Gross-Neveu model at large chemical potential[8] and in ferromagnetic superconductors.

Our results for the hysteresis suggest that the uniform-striped transition is of first order, while both disorder-order transitions are of second order. The formation of stripes was observed by introducing a momentum dependent order parameter. The problem in this direct consideration is that at our values of N usually just two stripes can be observed as manifestly stable. However, multi-stripe patterns are supposed to dominate the striped phase at large volumes. We will demonstrate this behavior in an indirect way in Sec. 5.

4 Dispersion Relation

The correlation functions with a spatial separation have a fast but non-exponential decay[2]; apparently it is distorted by the NC geometry. In Euclidean time direction, however, the decay of the correlator turned out to be exponential. At fixed spatial momenta $\vec{p} = (p_1, p_2)$, this allows us to determine the energy E and thus the dispersion relation $E^2(\vec{p}^{\,2})$. It is most instructive to look at it in the disordered phase – which does not suffer much from finite size effects – close to the ordering transition. Figure 2 shows examples close to the uniform order (on top) and close to the striped order (below). The former follows the familiar linear shape, whereas the latter has its energy minimum at non-zero momentum. Clearly for decreasing m^2, the minimal mode condenses, giving rise to a stripe pattern.

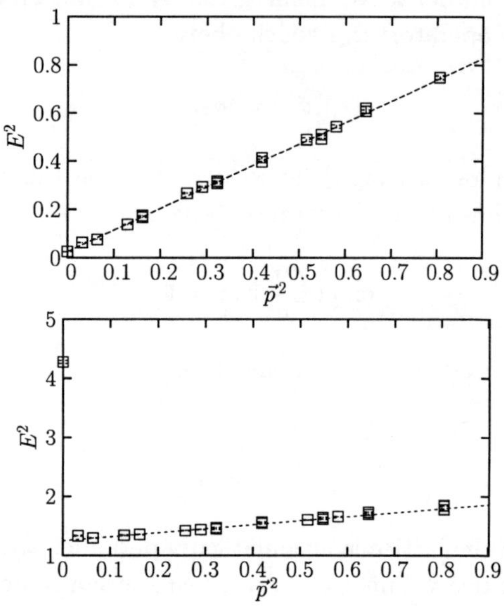

Figure 2. The dispersion relation in the disordered phase, close to ordering, at $N = 35$. For $\lambda = 0.06$ (on top), we obtain the usual linear dispersion, but for $\lambda = 100$ (below) the energy minimum moves to a finite momentum.

5 Continuum Limit

In order to obtain a "physical scale", we have to identify a dimensionful quantity. To this end, we consider the planar limit $N \to \infty$ at fixed λ and m^2. This pushes the IR jump in the dispersion towards zero, and we obtain (for all finite momenta) a linear dispersion of the form $E^2 = M_{\text{eff}}^2 + \vec{p}^{\,2}$. Measuring now M_{eff}^2 at fixed λ but varying m^2, we observed a linear dependence,

$$M_{\text{eff}}^2|_{\lambda = const.} = \mu^2 + \gamma m^2 \ , \tag{7}$$

which corresponds to the critical exponent $\nu = 1/2$. For instance, at $\lambda = 50$, we found the critical parameter $m_c^2 = -\mu^2/\gamma = -15.01(8)$.

The continuum limit is now taken such that the dimensionful effective mass, M_{eff}/a, remains constant (a being the lattice spacing). Hence the double scaling limit means $N \to \infty$ and $m^2 \to m_c^2$ such that $N(m^2 - m_c^2) = const.$ In our study, we chose the latter constant as 100, which implies $\theta = 100\gamma/\pi = 9.77(6)(a/M_{\text{eff}})^2$.

It turns out that the dispersion relation stabilizes in this double limit at all finite momenta, which demonstrates the nonperturbative renormalizability of this model. Figure 3 (on top) shows the energy minimum around $\vec{p}^{\,2}/a^2 \lesssim 0.1$. This translates into a dominant, finite stripe width, so after condensation we expect an infinite number of stripes in the double scaling limit to the continuum and infinite volume.

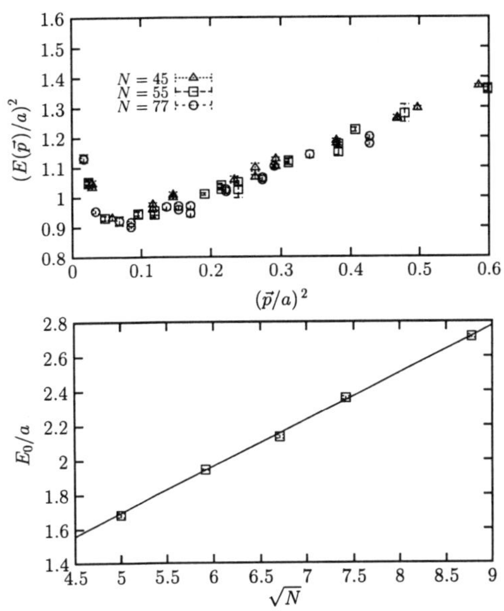

Figure 3. On top: The dispersion relation as we approach the double scaling limit. Its stabilization indicates nonperturbative renormalizability. The dip at finite $(\vec{p}/a)^2$ shows that the striped phase survives this limit. Below: E_0/a diverges linearly in $\sqrt{N} \propto 1/a$, in agreement with UV/IR mixing.

In this limit, the rest energy $E_0/a = E(\vec{p} = \vec{0})/a$ diverges linearly in $\sqrt{N} \propto 1/a$, see Fig. 3 (below). Also the UV divergence is linear in this model, hence our observation agrees perfectly with the concept of UV/IR mixing, which is known from perturbation theory[9]. Here this mixing is observed as a nonperturbative effect, so it belongs to the very nature of the system.

6 Conclusions

We have shown that the NC 3d $\lambda\phi^4$ model – with two NC spatial direction and a Euclidean time – is *nonperturbatively renormalizable*, and that its phase diagram includes a *striped phase*, which is there to stay in the continuum limit (more precisely: in the double scaling limit to zero lattice spacing and infinite volume, at a fixed noncommutativity parameter θ).

The striped phase implies the spontaneous breaking of translation symmetry. It also exists in the 2d version of this model (omitting the time direction)[4,2]; note that NC field theories are nonlocal, hence the Mermin-Wagner Theorem does not apply. At $\theta \to \infty$ perturbation theory suggests a commutative behavior (in this case the equivalence to a large N matrix field theory). However, in our phase diagram the uniform order does not return at large λ (which corresponds to large θ); in the case of spontaneous symmetry breaking the perturbative argument is not valid, because it does not capture an expansion around the striped ground state[10].

Acknowledgments

W.B. would like to thank the organizers of the XIth Regional Conference on Mathematical Physics in Tehran for their kind hospitality.

References

1. J. Ambjørn, Y.M. Makeenko, J. Nishimura and R.J. Szabo, *JHEP* **11**, 029 (1999); *Phys. Lett.* B **480**, 399 (2000); *JHEP* **05**, 023 (2000).
2. W. Bietenholz, F. Hofheinz and J. Nishimura, to appear in *JHEP*, arXiv: hep-th/0404020.
3. W. Bietenholz, F. Hofheinz and J. Nishimura, *Nucl. Phys.* B (Proc. Suppl.) **119**, 941 (2003); *Fortschr. Phys.* **51**, 745 (2003); *Nucl. Phys.* B

(Proc. Suppl.) **129**, 865 (2004); *Acta Phys. Polon.* B **34**, 4711 (2003). F. Hofheinz, Ph.D. thesis (2003), arXiv: hep-th/0403117.
4. J. Ambjørn and S. Catterall, *Phys. Lett.* B **549**, 253 (2002).
5. X. Martin, *Mod. Phys. Lett.* A **18**, 2389 (2003).
6. W. Bietenholz, F. Hofheinz and J. Nishimura, *JHEP* **09**, 009 (2002).
7. S.S. Gubser and S.L. Sondhi, *Nucl. Phys.* B **605**, 395 (2001); P. Castorina and D. Zappalà, *Phys. Rev.* D **68**, 065008 (2003).
8. O. Schnetz, M. Thies and K. Urlichs, arXiv: hep-th/0402014.
9. S. Minwalla, M. van Raamsdonk and N. Seiberg, *JHEP* **02**, 020 (2000).
10. W. Bietenholz, F. Hofheinz and J. Nishimura, *JHEP* **05**, 047 (2004).

HIGH-ENERGY BOUNDS ON SCATTERING AMPLITUDES IN QFT ON NONCOMMUTATIVE SPACE-TIME

M. CHAICHIAN and A. TUREANU

Department of Physical Sciences, University of Helsinki
Helsinki Institute of Physics
P.O. Box 64, 00014 Helsinki, Finland
E-mails: masud.chaichian, anca.tureanu@helsinki.fi

In the framework of Quantum Field Theory (QFT) on noncommutative (NC) space-time with $SO(1,1) \times SO(2)$ symmetry and space-space noncommutativity ($\theta_{0i} = 0$), we prove that, based on the causality condition usually taken in connection with this symmetry, it is merely impossible to draw any conclusion on the analyticity of the $2 \to 2$-scattering amplitude in $\cos\Theta$, Θ being the scattering angle. A physical choice of the causality condition rescues the situation and as a result an analog of Lehmann's ellipse as domain of analyticity in $\cos\Theta$ is obtained. However, the enlargement of this analyticity domain to Martin's ellipse and the derivation of the Froissart bound for the total cross section in NC QFT is possible *only* in the special case when the incoming momentum is orthogonal to the NC plane. This is the first example of a nonlocal theory in which the cross sections are subject to a high-energy bound.

1 Introduction

[a]The development of QFT on NC space-time, especially after the seminal work of Seiberg and Witten[1], which showed that the NC QFT arises from string theory, has triggered lately the interest also towards the formulation of an axiomatic approach to the subject. In the axiomatic approach to commutative QFT, one of the fundamental results consisted of the rigorous proof of the Froissart bound on the high-energy behavior of the scattering amplitude, based on its analyticity properties[2,3]. In Ref. [4], the aim was to obtain the analog of this bound when the space-time is NC. Such an achievement, besides being topical in itself, also proved fruitful in the conceptual understanding of subtle issues, such as causality, in nonlocal theories to which the NC QFT's belong.

In the following, we shall consider NC QFT on a space-time with the commutation relation

$$[x_\mu, x_\nu] = i\theta_{\mu\nu} \ , \qquad (1)$$

where $\theta_{\mu\nu}$ is an antisymmetric constant matrix. Such NC theories violate Lorentz invariance, but is invariant under the subgroup $SO(1,1) \times SO(2)$ of the Lorentz group, while translational invariance still holds. We shall consider the particular case of space-space noncommutativity, *i.e.* $\theta_{0i} = 0$, and choose the system of coordinates, such that $\theta_{13} = \theta_{23} = 0$ and $\theta_{12} = -\theta_{21} \equiv \theta$.

In the conventional (commutative) QFT, the Froissart bound was first obtained[2] using the conjectured Mandelstam representation (double dispersion relation), which assumes analyticity in the entire E and $\cos\Theta$ complex planes. The Froissart bound

$$\sigma_{tot}(E) \leq c \ \ln^2 \frac{E}{E_0} \ , \qquad (2)$$

expresses the upper limit of the total cross section σ_{tot} as a function of the CMS energy E, when $E \to \infty$. However, such an analyticity or equivalently the double dispersion relation has not been proven, while smaller domains of analyticity in $\cos\Theta$ were already known[5].

One of the main ingredients in rigorously obtaining the Froissart bound is the Jost-Lehmann-Dyson representation[6,7] of the Fourier transform of the matrix element of the commutator of currents, which is based on the causality as well as the spectral conditions. Based on this integral representation, one obtains the domain of analyticity of the scattering amplitude in $\cos\Theta$. This domain proves to be an ellipse – the so-called Lehmann's ellipse[5]. However, the domain of analyticity in $\cos\Theta$ can be enlarged to the so-called Martin's ellipse by using the dispersion relations satisfied by the scattering amplitude and the unitarity constraint on the partial-wave amplitudes. Using this larger domain of analyticity, the Froissart bound (2) was rigorously proven[3].

In Ref. [4], we followed the same prescription for the case of NC QFT with $\theta_{0i} = 0$ for the derivation of the high-energy bound on the scattering amplitude. The main points and peculiarities of the derivation

[a]Talk presented by A.T. at the XIth Regional Conference on Mathematical Physics, May 2004, Tehran, Iran.

are highlighted in the following.

2 Analyticity of the Scattering Amplitude in $\cos\Theta$. Lehmann's Ellipse

The process considered is the $2 \to 2$ scalar particles scattering, $k + p \to k' + p'$. In the center of mass system (CMS) and in a set in which the incoming particles are along the vector $\vec{\beta} = (0, 0, \theta)^b$, the scattering amplitude in NC QFT depends still on only two variables, the CM energy E and the cosine of the scattering angle, $\cos\Theta$.

For NC QFT with $SO(1,1) \times SO(2)$ symmetry, in Ref. [8] a new causality condition was proposed, involving (instead of the light-cone) the light-wedge corresponding to the coordinates x_0 and x_3, which form a two-dimensional space with the $SO(1,1)$ symmetry. Accordingly, we shall require the vanishing of the commutator of two currents (in general, observables) at space-like separations in the sense of $SO(1,1)$ as

$$[j_1(\frac{x}{2}), j_2(-\frac{x}{2})] = 0\ , \quad \text{for } \tilde{x}^2 \equiv x_0^2 - x_3^2 < 0\ . \quad (3)$$

The spectral condition compatible with (3) would require now that the physical momenta be in the forward light-wedge

$$\tilde{p}^2 \equiv p_0^2 - p_3^2 > 0 \text{ and } p_0 > 0\ . \quad (4)$$

In terms of the Jost-Lehmann-Dyson representation derived in Ref. [4], on the basis of the causal condition (3) and spectral condition (4), the scattering amplitude is written as

$$M(E, \cos\Theta) = i \int d^2\tilde{u} d\kappa^2 \frac{\phi(\tilde{u}, \kappa^2, k+p, (k'-p')_{1,2})}{\left[\frac{1}{2}(\tilde{k}' - \tilde{p}') + \tilde{u}\right]^2 - \kappa^2}\ , \quad (5)$$

where $\phi(\tilde{u}, \kappa^2, \cdots)$ is a function of its $SO(1,1)$- and $SO(2)$-invariant variables: $u_0^2 - u_3^2$, $(k_0+p_0)^2 - (k_3 - p_3)^2$, $(k_1+p_1)^2 + (k_2+p_2)^2$, $(k_1'-p_1')^2 + (k_2'-p_2')^2, \cdots$. The function ϕ is zero in a certain domain, determined by the causal and spectral conditions, but otherwise arbitrary.

For the discussion of analyticity of $M(E, \cos\Theta)$ in $\cos\Theta$, it is of crucial importance that all dependence on $\cos\Theta$ is contained in the denominator of (5). But, since the *arbitrary* function ϕ depends now on $(k' - p')_{1,2}$, it also depends on $\cos\Theta$. This makes impossible the mere consideration of any analyticity property of the scattering amplitude in $\cos\Theta$.

2.1 Causality in NC QFT

In the following, we shall challenge the causality condition

$$f(x) = 0\ , \quad \text{for } \tilde{x}^2 \equiv x_0^2 - x_3^2 < 0\ , \quad (6)$$

which takes into account *only* the variables connected with the $SO(1,1)$ symmetry. This causality condition would be suitable in the case when nonlocality in NC variables x_1 and x_2 is *infinite*, which is not the case on a space with the commutation relation $[x_1, x_2] = i\theta$, which implies $\Delta x_1 \Delta x_2 \geq \frac{\theta}{2}$. The fact that in the causality condition (6) the coordinates x_1 and x_2 do not enter means that the propagation of a signal in this plane is instantaneous, *i.e. no matter how far apart two events are in the noncommutative coordinates*, the allowed region for correlation is given only by the condition $x_0^2 - x_3^2 > 0$, which involves the propagation of a signal only in the x_3-direction, while the time for the propagation along x_1- and x_2-directions is totally ignored.

The uncertainty relation $\Delta x_1 \Delta x_2 \sim \theta$, which follows from the considered commutation relation of the coordinates, puts a lower bound on localization. Admitting that the scale of nonlocality in x_1 and x_2 is $l \sim \sqrt{\theta}$, then the propagation of interaction in the noncommutative coordinates is instantaneous *only within this distance* l. It follows then that two events are correlated, *i.e.* $f(x) \neq 0$, when $x_1^2 + x_2^2 \leq l^2$ (where $x_1^2 + x_2^2$ is the distance in the NC plane with $SO(2)$ symmetry), provided also that $x_0^2 - x_3^2 \geq 0$ (the events are time-like separated in the sense of $SO(1,1)$). Adding the two conditions, we obtain that

$$f(x) \neq 0\ , \text{ for } x_0^2 - x_3^2 - (x_1^2 + x_2^2 - l^2) \geq 0\ . \quad (7)$$

The negation of condition (7) leads to the conclusion that the locality condition should indeed be given by

$$f(x) = 0\ , \quad \text{for} \quad x_0^2 - x_3^2 - (x_1^2 + x_2^2 - l^2) < 0\ ,$$

or, equivalently,

$$f(x) = 0\ , \text{ for } x_0^2 - x_3^2 - (x_1^2 + x_2^2) < -l^2\ , \quad (8)$$

[b] The 'magnetic' vector $\vec{\beta}$ is defined as $\beta_i = \frac{1}{2}\epsilon_{ijk}\theta_{jk}$. The terminology stems from the antisymmetric background field $B_{\mu\nu}$ (analogous to $F_{\mu\nu}$ in QED), which gives rise to noncommutativity in string theory, with $\theta_{\mu\nu}$ essentially proportional to $B_{\mu\nu}$ (see, *e.g.*, Ref. [1]).

where l^2 is a constant proportional to NC parameter θ. When $l^2 \to 0$, (8) becomes the usual locality condition.

When $x_1^2 + x_2^2 > l^2$, for the propagation of a signal only the difference $x_1^2 + x_2^2 - l^2$ is time-consuming and thus in the locality condition, it is the quantity $x_0^2 - x_3^2 - (x_1^2 + x_2^2 - l^2)$ which will occur. Therefore, we shall have again the locality condition in the form

$$f(x) = 0 \ , \ \text{for} \ \ x_0^2 - x_3^2 - (x_1^2 + x_2^2 - l^2) < 0,$$

which is equivalent to (8). Correspondingly, the spectral condition will read as

$$p_0^2 - p_3^2 - (p_1^2 + p_2^2) \geq 0, \quad p_0 > 0 \ , \tag{9}$$

since there is no noncommutativity in momentum space. In fact, the consideration of nonlocal theories of the type (8) was initiated by Wightman[9], who asked the concrete question whether the vanishing of the commutator of fields (or observables), *i.e.* $f(x) = 0$, for $x_0^2 - x_1^2 - x_2^2 - x_3^2 < -l^2$ would imply its vanishing for $x_0^2 - x_1^2 - x_2^2 - x_3^2 < 0$. It was proven later[10,11,12] that, indeed, in a quantum field theory which satisfies the translational invariance and the spectral axiom (9), the nonlocal commutativity

$$f(x) = 0 \ , \quad \text{for} \quad x_0^2 - x_1^2 - x_2^2 - x_3^2 < -l^2,$$

implies the local commutativity

$$f(x) = 0 \ , \quad \text{for} \quad x_0^2 - x_1^2 - x_2^2 - x_3^2 < 0 \ . \tag{10}$$

This powerful theorem, which does not require Lorentz invariance, can be applied in the noncommutative case, since the hypotheses are fulfilled, with the conclusion that the causality properties of a QFT with space-space noncommutativity are physically identical to those of the corresponding commutative QFT. Consequently, the NC two-particle→two-particle scattering amplitude will have the same form as in the commutative case:

$$M(E, \cos\Theta) = i \int d^4u d\kappa^2 \frac{\phi(u, \kappa^2, k+p)}{\left[\frac{1}{2}(k' - p') + u\right]^2 - \kappa^2} \ . \tag{11}$$

This leads to the analyticity of the NC scattering amplitude in $\cos\Theta$ in the analog of the Lehmann ellipse, which behaves at high energies E the same way as in the commutative case, *i.e.* with the semi-major axis as

$$y_L = (\cos\Theta)_{max} = 1 + \frac{const}{E^4} \ . \tag{12}$$

The enlargement of the domain of analyticity in $\cos\Theta$ to Martin's ellipse requires, among other things, the use of the unitarity constraint on partial-wave amplitudes.

It was shown in Ref. [13], that a simple and manageable unitarity constraint can be obtained only for the configuration in which the incoming particle momentum is orthogonal to the NC plane. For this particular setting, $\vec{p} \parallel \vec{\beta}$, it is then straightforward, following the prescription developed for commutative QFT, to enlarge the analyticity domain of scattering amplitude to Martin's ellipse with the semi-major axis at high energies as $y_M = 1 + \frac{const}{E^2}$ and subsequently obtain the NC analog of the Froissart-Martin bound on the total cross section, in the CMS and for $\vec{p} \parallel \vec{\beta}$

$$\sigma_{tot}(E) \leq c \ \ln^2 \frac{E}{E_0} \ . \tag{13}$$

Acknowledgments

The support of the Academy of Finland under the Project no. 54023 is acknowledged.

References

1. N. Seiberg and E. Witten, *JHEP* **9909**, 032 (1999), arXiv: hep-th/9908142.
2. M. Froissart, *Phys. Rev.* **123**, 1053 (1961).
3. A. Martin, *Phys. Rev.* **129**, 1432 (1963); *Nuovo Cim.* **42**, 901 (1966).
4. M. Chaichian and A. Tureanu, arXiv: hep-th/0403032.
5. H. Lehmann, *Nuovo Cimento* **10**, 579 (1958).
6. R. Jost and H. Lehmann, *Nuovo Cimento* **5**, 1598 (1957).
7. F. Dyson, *Phys. Rev.* **110**, 1460 (1958).
8. L. Álvarez-Gaumé, J.L.F. Barbon and R. Zwicky, *JHEP* **0105**, 057 (2001), arXiv: hep-th/0103069.
9. A.S. Wightman, *Matematika* **6:4**, 96 (1962).
10. V.S. Vladimirov, *Sov. Math. Dokl.* **1**, 1039 (1960); in *Methods of the Theory of Functions of Several Complex Variables*, (Cambridge, Massachusetts, MIT Press, 1966).
11. D.Ya. Petrina, *Ukr. Mat. Zh.* 13, **No. 4**, 109 (1961) (in Russian).
12. A.S. Wightman, *J. Indian Math. Soc.* **24**, 625 (1960-61).
13. M. Chaichian, C. Montonen and A. Tureanu, *Phys. Lett.* B **566**, 263 (2003), arXiv: hep-th/0305243.

LORENTZ CONSERVING NONCOMMUTATIVE QED AND BJORKEN SCALING

M. HAGHIGHAT
Department of Physics, Isfahan University of Technology, Isfahan, Iran
E-mail: mansour@cc.iut.ac.ir

We consider the parton model in the noncommutative space-time at the lowest order. We show that Lorentz invariant noncommutative QED, even at the lowest order, violates the Bjorken scaling. Comparing the obtained results with the experimental data give an upper bound on noncommutativity parameter about $\frac{1}{\sqrt{\theta}} = 300GeV$.

1 Introduction

[a]Noncommutative field theories and its phenomenological aspects has been, recently, considered by many authors[1−7]. Such theories are mostly characterized on a noncommutative space-time with the noncommutativity parameter $\theta_{\mu\nu}$. In the canonical version of the noncommutative space-time one has

$$\theta^{\mu\nu} = -i\,[\hat{x}^\mu, \hat{x}^\nu]\,, \tag{1}$$

where "hat" indicates noncommutative coordinate and $\theta_{\mu\nu}$ is a real, constant and antisymmetric matrix. Obviously, the constant vectors θ_{0i} and θ_{ij} imply preferred directions in a given Lorentz frame which leads to violation of the Lorentz symmetry. Since the Lorentz symmetry is an almost exact symmetry of the nature, it is natural to explore the noncommutative (NC) field theories that are Lorentz invariant from the beginning. In this new class of NC theories the parameter of noncommutativity is not a constant quantity but is an operator which transforms as a Lorentz tensor[8−9]. Of course, in this way one needs to generalize the $\star$-product and operator trace for functions of both x^μ and $\theta_{\mu\nu}$, appropriately. However, in both cases experiment should confirm the theories. The obtained upper bounds for the violating Lorentz noncommutative field theory are twofolds: the first one comes from bound states such as the Hydrogen atom or the positronium[5,6] and the second one is obtained by scattering processes, for example, the electron-electron and the electron-photon scattering and so on[4,7]. In the case of Lorentz conserving noncommutative (LCNC) field theory, scattering process is only investigated. The dimensionful quantity θ in the noncommutative space implies new aspects in the parton model as well. In this paper, we explore parton model in the lowest order in which a virtual photon interacts with partons inside a nucleon in a LCNC-space. To this purpose, one should consider LCNCQED to find the effect of noncommutativity on the form factors.

2 Lorentz Conserving NCQED

To construct the noncommutative field theories that are Lorentz invariant, one needs to generalize the noncommutativity parameter. Now, we review the formalism of the Lorentz conserving NCQED introducing by Carlson, Carone and Zobin (CCZ)[8]. In the CCZ approach of NCQED $\hat{\theta}^{\mu\nu}$ is an operator and satisfies the following algebra

$$\begin{aligned}\left[\hat{x}^\mu, \hat{x}^\nu\right] &= i\hat{\theta}^{\mu\nu},\\ \left[\hat{\theta}^{\mu\nu}, \hat{x}^\lambda\right] &= 0,\\ \left[\hat{\theta}^{\mu\nu}, \hat{\theta}^{\alpha\beta}\right] &= 0,\end{aligned} \tag{2}$$

where $\hat{\theta}^{\mu\nu}$ is antisymmetric tensor that is not constant but transforms as a Lorentz tensor. The action for field theories on noncommutative spaces is then obtained using the Weyl-Moyal correspondence, according to that, in order to find the noncommutative action, the usual product of fields should be replaced by the $\star$-product

$$f \star g(x,\theta) = f(x,\theta)\exp(\frac{1}{2}\overleftarrow{\partial}_\mu \theta^{\mu\nu} \overrightarrow{\partial}_\nu) g(x,\theta). \tag{3}$$

It should be noted that here the mapping to c-number coordinates involves $\theta^{\mu\nu}$ as a c-number due to the presence of the operator $\hat{\theta}^{\mu\nu}$ in the Lorentz-conserving case. In this formulation, the operator trace that is a map from operator space to numbers, is defined as

$$Tr\hat{f} = \int d^4x\; d^6\theta\; W(\theta)\; f(x,\theta), \tag{4}$$

[a]Talk presented at the XIth Regional Conference on Mathematical Physics, May 2004, Tehran, Iran.

where $W(\theta)$ is a Lorentz invariant weight function and is assumed to be positive and even function of θ therefore one has

$$\int d^6\theta\ W(\theta)\ \theta^{\mu\nu} = 0. \quad (5)$$

Furthermore, the weight function is assumed to fall sufficiently fast, so that all integrals are well defined. Now, the properties of $W(\theta)$ and the definition of the operator trace (4) allows one to extract the interactions in the Lorentz conserving noncommutative field theory. To this end the action can be written as follows

$$S = \int d^4x\ d^6\theta\ W(\theta)\ \mathcal{L}(\phi, \partial\phi)_\star, \quad (6)$$

where $\mathcal{L}(\phi, \partial\phi)_\star$ depends on both x and θ and its subscript indicates the $\star$-product which is defined in (3). In LCNC field theory matter fields and gauge fields as well as the parameter of the gauge transformation $\Lambda(x, \theta)$ are functions of both x and θ. Therefore, using the same method as applied in the construction of $SU(N)$ noncommutative gauge theories[10], one should expand the fields as a power series in the variable θ, to extract the interactions. For instance, at the lowest order (*i.e.* tree level) of LCNC QED the correction to $qq\gamma$-vertex can be obtained for on shell fermions as[11]

$$-iq\left\{1 + \frac{\langle\theta^2\rangle}{384}k^4\right\}\gamma^\mu, \quad (7)$$

where k is the photon momentum. For the cross section of the process $e^-e^+ \longrightarrow \mu^+\mu^-$ such a correction results in

$$\frac{d\sigma}{d\cos\theta} = (\frac{d\sigma}{d\cos\theta})_{QED}(1 + \frac{\langle\theta^2\rangle}{96}s^2), \quad (8)$$

and for the cross process $e^-\mu^- \longrightarrow e^-\mu^-$ s should be replaced by t in (8).

3 Parton Model at the Lowest Order in LCNC-Space

In the canonical version of NCQED, the matter fields with charges 0 or ± 1 are allowed *i.e.* charged leptons and photon. But in the LCNCQED, the quarks as well as the leptons and photon, can be accommodated in the theory. Therefore we can examine the NC effects for the processes which contain quarks. For this purpose we consider inclusive inelastic electron-nucleon scattering. In this process the electron, at the lowest order of the parton model, interacts with free charged partons via one photon exchange therefore modification in the obtained results with respect to the usual space can be expected. To this end, we explore the differential cross section for the unpolarized e-N scattering as follows

$$\left.\frac{d^2\sigma}{dE'd\Omega}\right|_{eN} = \frac{E'\alpha^2}{EQ^4}L^{\mu\nu}W_{\mu\nu}, \quad (9)$$

where E, E', $\sqrt{-Q^2}$, $L^{\mu\nu}$ and $W_{\mu\nu}$ are initial and final energy of electron, the momentum transfer, the electron and the nucleon scattering tensor, respectively. The $ee\gamma$ vertex in the LCNCQED is given in (7) therefore $L^{\mu\nu}$ can be easily obtained as

$$L_{\mu\nu} = \frac{1}{2}(1+\frac{\langle\theta^2\rangle}{384}Q^4)^2 tr((\not{P}+m)\gamma_\mu(\not{P}+m)\gamma_\nu). \quad (10)$$

The inelastic nucleon scattering tensor is proportional to the absolute square of the nucleonic current, therefore we need the vertex function (Γ_μ) for the nucleonic current in the NC space. Since it is a Lorentz vector, therefore the most general form of Γ_μ can be written as

$$\Gamma_\mu = A\gamma_\mu + BP'_\mu + CP_\mu + iDP'^\nu\sigma_{\mu\nu} + iEP'^\nu\sigma_{\mu\nu} + FP^\nu\theta_{\mu\nu} + GP'^\nu\theta_{\mu\nu}, \quad (11)$$

where $A, B, \cdots, G$ depend on the Lorentz invariant quantity. Using the Gordon identity and gauge invariance lead to

$$W^{inel}_{\mu\nu} = (-g_{\mu\nu} + \frac{q_\mu q_\nu}{q^2})W_1(Q^2, \nu, \langle\theta^2\rangle) + (P_\mu - q_\mu\frac{P\cdot q}{q^2})(P_\nu - q_\nu\frac{P\cdot q}{q^2})\frac{W_2(Q^2, \nu, \langle\theta^2\rangle)}{M_N^2}, \quad (12)$$

where W_1 and W_2 are the structure functions and depend on the Lorentz invariant quantity such as Q^2, $\nu = P\cdot Q$ and $\langle\theta^2\rangle$. In the high energy limit the mass of the electron can be neglected and the differential cross section for the inclusive e-N scattering, in terms of Bjorken variable $x = \frac{Q^2}{2\nu}$ and the inelasticity parameter $y = \frac{E-E'}{E}$, can be cast into

$$\left.\frac{d^2\sigma}{dxdy}\right|_{eN} = \frac{4\pi s\alpha^2}{Q^4}(1 + \frac{\langle\theta^2\rangle}{384}Q^4)^2[xy^2F_1^{eN}(Q^2, x, \langle\theta^2\rangle) + (1 - y - \frac{xyM_N^2}{s})F_2^{eN}(Q^2, x, \langle\theta^2\rangle)], \quad (13)$$

where M_N is the nucleon mass, $F_1^{eN} = M_N W_1^{eN}$, $F_2^{eN} = \frac{\nu}{M_N}W_2^{eN}$ and s is the Mandelstam variable. In the parton model at the lowest order, one considers the elastic scattering of the electron off a free point charged parton with mass M_i, momentum P_i and charge $q_i e$. Therefore the cross section for this scattering can be easily constructed from the results for the electron-muon scattering (see (8)) as

$$\frac{d\sigma_i}{dQ^2} = (\frac{d\sigma_i}{dQ^2})_{QED}(1 + \frac{\langle\theta^2\rangle t_i^2}{96}), \quad (14)$$

where t_i is the Mandelstam variable for the parton i. If we neglect the electron and the partons masses in the Breit frame of reference and after a little algebra the cross section in terms of x and y variables becomes

$$\frac{d^2\sigma_i}{dxdy} = \frac{4\pi\alpha^2 q_i^2 x}{Q^2}(\frac{s^2+u^2}{2s^2})\delta(\xi_i - x)(1 + \frac{\langle\theta^2\rangle t^2}{96}), \quad (15)$$

where ξ_i is a fraction of the nucleon's total momentum carried by the i-th parton (*i.e.* $P_i^\mu = \xi_i P^\mu$) and the Mandelstam variables for i-th parton in terms of the Mandelstam variables for the whole nucleon are

$$\begin{aligned} s_i &= (p + P_i)^2 = \xi_i s, \\ t_i &= (P_i - P_i')^2 = t = -Q^2, \\ u_i &= (p' - P_i)^2 = \xi_i u. \end{aligned} \quad (16)$$

Various types of partons carry a different fraction of the parent nucleon's momentum. Therefore for the parton momentum distribution function $f_i(\xi_i)$ with the appropriate normalization

$$\int_0^1 d\xi_i \ f_i(\xi_i) = 1, \quad (17)$$

one has

$$\frac{d^2\sigma}{dxdy} = \frac{2\pi\alpha^2 s}{Q^4}[(y-1)^2+1]\sum_i f_i(x) q_i^2 x(1 + \frac{\langle\theta^2\rangle t^2}{96}). \quad (18)$$

Now comparing (18) with (13), where $M_N = 0$, we have

$$F_2^{eN}(Q^2, x, \langle\theta^2\rangle) = \sum_i f_i(x) q_i^2 x(1 + \frac{\langle\theta^2\rangle Q^4}{192}), \quad (19)$$

in other words the parton model at the lowest order in LCNCQED violates the Bjorken scaling but Callan-Gross relation still holds

$$F_2^{eN}(Q^2, x, \langle\theta^2\rangle) = 2xF_1^{eN}(Q^2, x, \langle\theta^2\rangle). \quad (20)$$

One can use the data on the deep inelastic e-N scattering to obtain the upper bound on the value of the parameter of noncommutativity in the LCNCQED, Λ_{LCNC}. Equation (19) shows that

$$\frac{F_2^{NC} - F_2^0}{F_2^0} = \frac{\langle\theta^2\rangle Q^4}{192} = \frac{Q^4}{16\Lambda_{LCNC}^4}, \quad (21)$$

where $\Lambda_{LCNC} = (\frac{12}{\langle\theta^2\rangle})^{\frac{1}{4}}$. The measurements of F_2 structure function in deep inelastic scattering can provide a test on the noncommutativity of space. For example, F_2 in positron-proton neutral current scattering has been measured with statistical and systematic uncertainties below 2%.[12] Therefore, as an estimation, one percent error in the experimental value of the structure function for $\sqrt{Q^2} = 200GeV$ results in $\Lambda_{LCNC} \sim 300GeV$.

Acknowledgements

The financial support of Isfahan University of Technology Research Council and IPM are acknowledged.

References

1. M. Hayakawa, *Phys. Lett.* B **478**, 394 (2000).
2. J. Madore, S. Schraml, P. Schupp and J. Wess, *Eur. Phys. J.* C **16**, 161 (2000).
3. M. Chaichian, P. Presnajder, M.M. Sheikh-Jabbari and A. Tureanu, *Phys. Lett.* B **526**, 132 (2002); arXiv: hep-th/0107055; I.F. Riad and M.M. Sheikh-Jabbari, *JHEP* **0008**, 045 (2000).
4. Y. Liao, *JHEP* **0111**, 067 (2001); H. Grosse and Y. Liao, *Phys. Rev.* D **64**, 115007 (2001).
5. M. Haghighat and F. Loran, *Mod. Phys. Lett.* A **16**, 1435 (2001); M. Haghighat, S.M. Zebarjad and F. Loran, *Phys. Rev.* D **66**, 016005 (2002); M. Haghighat and F. Loran, *Phys. Rev.* D **67**, 096003 (2003).
6. M. Chaichian, M.M. Sheikh-Jabbari and A. Tureanu, *Phys. Rev. Lett.* **86**, 2716 (2001).
7. J.L. Hewett, F.J. Petriello and T.G. Rizzo, *Phys. Rev.* D **64**, 075012 (2001); S.W. Baek, D.K. Ghosh, X.G. He and W.Y. Hwang, *Phys. Rev.* D **64**, 056001 (2001); Z. Guralnik, R. Jackiw, S.Y. Pi and A.P. Polychronakos, *Phys. Lett.* B **517**, 450 (2001); S.M. Carroll, J.A. Harvey, V.A. Kostelecky, C.D. Lane and T. Okamoto, *Phys. Rev. Lett.* **87**, 141601 (2001); S. Godfrey and M.A. Doncheski, *Phys. Rev.* D **65**, 015005 (2002).
8. C.E. Carlson, C.D. Carone and N. Zobin, *Phys. Rev.* D **66**, 075001 (2002).
9. K. Morita, *Prog. Theor. Phys.* **108**, 1099 (2003); H. Kase, K. Morita, Y. Okumura and E. Umezawa, *Prog. Theor. Phys.* **109**, 663 (2003).

10. B. Jurčo, L. Möller, S. Schraml, P. Schupp and J. Wess, *Eur. Phys. J.* C **21**, 383 (2001); X. Calmet, B. Jurčo, P. Schupp, J. Wess and M. Wohlgenannt, *Eur. Phys. J.* C **23**, 363 (2002).
11. J. M. Conroy, H. J. Kwee and V. Nazaryan, *Phys. Rev.* D **68**, 054004 (2003).
12. S. Chekanov *et al.*, (ZEUS Collab.) *Eur. Phys. J.* C **21**, 443 (2001).

DEFORMED INSTANTONS

A. IMAANPUR

Department of Physics, School of Sciences, Tarbiat Modares University
P.O. Box 14155-4838, Tehran, Iran
and
Institute for Studies in Theoretical Physics and Mathematics (IPM), School of Physics
P.O. Box 19395-5531, Tehran, Iran
Email: aimaanpu@theory.ipm.ac.ir

We study instantons of $\mathcal{N} = 1/2$ supersymmetric Yang-Mills theory. By deforming the instanton equation in the presence of fermions, we write down the zero modes equations. The solutions satisfy the equations of motion. The deformed instanton equations imply that the finite action solutions have $U(1)$ connections which are not flat anymore.

1 Introduction

[a]By studying D-branes in the background of graviphoton field[1,2,3], Seiberg[4] has shown that one can construct a super Yang-Mills Lagrangian, describing the low-energy dynamics on the brane, which preserves only half of the supersymmetries. The analysis of (anti)instantons in $\mathcal{N} = 1/2$ SYM model parallels the one in ordinary $\mathcal{N} = 1$ SYM theory, though, in the former case the equation of motion for chiral fermions Λ becomes more involved when $F^+ = 0$. In this background, there are antichiral $\overline{\Lambda}$ zero modes, which deform the equation for chiral fermions so that Λ cannot remain zero. Further, because of a quartic antichiral term in the action, the fermionic solutions to the equations of motion are not in general the zero modes of the action. This will have a further consequence that in the presence of fermions, instantons are not solutions to the equations of motion anymore. In this note, we would like to comment that in $\mathcal{N} = 1/2$ $U(2)$ SYM model the instanton equation should be deformed as follows,

$$F^+_{\mu\nu} + \frac{i}{2} C_{\mu\nu}\overline{\Lambda}\overline{\Lambda} = 0, \tag{1}$$

$$\not{D}\overline{\Lambda} = 0\,, \tag{2}$$

where $F^+_{\mu\nu}$ is the self-dual part of the field strength, and $C_{\mu\nu}$ being the deformation parameter. The above equations, as we will see shortly, are also the zero mode equation for $\overline{\Lambda}$. The solutions of the above equations, unlike instantons in the presence of fermionic solutions in $\mathcal{N} = 1/2$ theory, do satisfy the equations of motion. This happens partly because in this configuration $\Lambda = 0$ is still a solution. In what follows, we will see that for the $U(2)$ group only the $U(1)$ part of the instanton equation is deformed and finite action solutions will have non-flat $U(1)$ connection. The $SU(2)$ instanton equation and the corresponding Dirac equation for the adjoint fermions, on the other hand, remain undeformed.

2 Instantons and Zero Modes

Let us begin with assuming that the superspace coordinates θ^α are not anticommuting, and instead they satisfy the following anticommutation relation

$$\{\theta^\alpha, \theta^\beta\} = C^{\alpha\beta}\,, \tag{3}$$

where $C^{\alpha\beta}$ is a constant and symmetric 2×2 matrix. The anticommutation relation (3) will deform the supersymmetry algebra with $\overline{Q}^2$ proportional to the deformation parameter $C^{\alpha\beta}$. Seiberg[4] has considered the above deformation in $\mathcal{N} = 1$ supersymmetric model and has shown that half of the supersymmetries can be preserved. Indeed, if $W^\alpha = (A_\mu, \Lambda)$ denotes the usual $\mathcal{N} = 1$ gauge supermultiplet, then the Lagrangian of this $\mathcal{N} = 1/2$ model reads

$$\mathcal{L} = i\tau \int d^2\theta \mathrm{tr}\, W^\alpha W_\alpha - i\bar{\tau}\int d^2\bar{\theta}\mathrm{tr}\,\overline{W}^{\dot{\alpha}}\overline{W}_{\dot{\alpha}}$$
$$-(i\tau - i\bar{\tau})\mathrm{tr}\left(iC^{\mu\nu}F_{\mu\nu}\overline{\Lambda}\overline{\Lambda} - \frac{|C|^2}{4}(\overline{\Lambda}\overline{\Lambda})^2\right),$$

where

$$C^{\mu\nu} \equiv C^{\alpha\beta}\epsilon_{\beta\gamma}\sigma^{\mu\nu}_{\alpha}{}^{\gamma},$$

is a constant and antisymmetric self-dual matrix, and $|C|^2 = C_{\mu\nu}C^{\mu\nu}$.

[a]Talk presented at the XIth Regional Conference on Mathematical Physics, May 2004, Tehran, Iran.

The above Lagrangian is invariant under the following Q deformed supersymmetry transformations,

$$\delta\Lambda = i\epsilon D + \sigma^{\mu\nu}\epsilon\left(F_{\mu\nu} + \frac{i}{2}C_{\mu\nu}\overline{\Lambda\Lambda}\right), \quad (4)$$
$$\delta A_\mu = -i\overline{\Lambda}\bar{\sigma}_\mu\epsilon,$$
$$\delta D = -\epsilon\sigma^\mu D_\mu\overline{\Lambda},$$
$$\delta\overline{\Lambda} = 0,$$

whereas $\overline{Q}$ is broken. A supersymmetric state is invariant under the above supersymmetric transformations. Setting Λ and $\overline{\Lambda}$ to zero, with

$$F^+_{\mu\nu} = 0, \quad (5)$$

gives such a state preserving the whole unbroken supersymmetry. This is of course also a solution to the equations of motion. Moreover, in $\mathcal{N} = 1$ SYM theory, since in the background of instantons ($F^+ = 0$) there are antichiral zero modes with no chiral zero modes, instantons remain solutions to the equations of motion in the presence of fermionic solutions. But in $\mathcal{N} = 1/2$ SYM theory this is not the story.

As mentioned in Sec. 1, in the background of instantons, and because the action of $\mathcal{N} = 1/2$ SYM theory has a quartic antichiral fermionic term, the zero modes, in general, do not satisfy the equations of motion. Therefore, instantons will not remain solutions when there are fermions. To remedy this, we deform the instanton equation and show that they satisfy the equations of motion and like instantons have a finite action. The equations are

$$F^+_{\mu\nu} + \frac{i}{2}C_{\mu\nu}\overline{\Lambda\Lambda} = 0,$$
$$\not{D}\overline{\Lambda} = 0,$$
$$\Lambda = 0. \quad (6)$$

It is easy to see that a solution to (6) is also a solution to the equations of motion. In fact, if we set $\Lambda = 0$, the equation of motion for the gauge fields is satisfied

$$D_\mu\left(F^{\mu\nu} + iC^{\mu\nu}\overline{\Lambda\Lambda}\right)$$
$$= D_\mu\left(F^{\mu\nu-} + F^{\mu\nu+} + iC^{\mu\nu}\overline{\Lambda\Lambda}\right)$$
$$= D_\mu\left(2F^{\mu\nu+} + iC^{\mu\nu}\overline{\Lambda\Lambda}\right) = 0, \quad (7)$$

where in the last equality we used the Bianchi identity

$$D_\mu(F^{\mu\nu+} - F^{\mu\nu-}) = 0, \quad (8)$$

with

$$F^\pm_{\mu\nu} = \frac{1}{2}(F_{\mu\nu} \pm \frac{1}{2}\epsilon_{\mu\nu\rho\sigma}F^{\rho\sigma}). \quad (9)$$

The equations of motion for $\overline{\Lambda}$ and Λ read

$$\overline{\not{D}}\Lambda = -C^{\mu\nu}F^+_{\mu\nu}\overline{\Lambda} - i\frac{|C|^2}{2}(\overline{\Lambda\Lambda})\overline{\Lambda},$$
$$\not{D}\overline{\Lambda} = 0, \quad (10)$$

which are also satisfied by solutions of (6).

For $U(1)$ gauge group, since $\overline{\Lambda}$ is in the adjoint representation, the second equation of (6) reduces to

$$\not{\partial}\overline{\Lambda} = 0, \quad (11)$$

which has no normalizable solution on $\mathbf{R}^4$. Therefore, in the $U(1)$ case, (6) reduce to the Abelian instanton equations, which are known to have no nontrivial solutions on $\mathbf{R}^4$ except the flat ones. However, for gauge groups of higher rank, there might be nontrivial solutions different than instantons. Let us hence consider the $U(2)$ gauge group.

3 Analysis of Zero Mode Equations

In this section we analyze possible solutions to (6). We will see that in the background of fermionic solutions of Dirac equation, the $U(1)$ part of the connection can no longer be flat. Specially, we find solutions in the presence of adjoint fermions in the background of 't Hooft instantons. The $U(1)$ connections we find, have a zero instanton number which is consistent with the fact that the deformed instanton equations are also the equations for zero modes.

To begin with, let $T^a = (T^i = \frac{\sigma^i}{2}, T^4 = \frac{1}{2})$, $i = 1, 2, 3$ denote the generators of $SU(2)$ and $U(1)$ groups, respectively. Now, to isolate the $SU(2)$ and $U(1)$ parts of (6), we expand the quadratic term $\overline{\Lambda\Lambda}$ in that equation. So, for the $SU(2)$ part we have

$$F^{+i}_{\mu\nu} + \frac{i}{2}C_{\mu\nu}\overline{\Lambda}^i\overline{\Lambda}^4 = 0, \quad (12)$$
$$\not{D}\overline{\Lambda}^i = 0, \quad (13)$$

while the $U(1)$ part reads

$$F^{+4}_{\mu\nu} + \frac{i}{8}C_{\mu\nu}\overline{\Lambda}^i\overline{\Lambda}^i + \frac{i}{8}C_{\mu\nu}\overline{\Lambda}^4\overline{\Lambda}^4 = 0, \quad (14)$$
$$\not{\partial}\overline{\Lambda}^4 = 0. \quad (15)$$

The last equation has no normalizable solution and thus we set $\overline{\Lambda}^4 = 0$. This will reduce (12) and (13) to the ordinary $SU(2)$ instanton equation. Equation (14), however, reads

$$F^{+4}_{\mu\nu} + \frac{i}{8}C_{\mu\nu}\overline{\Lambda}^i\overline{\Lambda}^i = 0, \quad (16)$$

where $\overline{\Lambda}^i$ are solutions to (13). This is the equation we would like to study further.

A solution to Dirac equation (13) can in general be written in terms of ADHM data[5]. In particular, in the background of one instanton, one can write the explicit solutions. For the $SU(2)$ case, we quote the result for three different normalized solutions

$$\mathrm{tr}\,(\overline{\Lambda\Lambda}) = \frac{2}{\pi^2}\,\rho^2 f(x)^3, \tag{17}$$

$$\mathrm{tr}\,(\overline{\Lambda\Lambda}) = \frac{6}{\pi^2}\,\rho^4 f(x)^4, \tag{18}$$

$$\mathrm{tr}\,(\overline{\Lambda\Lambda}) = \frac{3}{\pi^2}\,\rho^2 r^2 f(x)^4\,, \tag{19}$$

where

$$f(x) = \frac{1}{r^2+\rho^2}\,, \tag{20}$$

and $r = \sqrt{\sum_{i=1}^4 x_i^2}$. Here, ρ is the instanton size, and we have set to zero the instanton position for simplicity. The analysis of (16) becomes easier if we make use of the Corrigan's identities to write it as

$$F^{+4}_{\mu\nu} = -\frac{i}{8}C_{\mu\nu}\partial^\alpha\partial_\alpha K(x)\,, \tag{21}$$

for $K(x)$ being any of the functions $\frac{2}{\pi^2}f(x)$, $\frac{6}{\pi^2}(r^2f(x)^2-2f(x))$ or $\frac{3}{\pi^2}(r^2f(x)^2+f(x))$. The solution to this equation – up to a gauge transformation – is then found to be

$$A_\mu(x) = \frac{i}{8}C_{\mu\nu}\partial_\nu K(x)\,, \tag{22}$$

which of course has a nonvanishing curvature. Therefore, what we have found is that in the presence of fermionic zero modes the $U(1)$ part of the connection cannot remain flat. This is in contrast with the supersymmetric solutions to the field equations (instantons) where one sets the fermions to zero, and for having a finite action solution one has to restrict to the flat part of the $U(1)$ connections.

For instantons of higher topological charge, as mentioned earlier, one can write the adjoint zero mode solutions in terms of the ADHM data. Interestingly, using the Corrigan's identity, it can be seen that the expression for $\mathrm{tr}\,(\overline{\Lambda\Lambda})$ is actually a total derivative for all topological charges k,

$$\mathrm{tr}\,(\overline{\Lambda\Lambda})_k = \partial^\alpha\partial_\alpha\mathcal{K}(x;k)\,, \tag{23}$$

for some function $\mathcal{K}(x;k)$, which in turn has an explicit expression in terms of ADHM data. Therefore, we conclude that the solutions to (16) for all values of k will have the general form of the solution we found for $k=-1$ in (22). Moreover, since $\mathcal{K}(x;k)$ has no singularity and at infinity goes like r^{-2}, the corresponding $U(1)$ gauge fields will all have zero instanton number.

References

1. J. de Boer, P. Grassi and P. van Nieuwenhuizen, arXiv: hep-th/0302078.
2. H. Ooguri and C. Vafa, arXiv: hep-th/0302109.
3. D. Klemm, S. Penati and L. Tamassia, *Class. Quant. Grav.* **20**, 2905 (2003), arXiv: hep-th/0104190.
4. N. Seiberg, *JHEP* **0306**, 010 (2003), arXiv: hep-th/0305248.
5. E. Corrigan, D.B. Fairlie, S. Templeton and P. Goddard, *Nucl. Phys.* B **140**, 31 (1978).

ADS INTERPRETATION OF TWO-POINT CORRELATION FUNCTION OF QED

S. MAMEDOV

Institute for Studies in Theoretical Physics and Mathematics (IPM), School of Physics
P.O. Box 19395-5531, Tehran, Iran
and
High Energy Physics Lab., Baku State University
Z. Khalilov Str. 23, Baku 370148, Azerbaijan
E-mail: sh_mamedov@yahoo.com & shahin@theory.ipm.ac.ir

The two-point correlation function of QED is considered in the worldline formalism. In the position space, it is presented in terms of heat kernel. This leads to introducing the $K_1(x_i)$ functions related to the bulk to boundary propagator of massless scalar field and to revealing the bulk to boundary propagator in the photon polarization operator.

1 Introuction

[a]The study of correlation functions is the one of interesting topics of *AdS*/CFT correspondence. It is indeed important to study the correlation functions of realistic models, such as scalar and spinor QED in the framework of *AdS*/CFT correspondence. The two-point correlation function of electromagnetic field in QED is the photon polarization operator. The worldline formalism, based on Schwinger parameter moduli space[1,2], turns out to be a useful tool for rewriting two- and three-point correlation functions of free scalar field theory in terms of heat kernel[3]. Following this approach, N-photon amplitudes written in terms of worldline formalism can be rewritten in terms of heat kernel and can be interpreted in the *AdS*/CFT correspondence language as well. In this paper, we will use this method to convert the photon polarization operator into an expression written in terms of bulk to boundary propagator in *AdS* space-time.

2 Photon Polarization Operator

We shall use the expression of photon polarization operator in scalar QED in Schwinger parametrization. It has coinciding expression obtained both from direct calculation and in the worldline formalism[1,2]

$$\Pi(k_1,k_2) = -\frac{\left(g\mu^{\frac{\epsilon}{2}}\right)^2}{(4\pi)^{2-\frac{\epsilon}{2}}}\int_0^\infty \frac{d\tau}{\tau^{\frac{d}{2}-1}}\int_0^1 d\alpha\, e^{\tau k_1 k_2\alpha(1-\alpha)}$$
$$\times\left[\frac{2}{\tau}(\delta(\alpha)-1)\,\epsilon_1\cdot\epsilon_2 + (1-2\alpha)^2(\epsilon_1\cdot k_2)(\epsilon_2\cdot k_1)\right]$$
$$\times\int_{-\infty}^{\infty}\frac{d^dz}{(2\pi)^d}e^{i(k_1+k_2)z}. \quad (1)$$

Here, d is the dimension of space-time, ϵ_1, ϵ_2 and k_1, k_2 are polarization vectors and momenta of incoming and outgoing photons and τ is the Schwinger parameter.

Note that (1) differs from two-point correlation function of free scalar field theory only by an additional square bracket[3]. In the position space, taking Gaussian integrals over the momenta and derivatives, we obtain the following expression of polarization operator

$$\Pi(x_1,x_2) = -C\int_0^\infty \frac{d\tau}{\tau^{\frac{d}{2}-1}}\int_0^1 d\alpha\int_0^1 d\beta\int_{-\infty}^{\infty} d^dz$$
$$\times\left(\frac{\pi}{\tau\beta\alpha(1-\alpha)}\right)^{\frac{d}{2}} e^{-\frac{(z-x_1)^2}{4\tau\beta\alpha(1-\alpha)}}$$
$$\times\left(\frac{\pi}{\tau(1-\beta)\alpha(1-\alpha)}\right)^{\frac{d}{2}} e^{-\frac{(z-x_2)^2}{4\tau(1-\beta)\alpha(1-\alpha)}}$$
$$\times\left[\frac{2}{\tau}(\delta(\alpha)-1)\,\epsilon_1\cdot\epsilon_2 - (1-2\alpha)^2\right.$$
$$\times\frac{1}{2\tau(1-\beta)\alpha(1-\alpha)}$$
$$\left.\times\ \epsilon_1\cdot(z-x_2)\frac{1}{2\tau\beta\alpha(1-\alpha)}\epsilon_2\cdot(z-x_1)\right]. \quad (2)$$

[a]Talk presented at the XIth Regional Conference on Mathematical Physics, May 2004, Tehran, Iran. This presentation is a collaboration with S. Parvizi (*IPM, Tehran, Iran*).

We can write (2) in terms of heat kernel

$$\left(\frac{1}{4\pi t}\right)^{\frac{d}{2}} e^{-\frac{(x-z)^2}{4t}} = \left\langle x \left| e^{t\partial^2} \right| z \right\rangle .$$

Making appropriate change of variables:

$$t = 4\tau\rho\beta(1-\beta)\alpha(1-\alpha),$$

$\rho_1 = \rho(1-\beta)$ and $\rho_2 = \beta\rho$, we arrive at

$$\Pi(x_1,x_2) = -C' \int_0^\infty \frac{dt}{t^{\frac{d}{2}-1}} \int_0^1 d\alpha \, [4\alpha(1-\alpha)]^{\frac{d}{2}-2}$$

$$\times \int_0^1 d\beta \, (\rho_1\rho_2)^{\frac{d}{2}-2} \rho^{-(\frac{d}{2}-2)} \int_{-\infty}^{\infty} d^d z \left\langle x_1 \left| e^{\frac{t}{4\rho_1}\partial^2} \right| z \right\rangle$$

$$\times \left\langle x_2 \left| e^{\frac{t}{4\rho_2}\partial^2} \right| z \right\rangle \left[\frac{8\rho_1\rho_2\alpha(1-\alpha)}{t\rho} (\delta(\alpha)-1)\,\epsilon_1\cdot\epsilon_2 \right.$$

$$\left. -(1-2\alpha)^2 \frac{4\rho_1\rho_2}{t^2}\epsilon_1\cdot(z-x_2)\,\epsilon_2\cdot(z-x_1) \right]. \quad (3)$$

Now, inserting $\Gamma(s)$ function representations for 1

$$1 = \frac{1}{\Gamma\left(\frac{d}{2}\right)} \int_0^\infty d\rho \rho^{\frac{d}{2}-1} e^{-\rho}, \quad (4)$$

into (3),[b] changing the integration variables ρ and β into ρ_1 and ρ_2 and using the equality $\int_0^\infty \rho d\rho \int_0^1 d\beta = \int_0^\infty d\rho_1 \int_0^\infty d\rho_2$, the two-point function $\Pi(x_1,x_2)$ can be given in a more symmetric form as

$$\Pi(x_1,x_2) = \int_0^\infty \frac{dt}{t^{\frac{d}{2}+1}} \int_{-\infty}^{\infty} d^d z \int_0^\infty d\rho_1 \rho_1^{\frac{d}{2}-1} e^{-\rho_1}$$

$$\times \left\langle x_1 \left| e^{\frac{t}{4\rho_1}\partial^2} \right| z \right\rangle \int_0^\infty d\rho_2 \rho_2^{\frac{d}{2}-1} e^{-\rho_2} \left\langle x_2 \left| e^{\frac{t}{4\rho_2}\partial^2} \right| z \right\rangle$$

$$\times \left[-\frac{C_1}{\Gamma\left(\frac{d}{2}+1\right)} t\, \epsilon_1\cdot\epsilon_2 + \frac{C_2}{\Gamma\left(\frac{d}{2}\right)} \epsilon_1\cdot(z-x_2) \right.$$

$$\left. \times \epsilon_2\cdot(z-x_1) \right]. \quad (5)$$

Here, $\rho_1+\rho_2=\rho$ is to be taken into account in the exponent. The integrals over the α-parameter are included in the constants $C_{1,2}$ defined by[c]

$$C_1 = 4^{\frac{d}{2}-1} C' \int_0^1 d\alpha \, [\alpha(1-\alpha)]^{\frac{d}{2}-1} (\delta(\alpha)-1),$$

$$C_2 = 4^{\frac{d}{2}-2} C' \int_0^1 d\alpha \, [\alpha(1-\alpha)]^{\frac{d}{2}-1} (1-2\alpha)^2 .$$

Following Ref. [3], the integrals over ρ_i, $i=1,2$ are to be separated and the resulting expressions are to be denoted by $K_1(x_i,z,t)$

$$K_1(x_i,z,t) = \int_0^\infty d\rho_i \, \rho_i^{\frac{d}{2}-1} e^{-\rho_i} \left\langle x_i \left| e^{\frac{t}{4\rho_i}\partial^2} \right| z \right\rangle . \quad (6)$$

As was shown in Ref. [3], identifying the variable t with the radius of a d-dimensional sphere z_0, $(t=z_0^2)$, the function $K_1(x_i,z,t)$ obeys the $d+1$-dimensional Klein-Gordon equation with zero mass

$$\left[-z_0^2\partial_{z_0}^2 + (d-1)\,z_0\partial_{z_0} - z_0^2\partial^2\right] K_1(x,z,t) = 0, \quad (7)$$

where ∂^2 is the d-dimensional Laplacian in the direction $\vec{z}$. This means that $K_1(x_i,z,t)$ is the bulk to boundary propagator of massless scalar or vector field in the $d+1$-dimensional *AdS* space-time. Now, we can write (5) in terms of this propagator in a form which is more suitable for the desired *AdS*/CFT interpretation

$$\Pi(x_1,x_2) = \int_0^\infty \frac{dt}{t^{\frac{d}{2}+1}} \int_{-\infty}^{\infty} d^d z \, K_1(x_1,z,t)\, K_1(x_2,z,t)$$

$$\times \left[-\frac{C_1}{\Gamma\left(\frac{d}{2}+1\right)}\epsilon_1\cdot\epsilon_2 t + C_2 \frac{1}{\Gamma\left(\frac{d}{2}\right)}\epsilon_1\cdot(z-x_2) \times \right.$$

$$\left. \epsilon_2\cdot(z-x_1) \right]. \quad (8)$$

Comparing now this expression for $\Pi(x_1,x_2)$ with the two-point correlation function $\Gamma(x_1,x_2)$ for free scalar field theory, we find the additional square bracket factor in our case, which should be replaced by t^3 for the last one. Thus, the photon polarization operator is expressed in terms of bulk to boundary propagator $K_1(x,z,t)$ of massless field. For the case of spinor QED, *i.e.*, when we have spinor particles in the loop, the photon polarization operator has a form similar to its Schwinger parameter expression (1) for scalar loop[1,2]

$$\Pi(k_1,k_2) = 2\frac{(g\mu^{\frac{\epsilon}{2}})^2}{(4\pi)^{2-\frac{\epsilon}{2}}} \int_0^\infty \frac{d\tau}{\tau^{\frac{d}{2}-1}} \int_0^1 d\alpha \, e^{\tau k_1 k_2 \alpha(1-\alpha)}$$

$$\times \left[\frac{2}{\tau}(\delta(\alpha)-1)\,\epsilon_1\cdot\epsilon_2 - \left[(1-2\alpha)^2-1\right](\epsilon_1\cdot k_2) \times \right.$$

$$\left. (\epsilon_2\cdot k_1) + (\epsilon_1\cdot\epsilon_2)(k_1\cdot k_2) \right] \int_{-\infty}^{\infty} \frac{d^d z}{(2\pi)^d} e^{i(k_1+k_2)z} . \quad (9)$$

[b] In the terms containing different degrees of ρ_l, $\Gamma(s)$ functions have to be introduced with different arguments.
[c] We suppose $d \geq 2$ for convergency of these integrals

This allows us to rewrite the formula (7) for spinor loop (8)

$$\Pi(x_1,x_2) = 2\int_0^\infty \frac{dt}{t^{\frac{d}{2}+1}} \int_{-\infty}^\infty d^d z\, K_1(x_1,z,t) \times K_1(x_2,z,t)\left[C_1 \frac{\epsilon_1\cdot\epsilon_2}{\Gamma\left(\frac{d}{2}+1\right)}t + \frac{1}{\Gamma\left(\frac{d}{2}\right)} \times \left[C_2\epsilon_1\cdot(z-x_2)\,\epsilon_2\cdot(z-x_1) - C_3\epsilon_1\cdot\epsilon_2\,(z-x_1)\cdot(z-x_2)\right]\right]. \quad (10)$$

We see therefore from (7) and (9) that the two-point correlation functions of vector field for scalar and spinor QED are expressed by the massless bulk to boundary propagator. Note that the two-point correlation function is studied here for massless vector field (photon field). The obtained result is therefore in terms of bulk to boundary propagator of this special field. If we multiply (7) with 2 and add it to (9), then the result agrees with the result obtained in Ref. [4]. This tells us that supersymmetry plays an important role for matching correlation functions in Field Theory with the *AdS* supergravity ones.

Acknowledgments

I thank Prof. F. Ardalan and IPM for the invitation to this institute and hospitality.

References

1. M.J. Strassler, *Nucl. Phys.* B **385**, 145 (1992), arXiv: hep-ph/9205205.
2. C. Schubert, *Phys. Rept.* **355**, 73 (2001), arXiv: hep-th/0101036.
3. R. Gopakumar, arXiv: hep-th/0308184.
4. G. Chalmers, H. Nastase, K. Schalm and R. Siebelnik, *Nucl. Phys.* B **540**, 247 (1999).

CHAPTER 7: PLASMA PHYSICS

SPOT SIZE EFFECTS ON THE LASER PLASMA INTERACTION FEATURES

H. ABBASI*,† and H. HAKIMI PAJOUH*

*,† *Faculty of Physics, Amir Kabir University of Technology*
P.O. Box 15875-4413, Tehran, Iran
and
* *Institute for Studies in Theoretical Physics and Mathematics (IPM), School of Physics*
P.O. Box 19395-5531, Tehran, Iran
E-mails: abbasi, hakimi@theory.ipm.ac.ir

A generalization of the paraxial approximation based on stationary transverse distribution of the laser pulse in laser-plasma interaction is addressed. The laser beam is assumed to have a nonuniform intensity distribution (along its wave front) in a plane transverse to the direction of propagation. The conditions are considered in such a way that the pulse variation in the perpendicular direction is negligible. Accordingly, the study is based on an approach that accounts for the transversal effects in an average manner. The formalism of the electron parametric instability is presented in this framework. The resulting equation is apparently similar to the equation that is obtained from the paraxial approximation but with averaged source term.

1 Introduction

[a]Ultrahigh intensity lasers[1] are being developed for a wide range of applications. Intense laser beam propagation in plasmas has applications in different areas as ultra-broadband radiation generation[2], optical harmonic generation[3], X-ray generation[4], inertial confinement fusion[5] and laser driven acceleration[6]. These and other applications provide a motivation for studying the physics of intense laser fields interacting with matters.

There are regimes in the interaction of an intense laser pulse with plasmas where the transverse characteristics of the laser beams, such as the laser spot size and the transverse distribution of the laser amplitude, do not significantly change in the time scales of interest. For instance, in the case when the relativistic self-focusing just balances the defocusing due to diffraction, the laser pulse can be self-guided. A necessary requirement for optically guiding is that the refractive index have a maximum on axis (*e.g.* Gaussian beam).

Consider the case when $\Delta r_\perp$, the scale of transverse variation, is much larger than Δz, the scale length in the propagation direction (Fig. 1). In this case, the variation of the transverse distribution is negligible.

In the case when the self-focusing is dominant the above approximation is broken. The relativistic self-focusing occurs when the laser power exceeds a critical power given by $P_c = 17(\omega_0/\omega_{p_e})^2$ GW, where ω_0 is the laser frequency and ω_{p_e} is the electron plasma frequency. However, for the range of parameters, $r_0 \propto (1\mu\text{m} - 1\text{mm}), \omega_0 \propto (10\omega_{p_e} - 100\omega_{p_e})$, there is an intensity in the interval $10^{13} - 10^{21}$ W/cm^2 for which the power of laser pulse is less than the critical power. As a result, the relativistic self-focusing does not occur and consequently the mentioned approximation is satisfied. Accordingly, the laser pulse propagates more than several Rayleigh length $Z_R = \pi r_0^2/\lambda_0$, where r_0 is the spot size and λ_0 is the laser wavelength without any significant change in the transversal size.

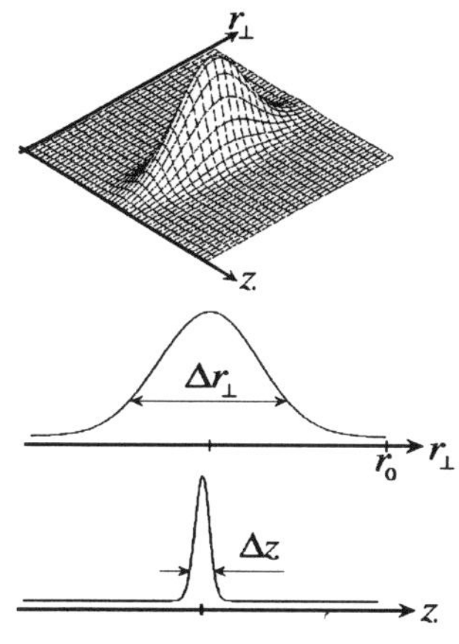

Figure 1. A typical laser pulse for which $\Delta r_\perp \gg \Delta z$ is fulfilled.

[a]Talk presented by H.A. at the XIth Regional Conference on Mathematical Physics, May 2004, Tehran, Iran.

Under the above conditions that the transversal dynamics is negligible, in most of the cases, under the paraxial approximation, the perpendicular dimension of the problem is completely ignored. Almost, all the analytical calculations in laser plasma interaction, such as growth rates, stationary solutions, and so on, are performed following this approximation. The simplification of the model due to the paraxial approximation is its advantage. However, as it is seen from Fig. 2, different transverse distribution are identical in the framework of this paraxial approximation. Since the area under the transverse distribution is somehow related to the total energy of the laser pulse, it has certain influence on different mechanisms in laser plasma interaction.

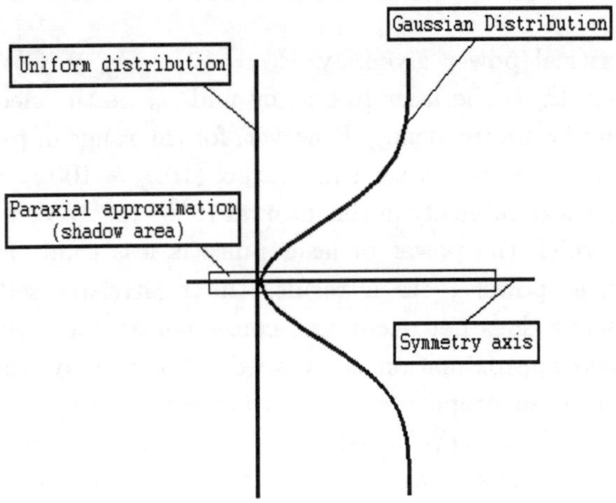

Figure 2. In the paraxial approximation, different transverse distributions are identical.

The main goal of this paper is to include the effect of the transverse distribution of the laser pulse in a logical manner. In other words, we are going to improve the paraxial approximation. The question is: How can the effect of stationary transverse distribution be considered?

It is clear that the areas near the pulse edge do not excite nonlinear effects (small amplitude areas), while the central part of the Gaussian distribution has the most contribution in deriving the nonlinear effects. This fact leads us to consider the transverse distribution as a probability density function. Therefore, all the governing equations, resulting from the paraxial approximation, might be averaged by this probability density function. In this way, appropriate contributions of the pulse pieces are taken into account.

For the case of weak pulses, nonlinear effects are negligible and averaging process is trivial, *i.e.*, averaging of linear terms can be canceled out. Therefore, the results of the paraxial approximation is still valid for the stationary transverse Gaussian profile in linear regime. However, for the intense laser pulses, the nonlinear nature of the interactions makes the averaging nontrivial.

In this paper, in the framework of generalized paraxial approximation, the relativistic modulational instability is considered as an example. The dispersion relation is obtained and compared with the case of paraxial approximation.

2 Formalism of Electron Parametric Instability

From hereon we try to explain the method that was introduced in the previous section. To this purpose, let us consider the formalism by which the electron parametric instability can be studied. To do that, we consider the purely one-dimensional case and assume that the light waves propagate along the x-direction in a cold plasma. The electromagnetic field is described by the vector potential $\mathbf{A}(x,t)$ directed in the y-z plane and the electrostatic field is described by the scalar potential Φ. Using the gauge condition $\nabla \cdot \mathbf{A} = 0$, one can write

$$\left(\frac{\partial^2}{\partial x^2} - \frac{1}{c^2}\frac{\partial^2}{\partial t^2}\right)\mathbf{A}(x,t) = -\mu_0 \mathbf{j}_\perp, \qquad (1)$$

and

$$\frac{\partial^2}{\partial x^2}\Phi = -\frac{1}{\epsilon_0}\rho, \qquad (2)$$

where $\mathbf{j}_\perp$ is the electric current perpendicular to the x-direction and ρ is the charge density. Due to the short time scales considered in this paper, we neglect the ion motion. On the other hand, the electron equation of motion can be put into the form

$$\frac{d}{dt}(\mathbf{p} - e\mathbf{A}) = e\left[\nabla\Phi - (\nabla \mathbf{A}.\mathbf{v})\right], \qquad (3)$$

where $\mathbf{p}$ is the electron momentum and $-e$ is the charge of an electron. In the perpendicular direction, we have simply $\mathbf{p}_\perp = e\mathbf{A}$ and (1) becomes

$$\left(\frac{\partial^2}{\partial x^2} - \frac{1}{c^2}\frac{\partial^2}{\partial t^2}\right)\mathbf{a} = \frac{\omega_p^2}{c^2}\frac{1}{\gamma}\mathbf{a}, \qquad (4)$$

where $\mathbf{a} = e\mathbf{A}/mc$, $\omega_p^2 = \left(ne^2/m\epsilon_0\right)$ is the plasma frequency based on the electron density n, m is the mass of an electron and γ is the Lorentz factor

$$\gamma = \left(1 + a^2\right)^{1/2}. \tag{5}$$

In the x-direction the equation of motion reads

$$\frac{\partial}{\partial t} p_x + v_x \frac{\partial}{\partial x} p_x = e\frac{\partial \Phi}{\partial x} - \frac{e^2}{m}(\nabla \mathbf{A}) \cdot \frac{\mathbf{A}}{\gamma} = e\frac{\partial \Phi}{\partial x} - \frac{e^2}{2m\gamma}\frac{\partial A^2}{\partial x}. \tag{6}$$

Together with the equation of conservation of the electron density

$$\frac{\partial}{\partial t} n + \frac{\partial}{\partial x} n v_x = 0, \tag{7}$$

the relations (2), (4) and (6) form the exact system of equations for the one-dimensional problem considered here.

We now consider the stability of a large amplitude circularly polarized wave with a stationary distribution perpendicular to the propagation direction

$$\mathbf{a} = \mathbf{e}_p a_\perp a_\parallel e^{i\alpha_0} + c.c., \tag{8}$$

where $\alpha_0 = (k_0 x - \omega_0 t)$, $\mathbf{e}_p = (\mathbf{e}_y \pm i\mathbf{e}_z)/2$ is the polarization vector and $a_\perp$ is the stationary transversal distribution, say $a_\perp = exp(-r_\perp^2/2r_0^2)$, where r_0 is the spot size.

Since in the study of the electron parametric instability we deal with the growth of perturbations over a nonlinear background caused by the strong pump field, it is convenient to use the following notation

$$n = n_0 + \delta n; \qquad \delta n \ll n_0, \tag{9}$$

$$\gamma = \gamma_0 + \delta\gamma; \qquad \delta\gamma \ll \gamma_0, \tag{10}$$

where δn, $\delta\gamma$ and γ_0 are defined in the following. Substituting (8) into (4), and using (9) and (10), we obtain

$$\left(\frac{\partial^2}{\partial x^2} - \frac{1}{c^2}\frac{\partial^2}{\partial t^2}\right) a_\perp a_\parallel e^{i\alpha_0} = \frac{\omega_0^2}{\gamma_0 c^2}\left(1 + \frac{\delta n}{n_0}\right)\left(1 - \frac{\delta\gamma}{\gamma_0}\right) a_\perp a_\parallel e^{i\alpha_0}. \tag{11}$$

Multiplying the above equation by $a_\perp 2\pi r_\perp dr_\perp$ and integrating over $r_\perp$, we finally have

$$\left(\frac{\partial^2}{\partial x^2} - \frac{1}{c^2}\frac{\partial^2}{\partial t^2}\right) a_\parallel e^{i\alpha_0} = \frac{\left[\int \frac{\omega_{p0}^2}{\gamma_0 c^2}\left(1 + \frac{\delta n}{n_0} - \frac{\delta\gamma}{\gamma_0}\right) a_\perp^2 2\pi r_\perp dr_\perp\right]}{\int a_\perp^2 2\pi r_\perp dr_\perp} a_\parallel e^{i\alpha_0}, \tag{12}$$

where the limits of integration is from zero to the spot size. $\delta n/n_0$ and $\delta\gamma/\gamma_0$ can be defined when $a_\parallel$ is divided into a pump wave of normalized amplitude a_0 (a_0 is assumed to be real) and two daughter waves a_+ and a_- propagating in the same direction

$$a_\parallel(x,t) = a_0 + a_+ e^{i\alpha} + a_- e^{-i\alpha}; \qquad a_+, a_- \ll a_0, \tag{13}$$

where $\alpha = (kx - \omega t)$.

Linearizing (2), (6) and (7) over a nonlinear background caused by the pump wave, and using (8) and (13) leads to

$$\frac{\delta n}{n_0} = \frac{k^2 c^2}{\omega^2 - \omega_{p0}^2/\gamma_0} \frac{a_\perp^2 a_0}{2\gamma_0^2}(a_+ + a_-)\left(e^{i\alpha} + e^{-i\alpha}\right). \tag{14}$$

In order to define $\delta\gamma$ and γ_0, we expand (5) using (8) and (13) to obtain

$$\gamma_0 = \left(1 + a_\perp^2 a_0^2\right)^{1/2}, \tag{15}$$

$$\delta\gamma = \frac{1}{2}\frac{a_\perp^2 a_0}{\gamma_0}(a_+ + a_-)\left(e^{i\alpha} + e^{-i\alpha}\right). \tag{16}$$

Equation (11) is the wave Equation in the paraxial approximation, and, as it was mentioned its average by the transverse distribution (12) gives a more precise form of the wave equation including the appropriate contributions of the pulse pieces. Equation (12) might be solved, in resonance condition, to obtain a dispersion relation by which the stimulated Raman and the modulational instabilities can be studied.

References

1. M.D. Perry and G.A. Mourou, *Science* **264**, 917 (1994); G.A. Mourou, C.P.J. Barty and M.D. Perry, *Phys. Today* **51** **(1)**, 22 (1998); E. Lemoff, G.Y. Yin, C.L. Gordon III, C.P.J. Barty and S.E. Harris, *Phys. Rev. Lett.* **74**, 1574 (1995).
2. *The Supercontinuum Laser Source*, ed. R.R. Alfano (Springer-Verlag, New York, 1989).
3. J. Zhou, J. Peatross, M.M. Murnane, H. Kapteyn and I.P. Christov, *Phys. Rev. Lett.* **76**, 752 (1996).
4. D.C. Eder, P. Amendt, L.B. DaSilva, R.A. London, B.J. MacGowan, D.L. Matthews, B.M. Penetrante, M.D. Rosen, S.C. Wilks, T.D. Donnelly, R.W. Falcone and G.L. Strobel, *Phys. Plasmas* **1**, 1744 (1994).

5. B. Deutsch, H. Furukawa, K. Mima, M. Murakami and K. Nishihara, Phys. Rev. Lett. **77**, 2483 (1996).
6. P. Sprangle, E. Esarey, A. Ting and G. Joyce, Appl. Phys. Lett. **53**, 2146 (1988); D. Umstadter, S.Y. Chen, A. Maksimchuk, G.A. Mourou and R. Wagner, *Science* **273**, 472 (1996); N.E. Andreev, L.M. Gorbunov, V.I. Kirsanov, K. Nakajima and A. Ogata, *Phys. Plasmas* **4**, 1145 (1997); C.D. Decker and W.B. Mori, *Phys. Rev. Lett.* **72**, 490 (1994); *Phys. Rev.* E **51**, 1364 (1995); D.P. Garuchava, I.G. Murusidze, G.I. Suramlishvili, N.L. Tsintsadze and D.D. Tskhakaya, *Phys. Rev.* E **56**, 4591 (1993).

ANALYSIS OF FREE-ELECTRON LASER WITH HELICAL WIGGLER AND AN ION-CHANNEL GUIDING BY RELATIVISTIC RAMAN BACKSCATTERING THEORY

A. A. KORDBACHEH and B. MARAGHECHI
Department of Physics, Amirkabir University of Technology, Tehran, Iran
and
Institute for Studies in Theoretical Physics and Mathematics (IPM), School of Physics
P.O. Box 19395-5531, Tehran, Iran
E-mails: amir@theory.ipm.ac.ir, behrouz@aut.ac.ir

H. AGHAHOSSEINI
Department of Physics, Amirkabir University of Technology, Tehran, Iran
E-mail: aghahos@aut.ac.ir

A one-dimensional relativistic theory of Raman backscatteing has been developed for a free-electron laser (FEL) with a helical wiggler and an ion channel guiding. The beam-frame linear dispersion equation and lab-frame spatial growth rate, which involves the coupling of space-charge and radiation waves by the wiggler field, have been derived. A numerical computation of the growth rate for ion channel guiding has been made.

[a]In a free-electron laser, the focusing mechanism of the intense electron beam in the transverse direction can be done by applying a solenoidal guide field. Takayama and Hiramatsu[1] proposed ion channel guiding of the electron beam as an alternative to the guide field in FEL. This technique involves the injection of a relativistic electron beam into a pre-ionized plasma channel, which leads to the formation of a positive ion core by expulsion of the plasma electrons. In both types of focusing mechanism, the growth rate may be enhanced by exploiting the resonance between guiding device and wiggler frequencies[2,3,4,5].

The purpose of the present study is to obtain the dispersion relation and growth rate for a helical wiggler FEL in the presence of ion channel using relativistic Raman backscattering theory. A relativistic, cold electron beam, moving along the z-axis, passes through an electrostatic helical wiggler, which is periodic along the z-direction, in the presence of a pre-ionized plasma channel with its axis coincident with wiggler axis. The physical explanation of FEL operation mechanism may be made by comprising the wiggler field as a propagating electromagnetic wave (pump wave) in the beam frame, which undergoes stimulated Raman backscattering.

This process is characterized by the parametric decay of the pump wave (ω_1, k_1) into a forward-scattered space-charge wave (ω_2, k_2) and a backscattered electromagnetic wave (ω_3, k_3). All the beam-frame quantities will be written as an unperturbed part (superscript 0) and small perturbation one (superscript 1)

$$\vec{E} = \vec{E}^{(0)} + \vec{E}^{(1)}, \tag{1}$$
$$\vec{B} = \vec{B}^{(0)} + \vec{B}^{(1)}, \tag{2}$$
$$\vec{V} = \vec{V}^{(0)} + \vec{V}^{(1)}, \tag{3}$$
$$N = n^{(0)} + n^{(1)}. \tag{4}$$

In the lab-frame (subscript L), the unperturbed state is characterized by the following equations:

$$\vec{E}_L^{(0)} = 2\pi e n_{iL}(\widehat{x}x_L + \widehat{y}y_L), \tag{5}$$
$$\vec{B}_L^{(0)} = \vec{B}_w(\widehat{x}cos(k_L z_L) + \widehat{y}sin(k_L z_L)), \tag{6}$$
$$\vec{V}_L^{(0)} = \vec{v}_w(\widehat{x}cos(k_L z_L) + \widehat{y}sin(k_L z_L)) + \widehat{z}v_{||}, \tag{7}$$

where

$$v_w = \frac{\Omega_{BL} k_L v_{||}^2}{(\Omega_{iL}^2 - \gamma_0 k_L^2 v_{||}^2)},$$
$$\Omega_{0L} = eB_0/mc, \;\; \Omega_{wL} = eB_w/mc,$$
$$n_L^{(0)} = n_0, \;\; k_L = k_w = 2\pi/\lambda_w,$$
$$\gamma_0 = (1 - {V_L^{(0)}}^2/c^2)^{-1/2}.$$

The equilibrium orbits of electrons in the combined wiggler magnetic field and ion-channel electrostatic field have been found in Ref. [3] where longitudinal

[a]Talk presented by B.M. at the XIth Regional Conference on Mathematical Physics, May 2004, Tehran, Iran.

components of $\vec{V}_L^{(0)}$ have been treated relativistically and the static electric and magnetic fields, arising from the electron beam, have been neglected. Thus, the unperturbed state in the beam-frame of reference is found as

$$\vec{E}^{(0)} = \frac{-\gamma_{||} B_w k_w^2 v_{||}^3}{2c(\omega_i^2 - k_w^2 v_{||}^2)}(\hat{x} - i\hat{y})e^{i(k_1 z + \omega_1 t)} + c.c., \tag{8}$$

$$\vec{B}^{(0)} = \frac{-iB_w(\omega_i^2 - \gamma_{||}^2 k_w^2 v_{||}^2)}{2\gamma_{||}(\omega_i^2 - k_w^2 v_{||}^2)}(\hat{x} - i\hat{y})e^{i(k_1 z + \omega_1 t)} + c.c., \tag{9}$$

$$\vec{V}^{(0)} = \frac{-\omega_w k_w v_{||}^2}{2(\omega_i^2 - k_w^2 v_{||}^2)}(\hat{x} - i\hat{y})e^{i(k_1 z + \omega_1 t)} + c.c., \tag{10}$$

where $\gamma_{||} = (1 - V_{||}{}^2/c^2)^{-1/2}$, and $k_1 = \gamma_{||} k$, $\omega_1 = \gamma_{||} k v_{||}$ are the beam frame wave number and angular frequency of the pump wave, respectively.

In order to obtain a dispersion equation describing the coupling of electromagnetic wave and space-charge wave by wiggler field, perturbation composed of a longitudinal plasma wave (ω_2, k_2), with variation as $\exp[i(k_2 z + \omega_2 t)]$, and of a transverse backscattered electromagnetic wave (ω_3, k_3), with variation as $\exp[i(k_3 z + \omega_3 t)]$, are considered.

The frequencies and wave numbers satisfy the phase-matching condition as

$$\omega_3 = \omega_2 - \omega_1, \tag{11}$$

$$k_3 = k_2 - k_1. \tag{12}$$

The nonlinear dispersion relation describing the Raman backscattering of an electromagnetic pump wave can be obtained by simultaneous solution of fluid equations together with Maxwell's equation for a cold electron beam. Neglecting the non-resonant terms, the result will be

$$\left(\omega^2 - \omega_p^2\right)\left(k_3^2 c_3^2 - \omega_3^2 + \omega_p^2\right) = -\frac{k_2 k_w^2 \omega_w^2 \omega_p^2 v_{||}^3}{2\left(\omega_i^2 - k_w^2 v_{||}^2\right)^2} \times \left[k_3 v_{||} - \frac{\omega_3(1 - a_1)\left(\omega_i^2 - k_1^2 v_{||}^2\right)}{k_1 v_{||}\left(\omega_3 - a_1 a_2\right)}\right]\left[1 - \frac{a_1 a_2}{\omega_3} - \frac{\omega_2 \omega_1^2}{k_1 k_2 \omega_3 v_{||}^2}\right], \tag{13}$$

where

$$a_1 = \frac{\omega_w^2 k_w^2 v_{||}^4}{2c^2\left(\omega_i^2 - k_w^2 v_{||}^2\right)}, \quad a_2 = \left(\gamma_b^2 - 1\right)\omega_1,$$

$$\omega_w = \gamma_{||} e B_w / \gamma_0 mc, \quad \omega_p = \left(4\pi e^2 n_0 / \gamma_0 m\right)^{-1/2},$$

$$\omega_i = \left(2\pi e^2 n_i / \gamma_0 m\right)^{-1/2}.$$

In the forgoing dispersion equation, the wiggler effects on the linear dispersion relations of the space-charge wave and radiation wave have been neglected. In addition, the variation of the energy exchange between the electron beam and the electromagnetic wave has been taken into account by taking γ as

$$\gamma = \gamma_0 \gamma_{||}^{-1} + c^{-2} \gamma_0^3 \gamma_{||}^{-3} \vec{V}^{(0)} \cdot \vec{V}^{(1)}. \tag{14}$$

Expressing the imaginary parts of the complex frequencies and wave numbers of the growing waves in terms of the lab frame growth rate Δ_L as

$$k_2 = Re\ k_2 - i\gamma_{||}\Delta_L, \tag{15}$$

$$\omega_2 = Re\ \omega_2 - i\gamma_{||} v_{||}\Delta_L, \tag{16}$$

$$k_3 = Re\ k_3 - i\gamma_{||}\Delta_L, \tag{17}$$

$$\omega_3 = Re\ \omega_3 - i\gamma_{||} v_{||}\Delta_L, \tag{18}$$

and substituting them into (13), the lab-frame spatial growth rate of the space-charge and radiation waves will be in the form

$$\Delta_L^2 = \frac{k_2 k_w^2 \omega_w^2 \omega_p^2 v_{||}^3}{8\gamma_{||}^2 \omega_2\left(\omega_i^2 - k_w^2 v_{||}^2\right)^2}\left[1 - \frac{a_1 a_2}{\omega_3} - \frac{\omega_2 \omega_1^2}{k_1 k_2 \omega_3 v_{||}^2}\right] \times \left[k_3 v_{||} - \frac{\omega_3(1 - a_1)\left(\omega_i^2 - k_1^2 v_{||}^2\right)}{k_1 v_{||}\left(\omega_3 - a_1 a_2\right)}\right]\left[k_3 c^2 - \omega_3 v_{||}\right]^{-1}.$$

The results of the numerical study of the lab-frame spatial growth rate Δ_L are illustrated in Fig. 1 in which lab frame values for the unperturbed quantities have been taken to be $n_{0L} = 10^{12}\ cm^{-1}$, $B_w = 500G$, $\lambda_W = 2\ cm$ and $\gamma_0 = 2.5$.

Figure 1 shows Δ_L as a function of dimensionless ion-channel frequency $\omega_i / k_w c$ for FEL with ion-channel resonance near $\omega_i / k_w c \approx 0.82$. Away from resonance region, however, Δ_L decreases for small and large values of B_0. In the present analysis, the effects of ion-channel on the equilibrium orbits is taken into account. The effects on the radiation, however, are neglected.

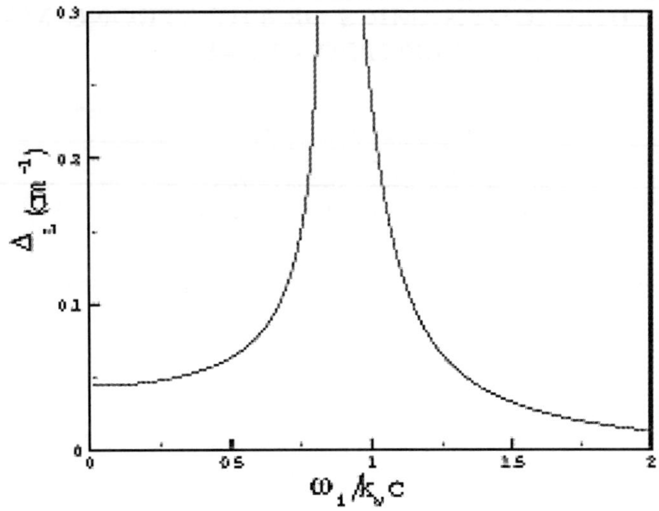

Figure 1. Lab-frame spatial growth rate Δ_L as a function of ion-channel frequency $\omega_i/k_w c$.

References

1. K. Takayama and S. Hiramatsu, *Phys. Rev.* A **37**, 730 (1988).
2. D. Whittum and A.M. Sessler, *Phys. Rev. Lett.* **64**, 25 (1990).
3. P. Jha and P. Kumar, *IEEE Trans. Plasma Sci.* **24**, 1359 (1996).
4. P. Jha and P. Kumar, *Phys. Rev.* E **57**, 2256 (1998).
5. M. Esmaeilzadeh, H. Mehdian and J.E. Willet, *Phys. Rev.* E **56**, 016501 (2001).

RELATIVISTIC THERMODYNAMICS OF THE STRONG MAGNETIZED DENSE ELECTRON PLASMA

N. L. TSINTSADZE
Tbilisi State University
Chavchavadze 3, Tbilisi, Georgia

Relativistic thermodynamics of a dense electron plasma, for an arbitrary temperature in the presence of a super strong magnetic field is investigated. A novel set of adiabatic equations are obtained.

1 Introduction

[a]The study of the problem of influence of a strong or super strong magnetic field on the thermodynamic properties of medium and the propagation of proper waves is important in supernovae and neutron stars, in the convective zone of the sun, as well as, the early pre-stellar period of the evolution of the universe.

Based on the astrophysical data, the surface magnetic field of a neutron star is $B \sim 10^{12} - 10^{13}G$, and the internal field can reach $B \sim 10^{15}G$ or even higher[1,2,3]. As was shown by Bisnovati-Kogan[4], the presence of rotation may increase the magnetic field by an additional factor of $10^3 - 10^4$, on the other hand, the surface density of the neutron star matter is $\rho \sim 1.6 \times 10^9 g/cm^3$ and near the center of the star is $\rho \sim 10^{14} g/cm^3$.

In such dense medium of a neutrino star, where the electrons are in a strongly degenerate state, the presence of a strong magnetic field will essentially change the thermodynamic quantities. This happens, when the characteristic energy of an electron on a Landau level reaches the size of the electron chemical potential $\mu \approx (|e|\hbar/2m_0c)B$, *i.e.* when

$$B \geq B_s(\mu/m_0c^2) \sim 10^{15} - 10^{16}G.$$

Here

$$B_s = m_0^2c^3/|e|\hbar = 4.4 \times 10^{13}G,$$

is the Schwinger's magnetic field, $2\pi\hbar$ is the Planck constant, m_0 is the electron rest mass, c is the speed of light in vacuum, and e is the magnitude of the electron charge. As it is well known, the strong magnetic field in the fermion gas leads to two magnetic effects[5]: There are, Pauli paramagnetism due to the spin of the electrons (positrons) and Landau diamagnetism due to the quantization of the orbital motion of the electrons (positrons).

2 Relativistic Thermodynamic Potential

In a strong magnetic field, the motion in a plane perpendicular to the magnetic field is quantized and the discrete energy values are called Landau levels. In the relativistic quantum theory[6] the electron energy levels $\varepsilon_{\ell\sigma}$ in a magnetic field B was determined as follows

$$\varepsilon_{\ell\sigma} = c\sqrt{P_z^2 + (2\ell + 1 + \sigma)\frac{|e|\hbar}{2c}B + m_0^2c^2}, \qquad (1)$$

where P_z is the momentum component along the magnetic field, $\sigma = \pm 1$ and $\ell = 0, 1, 2, \cdots$.

In expression (1), the term $(2\ell+1)(|e|\hbar/2m_0c)B$ corresponds to the diamagnetic effect, while

$$(|e|\hbar/2m_0c)\sigma = \sigma \cdot \mu_B \cdot B,$$

is the spin energy, and μ_B is the Bohr magneton.

The number of states in the interval dP_z for any given value of ℓ is

$$\frac{V|e|B}{(2\pi\hbar)^2c}\, dP_z. \qquad (2)$$

Now, we write down the thermodynamic potential of a relativistic electron gas with an arbitrary temperature

$$\Omega = -\frac{VT|e|B}{(2\pi\hbar)^3c}\sum_{\ell=0}^{\infty}\sum_{\sigma}\int_{-\infty}^{+\infty} dP_z \ln\left(1 + \exp\frac{\mu - \varepsilon_{\ell\sigma}}{T}\right) \qquad (3)$$

Unfortunately, the evaluation of (3) is nontrivial. Thus, we first want to restrict ourselves to the most simple limiting cases.

To this end, we consider the orders of magnitude of the characteristic energies ε_F, T and $\mu_B B$ of the system, where ε_F is the Fermi energy of the highest occupied state.

[a]Invited plenary talk presented at the XIth Regional Conference on Mathematical Physics, May 2004, Tehran, Iran.

Let us consider the case of a strong magnetic field satisfying $\sqrt{e\hbar c B} - \mu \gg T$, so that only the Landau ground level is filled ($\ell = 0$), *i.e.* only the Pauli paramagnetism and the self-energy of the particles are taken into account. Thus,

$$\varepsilon_{0,\sigma} = c\sqrt{P_z^2 + (1+\sigma)\frac{e\hbar B}{c} + m_0^2 c^2}. \tag{4}$$

3 Boltzmann gas

We now want to calculate the Boltzmann limit $T \gg \varepsilon_F$ of (3) first for an arbitrary magnetic field. In this case, $\exp(\mu/T) \ll 1$, and one may insert Boltzmann statistic, *i.e.*

$$\ln\left(1 + \exp\left(\frac{\mu - \varepsilon_{0,\sigma}}{T}\right)\right) \simeq \exp\left(\frac{\mu - \varepsilon_{0,\sigma}}{T}\right). \tag{5}$$

Substituting expression (5) into (3) and instead of $\int_{-\infty}^{+\infty} dP_z$, one may also write $2\int_0^{\infty} dP_z$, since only P_z^2 enters the integrand. We introduce the new variable $u = P_z/[(1+2\eta)\cdot(1+\sigma)]^{1/2}$, where $\eta = \beta_B B/m_0 c^2$ and $a_z = a_0\sqrt{1+2\eta(1+\sigma)}$ and $a_0 = m_0 c^2/T$. Now the thermodynamic potential has the form

$$\Omega = -\frac{VT(m_0c^2)^2}{\pi^2(\hbar c)^3}\beta_B \cdot B \sum_\sigma \sqrt{1+2\eta(1+\sigma)}$$

$$\int_0^\infty du \exp[a_\sigma(1 - \sqrt{1+u^2})]. \tag{6}$$

After integration, we obtain

$$\Omega = -\frac{VT^2}{\pi^2}\frac{\beta_B \cdot B}{(\hbar c)^3} m_0 c^2 \sum_\sigma e^{a_\sigma} a_\sigma K_1(a_\sigma), \tag{7}$$

where $K_1(x)$ is the McDonald function of the first order.

Let us now investigate (7) for different parameters a_σ ($a_2 = a_0\sqrt{1+4\eta}, a_0 = m_0c^2/T$):

1. First we suppose that $T \gg m_0 c^2, \beta_B B$. Under these conditions, we obtain

$$\Omega = -\frac{VT^2|e|B}{\pi^2(\hbar c)^2}. \tag{8}$$

The total energy of such electrons is defined by

$$E = -T^2 \frac{\partial}{\partial T}\left(\frac{\Omega}{T}\right)_{V,B} = -\Omega. \tag{9}$$

This is a novel relation in thermodynamics. The pressure is $P = T^2|e|B/\pi^2(\hbar c)^2$. For the entropy we have

$$S = -\left(\frac{\partial \Omega}{\partial T}\right)_{V,B} = \frac{2V|e|BT}{\pi^2(\hbar c)^2}, \tag{10}$$

and the entropy per particle is

$$\frac{S}{N} = \frac{2|e|}{\pi^2(\hbar c)^2}\frac{TB}{n}, \tag{11}$$

where $n = N/V$ is the density of electrons (positrons). Hence, for an adiabatic process $d\frac{S}{N} \equiv 0$, we obtain from (11) the adiabatic equation

$$\frac{TB}{n} = \text{constant}. \tag{12}$$

We note that because of the pressure $P = E/V$, the sound velocity is $c_s = c$, instead of $c/\sqrt{3}$ as isotropic electron (positron) gas.

2. We now consider the non-relativistic temperature $m_0 c^2 \gg T$ *i.e.* a_0 and $a_\sigma \gg 1$.

In this case Ω has the following form

$$\Omega = -\frac{VT^{3/2}|e|B(m_0c^2)^{1/2}}{(2\pi)^{3/2}(\hbar c)^2}\left[1 + \left(1 + 4\frac{\beta_B B}{m_0 c^2}\right)^{1/4}\right]. \tag{13}$$

In the limiting case of $\beta_B B/m_0 c^2 \ll 1$, for the adiabatic equation, we obtain the expression

$$\frac{T^{1/2}B}{n} = \text{constant}. \tag{14}$$

In the opposite case, *i.e.* $\beta_B B/m_0c^2 \gg 1$, we obtain then

$$\frac{T^{1/2}B^{5/4}}{n} = \text{constant}, \tag{15}$$

for the adiabatic equation. We can see, that all adiabatic equations contain three unknown quantities n, T and B, instead of two, as in the ordinary thermodynamics.

If we fix the magnetic field, in the non-relativistic limit, using Eqs. (13)-(15) and relation $P = -\Omega/V$, we can define the sound velocity by $c_s = (\partial P/\partial \rho)_S^{1/2}$ with $\rho = m_0 n$.

For the case of expression (14), we obtain

$$c_s^2 = \frac{3T_0}{2m_0}\left(\frac{n}{n_0}\right)^2\left(\frac{B_0}{B}\right)^2, \tag{16}$$

where n_0 and B_0 are initial values of the density and the magnetic field.

For the adiabatic equation (15) we have

$$c_s^2 = \frac{3T_0}{2m_0}\left(\frac{n}{n_0}\right)^2\left(\frac{B_0}{B}\right)^{1/4}. \tag{17}$$

For (16) and (17), it follows that the sound velocity depends very strongly on the magnetic field due to the Pauli magnetism and the self-energy of the particles.

4 A Fermi Gas

The study of the thermodynamics of a Fermi gas of sufficiently low temperatures and strong magnetic field is of fundamental significance. As the gas is compressed, the mean energy of the electrons increases; when it becomes comparable to, or larger than m_0c^2, relativistic effects become important. In this case, the temperature can be $10^8 - 10^9 K$ and the relativistic degenerate electron (positron) gas will be formed. For instance, a supernova explosion happens when the inner part of a star first collapses until very large densities $\rho \simeq 10^{14} g/cm^3$ comparable to atomic nuclei are obtained. In this case, the Fermi energy ε_F of electrons is $\varepsilon_F \approx 10^8 eV$, which is much more than the thermal energy, even for the temperature $T \sim 10^{10} K$.

To study the thermodynamic properties of the Fermi gas, we shall use the expression for Ω, (3), with the energy (4).

We now want to study the limiting case, when the thermal energy $k_B T$ (k_B is the Boltzmann coefficient) is much less than the Fermi energy ε_F. In this case, the Fermi distribution function n_F is, in a good approximation, described by the step function

$$n_F = \Theta(\varepsilon_F - \varepsilon) = \begin{cases} 1, \text{ if } \varepsilon \leq \varepsilon_F, \\ 0, \text{ if } \varepsilon > \varepsilon_F. \end{cases} \tag{18}$$

The chemical potential μ must be identical to the Fermi energy ε_F of the system. Taking into account the temperatures, which are much lower than the degeneracy temperature $T_F = \varepsilon_F / K_B$, the distribution function is appreciably different from unity or zero only in a narrow range of values of the energy ε close to the limiting energy ε_F. The width of this transition zone of the Fermi distribution is of the order of T. Thus, in this case the logarithm of the grand partition function (3) can be rewritten in the following form

$$\Omega = -\frac{V|e|B}{(2\pi\hbar)^2 c} \sum_\sigma \int_{-\infty}^{+\infty} dP_z P_z \frac{d\varepsilon_{0,\sigma}}{dP_z} \times \left\{ \Theta(\varepsilon_F - \varepsilon) + \left(\frac{1}{\exp\left(\frac{\varepsilon_{0,\sigma}-\mu}{T}\right) + 1} - \Theta(\varepsilon_F - \varepsilon) \right) \right\}. \tag{19}$$

The first integral

$$\Omega_0 = -\frac{V|e|B}{(2\pi\hbar)^2 c} \sum_\sigma \int_{-\infty}^{+\infty} dP_z P_z \frac{d\varepsilon_{0,\sigma}}{dP_z} \Theta(\varepsilon_F - \varepsilon), \tag{20}$$

denotes the value of Ω at absolute zero temperature. The expression (20) is the first term in the expansion of the corresponding quantity in powers of the small ratio T/T_F. After the same simple calculations, we obtain

$$\Omega = -\frac{VT^2 \beta_B B (m_0 c^2)^2}{\pi^2 (\hbar c)^3} \sum_\sigma a_\sigma^2 \frac{1}{2} \left[u_{F\sigma} \sqrt{1 + u_{F\sigma}^2} - \ln(u_{F\sigma} + \sqrt{1 + u_{F\sigma}^2}) \right] + \sum_\sigma \int_0^\infty dx \frac{[(a_{F\sigma} + x)^2 - a_\sigma^2]^{1/2} - [(a_{F\sigma} - x)^2 - a_\sigma^2]^{1/2}}{\exp(x) + 1}, \tag{21}$$

where we have introduced new symbols

$$a_{F\sigma} = a_0 \sqrt{1 + 2\eta(1+\sigma) + P_F^2 / m_0^2 c^2},$$

and

$$u_{F\sigma} = P_F / (m_0 c \sqrt{1 + 2\eta(1+\sigma)}).$$

We now consider (21) in the two limiting cases:

a) In a degenerate extreme relativistic electron gas, the energy of particles is large compared to $m_0\ c^2$. In this case, the relation between the energy and momentum of the electron is $\varepsilon = cP_z$ and in (21) we should take $u_{F\sigma}^2 \gg 1$. Then, we expand the expression (21) as a power series in $(1/u_{F\sigma})$ and retain only the two main terms, to obtain

$$\Omega = -\frac{V\beta_B B m_0 c^2 P_F^2 c^2}{\pi^2 (\hbar c)^3} - \frac{2VT^2 \beta_B B m_0 c^2}{3(\hbar c)^3}. \tag{22}$$

From (22), we define the entropy per electron

$$S = \frac{4}{3} \frac{T \beta_B B m_0 c^2}{(\hbar c)^3 n}, \tag{23}$$

and the adiabatic equation is

$$\frac{BT}{n} = \text{constant}. \tag{24}$$

It should be emphasized that the adiabatic equation (24) is the same as (12).

Here, we have considered the degenerate extreme relativistic gas of electrons in a strong magnetic field. In this case, there is a drastic difference between the thermodynamic relations without and with the magnetic field. For instance, the relation between Ω and the energy of particles E instead of being obtained from $\Omega = -1/3E|_{B=0}$, will be obtained from the following novel relation

$$-\Omega = E = 2\frac{V|e|B}{(2\pi\hbar)^2}\int_{-\infty}^{+\infty}\frac{dP_zP_z}{\exp(\frac{cP_z-\mu}{T})+1}, \quad (25)$$

and the sound velocity now equals the speed of light in vacuum.

b) In the non-relativistic limit, $u_{F\sigma}^2 \ll 1$, for the thermodynamic potential we get

$$\Omega = -\frac{2}{3}\frac{\pi^4 V\hbar^6 n^3}{(\beta_B B)^2 m_0^3} - \frac{VT^2(\beta_B B)^2 m_0^3\sqrt{1+4\eta}}{3\pi^2\hbar^6 n}. \quad (26)$$

From here, it follows the adiabatic equation

$$\frac{TB^2}{n^2}\sqrt{1+4\eta} = \text{constant}. \quad (27)$$

Using the expression for the thermodynamic potentials (22) and (26), we can obtain any thermodynamic quantity.

In this brief communication, we have developed the relativistic thermodynamics of the dense electron gas in the super strong magnetic field, taking into account the Pauli paramagnetism and the self-energy of the particles. We have discussed about the sound velocity and derived the adiabatic equations for a hot and a "cold" electron (positron) gas in the presence of a strong magnetic field.

References

1. I. Landstreet, *Phys. Rev.* **153**, 1372 (1967).
2. S.L. Shapiro and S.A. Teukolsky in *Black Holes, White Dwarfs and Neutron Stars*, (John Wiley and Sons, New York, 1981).
3. V.M. Lipunov in *Neutron Star Astrophysics*, (Nauka, Moscow, 1987).
4. G.S. Bisnovatyi-Kogan, *Astron. Zh.* **47**, 813 (1970).
5. L.D. Landau and E.M. Lifshitz in *Statistical Physics*, (Pergamon Press, Oxford, 1989).
6. V.B. Berestetskii, E.M. Lifshitz and L.P. Pitaevskii in *Quantum Electrodynamics*, (Pergamon Press, Oxford, 1982).